ICME-13 Monographs

Series editor

Each volume in the series presents state-of-the art research on a particular topic in mathematics education and reflects the international debate as broadly as possible, while also incorporating insights into lesser-known areas of the discussion. Each volume is based on the discussions and presentations during the ICME-13 conference and includes the best papers from one of the ICME-13 Topical Study Groups or Survey Teams.

More information about this series at http://www.springer.com/series/15585

Milton Rosa · Lawrence Shirley
Maria Elena Gavarrete · Wilfredo V. Alangui
Editors

Ethnomathematics and its Diverse Approaches for Mathematics Education

Springer

Editors
Milton Rosa
Centro de Educação Aberta e a Distância
Universidade Federal de Ouro Preto
Ouro Preto, Minas Gerais
Brazil

Lawrence Shirley
Department of Mathematics
Towson University
Towson, MD
USA

Maria Elena Gavarrete
Universidad Nacional de Costa Rica
Heredia
Costa Rica

Wilfredo V. Alangui
University of the Philippines Baguio
Baguio
Philippines

ISSN 2520-8322 ISSN 2520-8330 (electronic)
ICME-13 Monographs
ISBN 978-3-319-59219-0 ISBN 978-3-319-59220-6 (eBook)
DOI 10.1007/978-3-319-59220-6

Library of Congress Control Number: 2017940227

Printed on acid-free paper

This Springer imprint is published by Springer Nature
The registered company is Springer International Publishing AG
The registered company address is: Gewerbestrasse 11, 6330 Cham, Switzerland

Preface

Ethnomathematics as an important line of study and research in mathematics education investigates the roots of mathematical ideas, procedures, and practices, starting from the way individuals behave in different cultural groups. Over the past three decades, the amount of research, investigations, thesis, and dissertations that dealt with both the theoretical and practical aspects of ethnomathematics has expanded exponentially and worldwide.

In this regard, ethnomathematics has come to study our diverse cultural roots of mathematical knowledge starting from the various ways in which the members of distinct cultural groups mathematize. The study of ethnomathematics also considers the historical evolution of the mathematical knowledge with the acknowledgment of all social and cultural factors that mold this development.

Numerous articles, projects, chapters, and books have been written in many countries about the relation between culture, mathematics, and mathematics education. Over the same time, the early founders of ethnomathematics have found countless new voices that have gone on to make even more new discoveries, and offered even more new insights. As researchers matured, new studies involving ethnomathematics were discussed and debated in a succession of local, regional, national, and international meetings, seminars, conferences, and congresses as well as in numerous study groups around the world.

When the focus of a study is the pedagogy of mathematics, our attention must be centered both around legitimizing the students' knowledge, grown from experiences built in their own ways, and around the study of the pedagogical possibilities of how to work with the diversity of learning processes that occurs outside and inside of our school environments. Indeed, a discussion of the educational aspects of ethnomathematics helps teachers to establish cultural models of beliefs, thought, and behavior, in the sense of contemplating the potential of the pedagogical work that takes into account the previous knowledge of the students as well as a mathematical learning that is more meaningful and empowering.

In this book, we highlight the evolution of the field of ethnomathematics across the globe by acknowledging that Prof. D'Ambrosio is the most important theoretician in this field and offers encouragement, leadership, and dissemination of our many new ideas, concepts, and perspectives involved in ethnomathematics around the world and its applications to mathematics education. From his point of view, the *program ethnomathematics* is a way of generating, organizing, and diffusing knowledge developed by culturally identified groups that offer a possibility of meeting the challenges of proposing and keeping peace.

In order to elucidate and clarify, and perhaps to facilitate new discussions about ethnomathematics, we truly hope that readers will be able to capture the many authors' thoughts and concepts regarding ethnomathematics. From the authors' own particular vantage points, they each have done a great deal to add to a growing body of scientific discourse of this program as presented in the ICME13–TSG35 Ethnomathematics Study Group in Hamburg, Germany, in 2016. As well, we would like to state at the outset that there are other perspectives and diverse approaches for mathematics education as well as innovative views on ethnomathematics emerging from other researchers and in other knowledge fields. We hope that future books will consider them as well. We would also like to emphasize that the questions, answers, comments, conceptions, and discussions made in the chapters in this book are the authors' personal views on ethnomathematics and not necessarily those of the editors.

We are certain that not all educators, mathematicians, and philosophers will agree upon these views and conceptions on ethnomathematics. As well, we are confident that, in some cases, the approaches, perspectives, and innovations presented here may be in discordance with views of other ethnomathematicians. Thus, we are pleased that this book will illustrate what happens within a research field as it continues to evolve and has spread itself worldwide to include a diversity of schools, colleges, universities, and local communities, in a relatively short period of time.

We have no doubt that ethnomathematics is alive, and it is evolving as more and more research is uncovered worldwide. We also understand that it will continue this growth process. It is a research field that has not yet crystallized and that to us is very, very exciting! As it stands currently, it seeks to document and understand widely diverse mathematical ideas, procedures, and practices as distinct cultural group members gain voice and present their perspectives in order to become empowered and value their previous knowledge. As this diversity of voices begins to speak, they have remarkably similar, yet different points of view.

From the discussions provided by the authors who presented at ICME13–TSG35 Ethnomathematics Study Group in Hamburg and who wrote chapters for this book, we can safely conclude that mathematical knowledge, as we currently experience it, is constructed by the development of different ideas, procedures, and practices that are common to the members of all our diverse sociocultural groups. These processes enable us to elaborate and use our abilities and competencies, which include the universal mechanisms of *counting, locating, measuring, drawing, representing, playing, understanding, comprehending, explaining*, and *modeling*. Today, ethnomathematics investigates the roots of mathematical ideas and practices, starting from the way diverse individuals behave in different cultural groups.

In other words, many of the ethnomathematical studies as presented in this book identify the mathematical practices that begin with the knowledge of the *others* in their own terms and rationality. To know and understand the value of the plurality of the nature of our diverse social, cultural, economic, and political realities is a necessity in order to take a firm stand against prejudices based on cultural differences, social classes, beliefs, gender, sexual orientation, ethnics, or other social, cultural, political, and individual characteristics.

The authors in this book have shared and debated the necessity of issues regarding research, mathematics education, classroom practices, philosophies, and the knowledge of the members of specific cultural groups. Ethnomathematics clearly has a role in helping us to clarify the nature of mathematical knowledge and of knowledge in general. An important objective of this book is to show that in a globalized and interdependent world, it is fundamental for researchers and educators to understand that the diversity of ideas and thoughts that they come into contact with, are greatly influenced by members of distinct cultural groups and their unique mathematics. This pedagogical approach is not often reflected in traditional mathematics classrooms, yet high equitable expectations along with personalized connections in mathematics instruction are essential for success for all our students.

It is important to keep in mind that ethnomathematics grew out of the history of mathematics, mathematics education, and issues of mathematics in anthropology, sociology, economic, environmental issues, and political science as well as that it recognizes that all cultural groups do activities that involve mathematical thinking, even if the mathematics and it may not look like traditional Eurocentric academic mathematics that students learn in schools and universities. In its insubordinative and creative way, ethnomathematics is considered as basic to the counting terms in various languages or the use of symmetries in craft products, or as complex and controversial as oppressed societies using mathematics to encourage open-minded thinking to challenge power relations, which is the main purpose of this book.

The chapters discussed and debated in this book demonstrate the universal concern regarding mathematics education, classroom knowledge, and knowledge of cultural groups. Because ethnomathematics has a role in helping us to clarify the nature of mathematical knowledge and of knowledge in general, the discussions surrounding these issues do not imply that ethnomathematics is only an instrument to improve mathematical education. It is necessary to shift the research from theoretical issues toward practical issues that help educators and students to access their full potential by searching innovative forms and diverse approaches for mathematics education. Again, ethnomathematics is a young field of research, it is alive, and it is dynamic and evolving as more and more voices are added to the discussion. It is a privilege to be part of this movement!

Finally, the editors of this book wish to thank the ICME13–TSG35 Ethnomathematics Study Group for the fine contributions by the authors of the chapters in this book.

Ouro Preto, Brazil — Milton Rosa
Towson, USA — Lawrence Shirley
Heredia, Costa Rica — Maria Elena Gavarrete
Baguio, Philippines — Wilfredo V. Alangui

Contents

Part I
Introduction

Chapter 1
An Ethnomathematics Overview: An Introduction

Milton Rosa and Maria Elena Gavarrete

Abstract After three decades since the emergence of the ethnomathematics as a program, many investigators and educators in many countries are not comfortable about the role of ethnomathematics in mathematics education. This particular pedagogical action underscores the importance of doing the ethnomathematical work first in order to come to a good understanding of the mathematical aspects of culture by having a clear purpose in regards to educational activities. Both the implantation and implementation of ethnomathematical perspectives in classrooms must be preceded by investigations of the mathematical ideas, procedures, and practices developed by the members of diverse cultural groups. Ethnomathematics helps to establish a meta-awareness of the role of mathematical knowledge in the society and cultural context of mathematics. Hence, ethnomathematics is a reciprocal program as it is possible to think of traditional academic mathematics and its role within its host cultural group. This reciprocity is a vital aspect of ethnomathematics.

Keywords Challenges · Cultural features · Ethnomathematics · Mathematics education · Opportunities

1.1 Preliminary Thoughts…

Over time, there has been an increased interest in how we, as researchers in ethnomathematics, incorporate new ideas and technologies in novel and creative ways and how these interactions are increasingly affecting our thinking and learning processes. Our own culture and society considerably influences the way in which

M. Rosa (✉)
Universidade Federal de Ouro Preto, Ouro Preto, Brazil
e-mail: milton@cead.ufop.br

M.E. Gavarrete
Universidad Nacional de Costa Rica, Heredia, Costa Rica
e-mail: marielgavarrete@gmail.com

M. Rosa et al. (eds.), *Ethnomathematics and its Diverse Approaches for Mathematics Education*, ICME-13 Monographs,
DOI 10.1007/978-3-319-59220-6_1

we understand mathematical ideas, procedures, and practices. In this regard, we need to make an effort to open our eyes to *alternative* narratives, understandings, histories, and technological sophistication in relation to local non-western cultures and their mathematics.

Hence, the core of ethnomathematics research increasingly demonstrates how mathematics is made of many historically rich, diverse, and distinct traditions. The members of distinct cultural groups have developed mathematical concepts that are rooted in the universal human endowments of curiosity, ability, and transcendence. They characterize our very humanness. Awareness and appreciation of cultural diversity that can be seen in clothing, methods of discourse, religious views, morals, and our own unique worldview combine to allow us to understand each aspect of our daily life.

This context allowed D'Ambrosio (1985) to state that all people have developed unique, often distinct ways of mathematical knowledge that frequently were incorporated into cultural systems as diverse people interacted, immigrated and created new contexts. This is most obvious in ways that diverse groups order, quantify, use numbers, incorporate geometric forms and relationships, measure, and classify objects. Consequently, mathematical thinking is influenced by a wide diversity of human environments, which include language, religion, morals, economics, social, and political activities.

For example, language plays a central role in ethnomathematics research. The research related to the use of local languages empowers the knowledge of the *others*. Language implies a logical order and a cognitive structure that in turn carries implicit features of the worldview of a culture. It is through the knowledge of the language as well as its signs and meanings that we are able to understand and delve into the logic of differentiated groups. Thus, it is essential that investigators familiarize themselves with local languages in order to access distinct forms of expressions associated with mathematical knowledge of members of other cultures.

Within this context, Meaney et al. (2008) state that, "specific cultural practices such as ethnomathematical ones are embedded within the language of the culture" (p. 53). Along with language, the members of distinct cultural groups have come to develop logical processes related to quantification, measurement, and modelling in order to understand and explain their social, cultural, and historical contexts (Rosa and Orey 2010).

These processes enable the members of each cultural group to develop their own way to *mathematize*[1] their reality in order to examine how mathematical ideas and practices are processed and used in daily activities. These tools allow for the identification and integration of specific mathematical ideas, notions, procedures, and practices by schematizing, formulating, and visualizing a problem in different ways, discovering relations, patterns, and regularities, and transferring a real

[1]Mathematization is a process in which members of a distinct cultural group develop specific mathematical tools that can help them to organize, analyze, comprehend, understand, and solve concrete problems located in the context of their real-life situations (Rosa and Orey 2010).

world-situation into mathematical ideas through the mathematization process (Rosa and Orey 2012).

Inclusion of a diversity of ideas brought by people from diverse cultural contexts gives confidence and dignity to students, while allowing them to perceive a variety of perspectives in order to provide a base in which they are able to learn mathematics (Rosa and Orey 2015). Equally important is the search for alternative methodological approaches needed to reach this goal. On this matter, as traditional Western science-based mathematical practices were accepted worldwide, it is even more vital that we record historical, diverse, and alternate forms of mathematical ideas, notions, and procedures before many of these local practices are lost to us. In this way, students can see how mathematics has developed and realize how individuals help the overall evolution of mathematical knowledge.

Therefore, the development of mathematical ideas, procedures, and practices serves as a vehicle to transfer meanings and values from the culturally constituted to that of the constituted academic world. The communication of these ideas is represented best in the moderator process as the effect of culture on the mathematics concepts developed by the members of distinct cultural groups. When considering this process, it is worth noting that from an emic perspective, culture may not be seen as a construct apart from, and causing the development of, mathematical practices. We depict culture as causing the development of mathematical ideas, procedures, and practices in order to develop a framework that individuals can easily implement to compare mathematical practices developed by different cultures and isolate the cultural causes of the distinction of these practices.

In this regard, emic researchers view culture as inseparable from the individual, as an inherent quality (Geertz 1973). Thus, the emic approach focuses on the meaning of objects such as mathematical artifacts in the lives of the individual has also applied values theory to explain how people organize information in their own environment. An etic approach understands the mathematical phenomenon more cross culturally rather than cultural specific meanings. Studying culture according to pre-established etic procedures impedes the discovery of cultural diversity, whereas *emic* analysis broadens this view (Headland et al. 1990).

According to this context, Rosa and Orey (2011) argue that ethnomathematics attempts to establish relations between the mathematical ideas and procedures embedded in local practices (emic) and academic conceptual frameworks (etic). The goal of ethnomathematics research is about the acquisition of both emic and etic approaches because an emic approach is essential for an intuitive and empathic understanding of a culture, and it is essential for conducting effective ethnographic fieldwork while an etic approach is essential for cross-cultural comparisons, and forms an essential component of ethnology since such comparisons necessarily demand standard units and categories. Both etic and emic approaches refer to similar constructs but from different perspectives, that is, *between-cultures* versus *within-cultures*. The notion of values, or at least some variants of it, is a central component to most views of culture.

1.2 Cultural Features of Ethnomathematics

Ethnomathematics is a research program incorporating history, anthropology, pedagogy, linguistics, and philosophy of mathematics with pedagogical implications that focus on the techniques of explaining, understanding and coping with different sociocultural environments. According to D'Ambrosio (1985), the etymology of ethnomathematics, the prefix *ethno* refers to sociocultural contexts and, therefore, includes language, jargon, and codes of behavior, myths, and symbols. The derivation of *mathema* means to explain, to know, to understand, and to perform activities such as ciphering, measuring, classifying, ordering, inferring, and modelling. The suffix *tics* is derived from *techné* and has the same root as art and technique. In this case, *ethno* relates to the members of distinct cultural groups who identified themselves by their cultural traditions, codes, symbols, myths, and specific ways of reasoning and inferring.

It is the application of mathematical ideas, procedures, and practices developed and applied by members of a specific cultural group in distinct contexts, which are often used currently in present day contexts (D'Ambrosio 1985). The main objective of this program is to offer an innovative theoretical basis composed by philosophical, political, and epistemological dimensions of the development of mathematical knowledge as well as the comprehension of human behavior by making sense of the mathematical ideas and procedures practiced by humanity. This means that the study of the history of mathematics attempts to identify the cultural and mathematical contributions of different cultures across the world.

Much of what we call *modern mathematics* came about as diverse European groups sought to resolve unique problems related to colonization, commerce, art, religion, exploration, colonization and communications, the construction of railroads, census data, space travel, and other problem-solving techniques that arose from particular communities. For example, the Mayans invented the number zero and the positional value that are often attributed to the Hindus around the 9th century (Rosa and Orey 2005). These concepts were transmitted to the Arabs from the Hindus by means of exchanges and commercial activities. And the most common form of quantification is based on a Hindu-Arabic numeral system, which has resulted from a historical relationship between two distinct cultural groups that developed their own mathematical knowledge bases.

While making use of *modern mathematics* and science, ethnomathematics also embraces the mathematical ideas, thoughts, notions, and practices as developed by all cultures across time and space. From this perspective, a body of anthropological research has come to focus on both intuitive mathematical thinking and the cognitive processes largely developed in minority[2] cultural groups (Barton 1996). It is a

[2]One of the main characteristics people of minority groups have in common is that they often face discrimination, marginalization and exclusion from society. International human rights laws, which focus on the principle of equality, guarantee the educational right to all people. However, people of many minority groups are likely to be denied their right to education. In this context, the

program, which seeks to study how diverse groups of people understand, comprehend, articulate, process, and use mathematical ideas, procedures, and practices in order to solve problems related to their daily lives.

These basic principles of ethnomathematics define its philosophical and ideological postures, which are the roots of a *holistic theory* whose focus consists essentially of a critical analysis of the generation (creativity) and production of mathematical knowledge and its intellectual processes, as well as its institutionalization (academics) and diffusion through educational process (D'Ambrosio 2006). This *holistic* context includes diverse perspectives, patterns of thought, and histories, the study of the *systems*[3] taken from reality in order to help students to reflect, understand, and comprehend extant relations among all of the components of the system (Rosa and Orey 2010).

Cultural variables have strongly influenced how students come to understand their world and interpret both their own and others people's experiences (D'Ambrosio 1995). In attempting to create and integrate mathematical materials related to different cultures and that draw on students' own experiences in an instructional mathematics curriculum, it is possible to apply ethnomathematical strategies in teaching and learning mathematics. These strategies include, but are not limited to the historical development of mathematics in different cultures that:

> (…) use mathematics (e.g. an African-American biologist, an Asian-American athlete). Mathematical applications can be made in cultural contexts (e.g. using fractions in food recipes from different cultures). Social issues can be addressed via mathematics applications (e.g. use statistics to analyze demographic data) (Scott 1992, pp. 3–4).

Similarly, it is important to show how ethnomathematics describes the ideas and procedures that are implicit in the procedures and practices developed locally by members of distinct cultural groups because their mathematical thinking has been influenced by the vast diversity of human characteristics such as languages, religions, morals, and economical-social-political activities. In concert with these factors, throughout history, D'Ambrosio (2006) argues that humanity has developed logical processes related to universal needs regarding quantification, measurement, modelling, and explanation, all shaped and operating within different social and historical contexts in order to favor respect for solidarity and cooperation with the *others* in school practices.

(Footnote 2 continued)

majority of children who are members of non-traditional, marginalized or out-of-school populations and/or are also deprived of access to formal education that is relevant and responsive to their specific context and needs (UNICEF 2009). For example, globally speaking, there are some educational institutions in which a particular ethnic, racial or cultural group is a majority, thus, the experiences of students of a minority group such as Indigenous, English Language Learners or Special Education are not reflected in the mainstream cultural and educational materials of the broader national sense.

[3]Systems are sets composed of elements taken from reality. The analysis of the interrelationship among these elements seeks to develop reflection, understanding, and comprehension of phenomenon that are part of reality (Rosa and Orey 2016).

The main goal of ethnomathematics is building up a civilization free from truculence, arrogance, intolerance, discrimination, inequity, bigotry, and hatred of the others. In this regard, Western scientific arrogance is a disrespect of and outright refusal to acknowledge cultural identities by scientists and mathematicians that puts all processes of understanding and comprehension of many non-Western cultural systems at risk. These particularities should not be ignored and they should be respected when individuals attend school because this aspect gives confidence and dignity to students when their previous knowledge is acknowledged (D'Ambrosio 1985).

According to this context, Rosa and Orey (2015) state that ethnomathematics program can be considered as a creative insubordination movement in mathematics education because it caused the disruption of the existing order in academic mathematics by developing the study of ideas, procedures, and mathematical practices that are found in various and specific cultural contexts outside of mainstream science and academia. In this regard, this program broke the norms and bureaucratic rules of academic mathematics in order to recognize different ways and value diverse modes of producing mathematics in other cultures. The introduction of ethnomathematics by D'Ambrosio has played a large part in creating an environment that was receptive to the social turn (Lerman 2000, as cited in Alangui and Rosa 2016), which allowed mathematics educators to ask questions, discuss, and create views that challenge conventional notions about the nature of mathematics.

Since ethnomathematics is associated with the pursuit of peace, the challenge that many communities and school systems face today is in determining how to shape a new open, modern, international culture, which integrates and respects new and alternative ideas, and where diverse ideas coexist in balance with those of western science. This also includes an increased cultural, ethnic, and racial diversity. In this respect, D'Ambrosio (2009) states that:

> Education is a practice present in every culturally identified group. The major aims of education are to convey to new generations the shared knowledge and behavior and supporting values of the group, and, at the same time, to stimulate and enhance creativity and progress (p. 242).

Indeed, the most creative, dynamic, and productive societies develop these features accordingly (Florida 2004). The inclusion of moral consequences into mathematical-scientific thinking, ideas, procedures, and experiences as we explore the mathematics found in different cultural contexts is vital. Thereby, Rosa and Orey (2016) argue that it is important to acknowledge contributions that members from diverse cultural groups make to mathematical understanding, the recognition and identification of diverse mathematical practices in varied contexts, and the link between academic mathematics and student experiences should all become central ingredients to a complete study of mathematics. This is one of the most important objectives of an ethnomathematics perspective in the mathematics curriculum development.

This perspective is crucial in giving students a sense of cultural ownership of mathematics, rather than a mere gesture toward inclusiveness (Rosa and Orey 2010). An essential aspect of this program includes an ongoing critical analysis of the generation of mathematical knowledge as well as the intellectual processes of this production that seeks to explain, understand, and comprehend mathematical procedures, techniques and abilities through a deeper investigation and critical analyses of students' own cultures (D'Ambrosio 2016).

For example, in Brazil, the use of designs found in *Bakairi* body painting in indigenous schools has facilitated the comprehension of spatial relations such as form, texture, and symmetry, which enables the construction and systematization of students' geometrical knowledge (Rosa 2005). In this regard, ethnomathematics rescues the ancestral knowledge of the environment in order to generate models related to the environmental balance such as the patterns used for sowing or fishing. Therefore, ethnomathematics attends to the sociological and anthropological dynamics of the differentiated groups since it contributes to the raising of awareness about the elements of the socio-environmental context that is part of the identity of the peoples.

This approach allows students to experience mathematical language and learn geometric concepts through the study of cultural artifacts (Rosa 2005) that can be considered as the "physical or expressions of a specific culture and they include but are not limited to food, clothing, tools, art, and architecture" (Rosa and Orey 2012, p. 194). Therefore, cultural artifacts are viewed as inventions and dependent entities that exist a priori waiting to be discovered. Since mathematics is considered a social and cultural product, these artifacts are embedded in and embody different worldviews (Alangui and Rosa 2016).

Consequently, it is important to uncover tacit mathematical knowledge implied in the creation of cultural artifacts, which reveal the development of daily activities since these objects are "created by the members of cultural groups, which inherently give cultural clues and information about the culture of its creators and users" (Rosa and Orey 2012, p. 194). The use of cultural artifacts from diverse groups in educational settings raises students' self-confidence, enhances and stimulates creativity, creates a sense of connection, and promotes cultural dignity (D'Ambrosio 2006).

It is necessary to connect mathematics with students' own communities. Ethnomathematics studies multifaceted strategies and processes applied in these cultural artifacts in order to reveal techniques and identify cognitive processes that are in unremitting dynamism with the nature of mathematics. Studies have shown sophisticated mathematical ideas and practices that include, for example, geometric principles in craftwork, architectural concepts, and practices in the activities and artifacts of many indigenous, local, and vernacular cultures (Eglash et al. 2006; Orey 2000; Urton 1997).

Ethnomathematics reveals the importance of mathematical ideas and practices in developing local cultures because "decorative and symmetrical patterns displayed on these cultural artifacts are also expressions of beliefs, values, taboos, and religion of the identifiable people whose culture they represent" (Rosa and Orey 2012, p. 195). Therefore, it is important to highlight that the conduction of ethnomathematical investigations makes possible to describe how mathematics has been used to create cultural artifacts (Barton 1996). Mathematical concepts related to a variety of mathematical procedures used in cultural artifacts form part of the numeric relations found in universal actions of measuring, calculation, games, divination, navigation, astronomy, and modelling (Eglash et al. 2006).

This program takes into consideration the diverse processes that help in the construction and development of scientific and mathematical knowledge that includes collectivity, and the overall sense of and value for creative and new inventions and ideas. Accordingly, Rosa and Orey (2015) argue that ethnomathematics is a body of knowledge often built up by the members of distinct cultural groups over time and across generations of living who are in close contact with their own historical, social, cultural, and natural environment.

From this perspective, it is necessary to focus on both the intuitive mathematical thinking and the cognitive process that are largely developed in local cultures in order to study how students have come to understand, comprehend, articulate, process, and ultimately use mathematical ideas, concepts, and practices that they use to solve problems faced in their own contexts.

Ethnomathematics allows us to understand mathematics as a social science that contributes to the sociocultural development of communities, to the generation and enrichment of knowledge, and to the development of dialogue between academia and society. Likewise, it can enhance the identity traits of peoples by its incorporation into curricular innovation through its didactic and pedagogical action. This approach helps to enrich cultural autonomy and idiosyncrasy in order to promote respect for diversity since mathematics is perceived as a social phenomenon and as a human activity.

1.3 Discussing the Role of Ethnomathematics in Mathematics Education

The world's economy is globalized, yet traditional academic mathematics curricula neglects, indeed often rejects, the truly diverse contributions made by members of colonized and non-dominant cultures. Consequently, the role of ethnomathematics in mathematics education is primeval in recognizing the emergence of perceptions of space, time, and techniques of observing, comparing, classifying, ordering, measuring, quantifying, inferring, and modelling that are different styles of abstract thinking in the school curricula.

Ethnomathematics as a line of study and research of mathematics education, which investigates the roots of mathematical ideas and practices, starting from the way individuals behave in different cultural groups. It is a contemporary pedagogical trend in education that offers more inclusive, new and greatly expanded definitions of a given group's particular mathematical-scientific contributions. It attempts to identify mathematical practices that begin in the knowledge of the *others* as well as in their own rationality and terms.

This program promotes the integration of socially vulnerable sectors of society and encourages sustainable academic practices. It also recognizes that members of distinct cultures and different groups develop unique mathematical techniques, methods, and explanations that allow them to develop alternative understandings and social transformations in order to contribute to the achievement of social justice, peace, and dignity for all.

Pedagogically, ethnomathematics enables school mathematics to be seen as the "process of inducting young people into mathematical aspects of their culture" (Gilmer 1990, p. 4). An ethnomathematics perspective has reshaped cultural identity in a positive way by requiring the inclusion of a greater representation of practices and problems of a student's own community. The application of ethnomathematics as pedagogical action restores a sense of enjoyment or engagement and can enhance creativity in doing of mathematics (Rosa and Orey 2015). An ethnomathematics program helps both educators and students alike to understand mathematics in the context of ideas, procedures, and practices used in the day-to-day life. It further encourages an understanding of professional practitioners, workers, and academic or school mathematics. Such depth is accomplished by taking into account historical evolution and the recognition of natural, social and cultural factors that shape human development (D'Ambrosio 2006).

One of the most relevant reasons for teaching mathematics involves the consideration of mathematics as an expression of human development, culture, and thought, which is an integral part of the cultural heritage of humankind (D'Ambrosio 1995). Therefore, it is necessary to start by using sociocultural contexts, realities, and the many interests and needs of students and not mere enforcement of a rigid set of external values or often-decontextualized curricular activities. An ethnomathematics perspective in mathematics education enables educators to rethink about the nature of mathematics in order to acknowledge that diverse people, despite their formal schooling experiences, actually come to measure, classify, order, organize, infer, model, and reason with numbers, algebra, and visuospatially. These mathematical activities occurs in everyday living, which are important aspects of diverse modes of teaching and learning of mathematics.

This pedagogical approach helps us to identify mathematical elements of students' daily, which can help to improve students' self-perception and motivation towards mathematics. It is important that educators are concerned about respecting students' cultural background. In order to do so, they must establish respectful dialogues with students (Shirley and Palhares 2016). The importance of integrating an ethnomathematics perspective into the mathematics curriculum is related to its social impact in relation to the changes in the students' attitudes towards to their own cultural backgrounds. For this reason, the role of educators has become even more central in this highly multiethnic society increasingly integrated to a globalized world.

Mathematical ideas, procedures, and practices developed by the members of specific groups that occur in daily life strengthen their own cultural identity in relation to other cultures, which allows them to reflect on the role of mathematics in determining cultural otherness. These ideas strengthen cultural autonomy in different groups because they address the challenges posed by UNESCO (2012) in respect to the attention to diversity, to the avoidance of exclusion and isolation, to help students understand others' cultural contributions; and finally to show mathematics as a human activity.

When the focus of a study is the pedagogy of mathematics, the attention has been centered both around legitimizing the students' knowledge through the use of culturally relevant activities, grown from experiences built in their own ways and around the study of the possibilities of how to work with the learning of the ones outside the school and the ones inside the school. This means that "Curricular activities developed according to principles of culturally relevant pedagogy focus on the role of mathematics in sociocultural contexts. These activities involve ideas and procedures associated with ethnomathematical perspectives to solve problems" (Rosa and Gavarrete 2016).

Indeed, with a discussion of ethnomathematics helps educators to establish cultural models of beliefs, thought and behavior, in the sense of contemplating not only the potential of the pedagogic work that takes into account the knowledge of the students, but also a learning the school, which is more meaningful and empowering. It is important to understand how to make mathematics meaningful for students by using cultural artifacts found in distinct cultures. Therefore, it is necessary to develop contextualized mathematical activities in order to improve mathematical performance of the students by applying culturally relevant pedagogy into the mathematics curriculum.

The introduction of abilities and competencies into the mathematics curriculum is part of a worldwide trend. A focus on abilities and competencies emphasizes the individuals and tends to remove mathematics from its context. Ethnomathematics, on the other hand, emphasizes the communal and tends to connect mathematics with its own contexts. An ethnomathematical aspect may therefore provide a necessary balance to the new curriculum. If these two components are to be brought together, then we need to conceive ethnomathematics as an overarching aspect of this curriculum by humanizing mathematics. However, we are unsure exactly what this might mean, but perhaps ethnomathematics can be considered as a philosophical approach to the mathematics curriculum, or a context for it, or perhaps an affective or attitudinal response to the educations demands of the students.

It is necessary to propose a discussion about cultural relevance into the mathematics curriculum in order to help educators to acknowledge the relationship between cultural and school mathematical knowledge (Rosa and Gavarrete 2016). This approach fosters a critical and reflective attitude about the universality and contextualization of mathematical knowledge, since pedagogical work with ethnomathematics promotes educators' creativity when developing a mathematics curriculum that is connected to the social and cultural environments of the students.

Recent trends in mathematics education offer some hope for changing the role of mathematics that respects distinct and diverse contexts. In particular, the study of ethnomathematics can help educators to connect school mathematics to the students and their communities because culture influences the development of students' mathematical knowledge. Various educational initiatives brought to light the continuing need to develop mathematics lessons that are culturally relevant for students. Therefore, it is necessary to reiterate the importance of promoting sociocultural approaches in the mathematics curriculum to combat curricular *decontextualization*[4] resulting from its monocultural view (Rosa and Gavarrete 2016).

This pedagogical action can be achieved by the application of *innovative approaches*[5] to ethnomathematics such as the trivium curriculum and ethnomodelling, which need more investigations to address the pedagogical purposes of the ethnomathematics program:

(1) The Trivium curriculum is composed of *literacy*, *matheracy*, and *technoracy* that allows for the development of school activities based on an ethnomathematics foundation. Literacy is the capacity students have to process information present in their daily lives; matheracy is the capacity students have to interpret and analyze signs and codes in order to propose models to find solutions for problems faced daily; and technoracy is the capacity students have to use and combine different instruments in order to help them to solve these problems (Rosa and Orey 2016).

(2) In the ethnomodelling approach, the use of ethnomathematics assumptions and the application of tools and techniques of mathematical modelling allow us to perceive reality by using different lenses, which gives us insight into mathematics performed in a holistic way. Ethnomodelling is a research paradigm related to critical-reflective dimensions of learning that allows learners the opportunity to develop a sense of purpose and their own potential by using mathematics to examine and solve problems they themselves choose and deem important (Rosa and Orey 2016).

Lastly, it is important to emphasize the in an increasingly *glocalized*[6] and interdependent world, it is fundamental that educators are given experiences that allow them to understand that the diversity of ideas and thoughts that come into contact either through communications, business, education, and science are greatly influenced by the way in which individuals who belong to different cultural groups learn mathematics. This pedagogical approach is not often reflected in the traditional mathematics classroom, yet high equitable expectations along with personalized connections in mathematics instruction are essential for success for all students.

In this way, the pedagogical innovation proposals raised from the ethnomathematics program can encompass the incorporation of sociocultural components in teacher training programs, which envisions mathematics as a human activity in all cultures as well as a social phenomenon, thus, contributing to a functional vision of

---

[4]Decontextualization is a consciously or subconsciously process of examining or interpreting mathematical ideas, procedures, and practices separated from the sociocultural context in which they are embedded.

[5]Other innovative approaches of ethnomathematics are: social justice, civil rights, indigenous education, professional contexts, game playing, urban and rural contexts, ethnotransdisciplinarity, ethnopedagogy, ethnomethodology, and ethnocomputing (Rosa and Orey 2016).

[6]*Glocalization* is the acceleration and intensification of interaction and integration among members of distinct cultural groups (Rosa and Orey 2016).

the notion of mathematical ideas and procedures (D'Ambrosio 2006) and, thereby, allowing the value of mathematical practices to be strengthened (Bishop 1988).

Accordingly, it is necessary to debate about issues regarding mathematics education and mathematical knowledge developed by the members of a specific cultural group. However, the discussions surrounding these issues do not imply that ethnomathematics is only an instrument to improve mathematical education because it also has a role in helping us to clarify the nature of mathematical knowledge and of knowledge in general. In so doing, we recommend to shift ethnomathematical research from theoretical issues toward educational and practical affairs.

1.4 Acknowledging Some Opportunities and Challenges…

After three decades since the emergence of the ethnomathematics as a program, some investigators, philosophers, and educators are still not comfortable about the role of ethnomathematics in mathematics education. Furthermore, they must be sure about some ethnomathematical features that are unrelated to the mathematics classrooms because:

1. Ethnomathematics is not a cultural content, thus, it is necessary to question: Which culture(s) should be in the mathematics curriculum? How would we know whether or not the educators and students make the link between culture and mathematics?
2. Ethnomathematics is not an imperative to take an account of the sociocultural contexts in schools, which has been known for many years and discussed in many different ways. Hence, how is it possible to develop an ethnomathematical program in the context of mathematical education that involves the characteristics of the whole school system?
3. Ethnomathematics should not be confused with ethnic-mathematics since its ethnic component is the *ethno*-graphic study of mathematical ideas, procedures, and practices developed by the members of distinct cultural groups and it is based on gathering empirical data on the form of mathematics practiced in diverse cultures. How can we know the extent of the ethnographic studies? What should we look for in the data?
4. Ethnomathematics cannot be considered as a discipline because it is not just a process of teaching a set of frozen mathematical theories. It proposes a lively and dynamic pedagogical action that deal with environmental, social, political, cultural, an economic contexts, which have important mathematics components. How can an ethnomathematical attitude help develop a critical and reflective understanding of the mathematics curriculum?

This kind of investigations and pedagogical actions underscore the importance of doing the ethnomathematical work first in order to come to a good understanding of the mathematical aspects of culture by having a clear purpose to the educational activities. For this reason, the implantation and implementation an ethnomathematical perspective in classrooms must be preceded by a full investigation on the mathematical ideas, procedures, and practices developed by the members of the diverse cultural groups. Consequently, the ethnomathematical perspective must be clearly situated within the existing school curriculum and is intended to enhance the learning of mathematics.

It is necessary to understand the pedagogical action of ethnomathematics in schools. Classroom investigations using an ethnomathematics perspective is important because they are where D'Ambrosio's (1985) vision must be implemented. If starting with the students' own realities does not work, for whatever reason, it is necessary to think about how further investigations can help us to understand about the problems in using this approach. There are many arguments, but it is important to focus on the reasons to implement this program in the classrooms: (a) to be an effective path to traditional mathematics, (b) to be a way to develop intercultural classrooms, and (c) to be a way to transform the relationship between mathematics and society. Meanwhile, whatever the reason is, it is necessary to communicate this purpose in order to obtain the *buy-in* from the students. The critical education ideas of Freire (1973) are one way to develop this approach.

The suggestion of starting with the student's community sociocultural reality is another way to develop this approach, but they may refuse to study their own reality because it is oppressive. They do not identify this context as mathematics, and they may already have a grounded mathematical conception. Perhaps, in this case, educators should start with students' existing mathematical conceptions, even though if they are traditional, provided that what follows is their critical and reflective examination. The consequence of this approach for teacher education is significant. It means that educators must know more about mathematics and additional pedagogical skills in order to help students to undertake a critical and reflective examination of these mathematical conceptions. In this way, ethnomathematics is a tool to generate social awareness and promote processes of emancipation or empowerment of communities for justice and peace.

As a consequence, ethnomathematics becomes a high order task in this pedagogical action. The consequence of these ideas is that ethnomathematical work in the schools is not a simplistic presentation of cultural examples, nor situating mathematics in cultural contexts. Rather, it requires considerable background work, complete understanding, and pedagogical sophistication. This is a complex task; it takes time, and is difficult to access all of it is possibilities. According to Shirley (2015), one possible way to avoid this problem in order to bring the goals of ethnomathematics even more directly to students, is to encourage them to study problems taken from their own individual cultures, heritage, and personal interests. Perhaps, the importance of an ethnomathematical perspective in the schools is that it alerts us to the way in which cultural information can be used in the classrooms.

In this environment, ethnomathematics is considered as a way to contextualize mathematical ideas since it is related to the techniques developed as a study of mathematical procedures practiced by the members of distinct cultural groups. The idea of mathematics in cultural practices involves designing tasks that are contextualized in the cultural heritage based on different ways of knowing in order to help us to reflect on certain mathematical notions as well as on the nature of mathematical knowledge. However, how this pedagogical work is realized in the classrooms may be problematic in regards to misconceptions related to ethnomathematics as a program. Under this circumstance, a modified mathematics curriculum is necessary because it admits a wider possibility for mathematical thinking and investigation that is based on diverse cultural practices. Yet, the challenge is how to enlist this cultural strength in schools.

It is recommended that investigators conduct studies that help us to understand know how ethnomathematics can contribute to the development of contextualized activities in classrooms. Thus, it is important to recognize how investigators identify ethnomathematical forms of mathematics, but there is no acknowledgement of the pedagogical actions linked to the educational environment. Centered on this possibility and questioned whether it is possible to infer the causes for the difficulties of the implementation of an ethnomathematical perspective into the mathematics curriculum, it is necessary to identify if educators are locating the problem outside the schools. The answers may be responded in many ways, one of which may come from the teacher's point of view of trying to observe and understand students' own reality.

However, in order to achieve this goal, it is necessary to promote a change in teacher education programs and bring prospective teachers to understand sociocultural realities in which they are professionally involved. This is because reflections on mathematics, culture, education, and society as well as on the relations maintained among them can be oriented towards inclusive pedagogical practices in which learning mathematics addresses deeper notions of equity (Gavarrete 2015). Therefore, the idea of incorporating the pedagogical action of ethnomathematics in educational programs should be reinforced because it poses new challenges for research in mathematics education. In this context, ethnomathematics constitutes a prospective vision of an innovative research paradigm and at the same time contributes to determining the role of this program in mathematics education.

For example, in mathematics education, if the focus is the pedagogy of mathematics, the attention must be centered both around legitimizing the students' knowledge, grown from experiences built in their own ways and around the study of the pedagogical possibilities of how to work with the learning process that occurs outside and inside of the school environments. A discussion of the educational aspects of ethnomathematics helps educators to establish cultural models of beliefs, thought and behavior, in the sense of contemplating the potential of the pedagogical work that takes into account the tacit knowledge of the students as well as the development of mathematical learning process that is more meaningful and empowering. Making the practical from the theoretical happen in classrooms is one of the fundamental principles of the pedagogical action of the ethnomathematics as a program.

This debate alerts us that there is a need to for continued dialogue and development of ethnomathematical curricular activities in order to rescue their cultural heritage and support learning as well as linguistic and social inclusion in schools. Ethnomathematics is a response to this important social justice issue. However, it is necessary that educators understand: (a) how they can help students in their fight against social injustices, (b) how teacher education can be part of the fight for social justice, and (c) how to prevent a domination of Western mathematics in the school curricula. Hence, it is necessary to consider the danger of assimilationist experiences by the students. In this context, whatever may happens, the members of each cultural group need to be in control of their own willingness to further develop this approach.

In continuing this theoretical debate, and what is often difficult for educators, is how to connect to what Freire (1985) would consider most fundamental, the community, to what has become almost universal, the formalized academic mathematics. Hence, another principle of ethnomathematics is to stretch the limits of what is perceived as mathematics and its related thinking in order to link it to the *greater* mathematical knowledge. In this case, ethnomathematics reminds us to look at the larger mathematical ideas and conceptions not just to search for isolated procedures and practices.

In closing, ethnomathematics helps us to establish a meta-awareness of the role of mathematical knowledge in the society and cultural context in which it manifests itself. In this respect, ethnomathematics is also reciprocal because it is possible to think of conventional mathematics and its role within its host cultural group. Indeed, this reciprocity is a vital aspect of ethnomathematics.

References

Alangui, W. V., & Rosa, M. (2016). Role of ethnomathematics in mathematics education. In M. Rosa, U. D'Ambrosio, D. C. Orey, L. Shirley, W. V. Alangui, P. Palhares, & M. E. Gavarrete (Eds.), *Current and future perspectives of ethnomathematics as a program* (pp. 31–37). ICME13 Topical Surveys. London, England: SpringerOpen.

Barton, B. (1996). Making sense of ethnomathematics: Ethnomathematics is making sense. *Educational Studies in Mathematics, 31*(1–2), 201–233.

Bishop, A. J. (1988). *Mathematical enculturation*. Dordrecht, The Netherlands: Reidel.

D'Ambrosio, U. (1985). Ethnomathematics and its place in the history and pedagogy of mathematics. *For the Learning of Mathematics, 5*(1), 44–48.

D'Ambrosio, U. (1995). Multiculturalism and mathematics education. *International Journal on Mathematics Science, and Technology Education, 26*(3), 337–346.

D'Ambrosio, U. (2006). *Ethnomathematics: Link between traditions and modernity*. Roterdam, The Netherlands: Sense Publishers.

D'Ambrosio, U. (2009). A nonkilling mathematics? In J. E. Pim (Ed.), *Toward a nonkilling paradigm* (pp. 241–268). Hawaii, Honolulu, HA: Center for Global Nonkilling.

D'Ambrosio, U. (2016). An overview of the history of ethnomathematics. In M. Rosa, U. D'Ambrosio, D. C. Orey, L. Shirley, W. V. Alangui, P. Palhares, & M. E. Gavarrete (Eds.), *Current and future perspectives of ethnomathematics as a program* (pp. 5–10). ICME13 Topical Surveys. London, England: SpringerOpen.

Eglash, R., Bennett, A., O'Donnell, C., Jennings, S., & Cintorino, M. (2006). Culturally situated designed tools: Ethnocomputing from field site to classroom. *American Anthropologist, 108*(2), 347–362.

Florida, R. (2004). *The rise of the creative class and how it is transforming work, leisure, community and everyday life*. New York, NY: Basic Books.

Freire, P. (1973). *Education as the practice of freedom in education for critical consciousness*. New York, NY: Continuum.

Freire, P. (1985). *The politics of education: Culture, power, and liberation* (D. Macedo, Trans.). New York, NY: Bergin & Garvey.

Gavarrete, M. E. (2015). The challenges of mathematics education for Indigenous teacher training. *Journal of Intercultural Education, 26*(4), 326–337.

Geertz, C. J. (1973). Thick description: Toward an interpretative theory of culture. In C. Geertz (Ed.), *The interpretation of culture: Selected essays* (pp. 3–30). New York, NY: Basic Books.

Gilmer, G. (1990). An ethnomath approach to curriculum development. *ISGEm Newsletter, 5*(2), 4–5.

Headland, T. N., Pike, K. L., & Harris, M. (1990). *Emics and etics: The insider/outsider debate*. Newbury Park, CA: Sage Publications.

Meaney, T., Fairhall, U., & Trinick, T. (2008). The role of language in ethnomathematics. *The Journal of Mathematics and Culture, 3*(1), 52–65.

Orey, D. C. (2000). The ethnomathematics of the Sioux tipi and cone. In H. Selin (Ed.), *Mathematics across culture: The history of non-western mathematics* (pp. 239–252). Dordrecht, The Netherlands: Kluwer Academic Publishers.

Rosa, M. (2005). The ethnomathematics of Bakairi body painting. In *NCTM Annual Meeting and Exposition: Embracing Mathematical Diversity*. Anaheim, CA: NCTM.

Rosa, M., & Gavarrete, M. E. (2016). Polysemic interactions between ethnomathematics and culturally relevant pedagogy. In M. Rosa, U. D'Ambrosio, D. C. Orey, L. Shirley, W. V. Alangui, P. Palhares, & M. E. Gavarrete (Eds.), *Current and future perspectives of ethnomathematics as a program* (pp. 23–30). ICME13 Topical Surveys. London, England: SpringerOpen.

Rosa, M., & Orey, D. C. (2005). Las raízes históricas del programa etnomatemáticas [Historical roots of the ethnomathematics program]. *RELIME, 8*(3), 363–377.

Rosa, M., & Orey, D. C. (2010). Ethnomodelling: A pedagogical action for uncovering ethnomathematical practices. *Journal of Mathematical Modelling and Application, 1*(3), 58–67.

Rosa, M., & Orey, D. C. (2011). Ethnomathematics: The cultural aspects of mathematics. *Revista Latinoamericana de Etnomatemática, 4*(2), 32–54.

Rosa, M., & Orey, D. C. (2012). An ethnomathematical study of the symmetrical freedom quilts. *Symmetry: Culture and Science, 23*(2), 191–220.

Rosa, M., & Orey, D. C. (2015). A trivium curriculum for mathematics based on literacy, matheracy, and technoracy: An ethnomathematics perspective. *ZDM Mathematics Education, 47*(4), 587–598.

Rosa, M., & Orey, D. C. (2016). Innovative approaches in ethnomathematics. In M. Rosa, U. D'Ambrosio, D. C. Orey, L. Shirley, W. V. Alangui, P. Palhares, & M. E. Gavarrete (Eds.), *Current and future perspectives of ethnomathematics as a program* (pp. 18–23). ICME13 Topical Surveys. London, England: SpringerOpen.

Scott, P. J. (1992). Curriculum, resources, and materials for multicultural/multilingual classrooms. *ISGEm Newsletter, 8*(1), 3–5.

Shirley, L. (2015). Mathematics of students' culture: A goal of localized ethnomathematics. *Revista Latinoamericana de Etnomatemática, 8*(2), 316–325.

Shirley, L., & Palhares, P. (2016). Ethnomathematics and is diverse pedagogical approaches. In M. Rosa, U. D'Ambrosio, D. C. Orey, L. Shirley, W. V. Alangui, P. Palhares, & M. E. Gavarrete (Eds.), *Current and future perspectives of ethnomathematics as a program* (pp. 13–17). ICME13 Topical Surveys. London, England: SpringerOpen.

UNESCO. (2012). *Challenges in basic mathematics education*. Paris, France: United Nations Educational.

UNICEF. (2009). *State of the world's minorities and indigenous peoples*. Minority Rights Group International. London, England: MRGI.

Urton, G. (1997). *The social life of numbers: A Quechua ontology of numbers and philosophy of arithmetic*. Austin, TX: University of Texas Press.

Part II
Research Approaches on Ethnomathematics: Collection of Field Data

As a research program, ethnomathematics seeks to understand the many diverse processes of thinking and explaining. Because it is critical of formal school orientations not related to sociocultural and political aspects of mathematics, it is useful to understand how current research in ethnomathematics might offer mathematics educators innovative methodological alternatives. One of the motivating aspects of the work produced by researchers is to understand mathematical *knowing/doing* throughout the history of humanity in distinct contexts such as found in our communities. It is also important to emphasize the dynamic character of this research program since it is open to new foci, methodologies, and worldviews.

Chapter 2
Weaving Culture and Mathematics in the Classroom: The Case of Bedouin Ethnomathematics

Miriam Amit and Fouze Abu Qouder

Abstract Our study attempted to address young Bedouin students' persistent difficulties with mathematics by integrating ethnomathematics into a standard curriculum. First, we conducted extensive interviews with 35 Bedouin elders to identify the mathematical elements of their daily lives—particularly traditional units of length and weight. We then combined these with the standard curriculum to make an integrated 30-hour 7th grade teaching unit that was implemented in two Bedouin schools. Comparisons between the experimental group (75) and the control group (70) showed that studying the integrated curriculum improved the students' self-perception and motivation, but had almost no effect on achievements in school tests that were conducted immediately after the experiment. The experiment had an extra social impact, changing students' attitudes to their own culture and the tribe's older generation.

Keywords Ethnomathematics · Integrated curriculum · Motivation and self-esteem · Multicultural mathematics · Non-standard units · Teaching experiment

2.1 Introduction

In the past few decades, the effectiveness of mathematics education has been hotly debated in many countries around the world (Keitel et al. 1989). Mathematics, more than any other topic, is perceived as being free of the influence of culture and values, so many mathematics educators believe that it does not have to take the diversity of the students who learn it into account. The fact that this view is

M. Amit (✉) · F. Abu Qouder
Ben-Gurion University, Beer-Sheva, Israel
e-mail: amit@bgu.ac.il

F. Abu Qouder
e-mail: abuceoud@gmail.com

M. Rosa et al. (eds.), *Ethnomathematics and its Diverse Approaches for Mathematics Education*, ICME-13 Monographs,
DOI 10.1007/978-3-319-59220-6_2

problematic, and that it has contributed to the varying effectiveness of the education system, has been most definitively proven (Bishop 1988; Gilmer 1990; Powell and Frankenstein 1997). In countries with diverse populations, like Brazil (D'Ambrosio 1985), educators have called for the recognition of the fact that mathematics are a cultural product, and that elements like the students' ethnicity can be a factor in how it is learnt (Presmeg 1998).

Does teaching mathematics in combination with traditional cultural elements and values help students understand various mathematical topics? Can including elements from students' everyday lives in their mathematics education encourage more meaningful learning and more effective achievements? Can it increase students' motivation to learn? The study presented here addresses these questions in the context of a particular minority group that is currently struggling with the study of mathematics, the Bedouin population of Israel's Negev Region.

Surveys and national tests conducted by the Ministry of Education over the past 20 years have shown that Bedouin students tend to score very low in mathematics at all ages. Various studies have been conducted in an attempt to determine the causes of their difficulty, and to come up with an effective means of overcoming it (Amit et al. 2008). Though in the last few decades many changes have been made to the country's national mathematics curriculum in an effort to develop teaching strategies that promote the mathematics education of all students, teachers must still contend with the extensive difficulty of conveying the material so that students are able to understand it, and with their students' low levels of motivation to learn.

These problems, still prevalent throughout the educational system, are particularly acute amongst the Bedouin students. Our study describes an attempt to address these difficulties by means of an ethnomathematical teaching unit, which is designed to increase Bedouin students' interest in and motivation to learn mathematics by highlighting the presence of mathematics in their own culture and emphasizing its relevance to their daily lives. The unit is designed not to *replace* the contents of the standard mathematics curriculum, but rather to *supplement* them with content drawn from the students' own culture.

The development of the unit therefore began with the gathering of information regarding mathematical concepts traditionally used by the Bedouins (in this case, concepts related to measuring length and weight). These concepts were then incorporated into the standard 7th grade teaching unit to create an integrated unit. The unit was taught to an experimental group of 70 students, after which its effect on their motivation, self-perception and achievements was assessed.

2.2 Context

Ethnomathematics is a term that arose in the 1980s out of the awareness of the connection and mutual interaction that exists between mathematical culture and political culture. It has been defined as the connection between various branches of mathematics, various fields in which mathematics are used, the historical roots of mathematical content and the connections between the *real world* and the world of mathematical work (Powell and Frankenstein 1997). These connections are being acknowledged and studied around the world in projects like *Realistic Mathematics* in the Netherlands (Presmeg 1998).

2.2.1 What is Ethnomathematics?

The *ethno* part of ethnomathematics consists of the language and vocabulary, the behavioral norms and the symbols of certain groups. It is dependent on the culture of a particular group, and is influenced by its historical development and by its accumulated mathematical experience. The *mathematics* part refers to the opinions, understanding, explanation and execution of actions like coding, measurement, sorting, organization, deduction and modeling. Ascher defined ethnomathematics as learning the opinions of the educated people, claiming that, though mathematical opinions are present in all cultures, they differ from society to society in how they manifest and in their cultural content (Ascher and Ascher 1986; Ascher 2000). Anderson (1990) also connects ethnomathematics with mathematical opinions, which can be expressed in different ways: in writing, acts, or speech.

D'Ambrosio, a leading researcher in the field who developed an ethnomathematical study program, claimed that the belief in the universality of mathematics can prevent individuals from examining and identifying different aspects of thought and culture that could lead to different mathematical structures, even at levels as basic as counting, sorting, measuring, deduction, categorization and modeling (1985). He claimed that only by deconstructing Eurocentric assumptions of universality will we be able to achieve anthropological awareness of the fact that different cultures are capable of generating unique mathematical products, and that mathematical culture is susceptible to change over time.

Mathematicians work on local ethnomathematics out of the awareness that every culture constitutes a complex, valuable knowledge system that may have something to teach the rest of the world about its own alternative knowledge forms (Adam et al. 2003). The purpose of ethnomathematics is therefore not to *replace* the so-called *pure*, formal mathematics, but rather to encourage educators to reflect upon the mathematics that they teach, and to be aware that disregarding the cultural context of mathematics completely is not a neutral choice, but one which may be discriminating against certain students and perpetuating a culture of intolerance and inequality (D'Ambrosio 2002).

Some researchers differentiate between two types of mathematics: the *academic* mathematics taught and practiced in schools, and the practical *ethnomathematics* practiced by different cultural sub-groups like national tribes, worker organizations, and even students in the same age group (D'Ambrosio 1994). In this context, ethnomathematics suggests that mathematical experience outside of school should also be addressed as legitimate and pertinent knowledge (Gerdes 1990; Ascher 1991).

Researchers of ethnomathematics have shown that there is a gap between the mathematics learned in school and the mathematics that is relevant to students' everyday lives, and that struggling with the former does not necessarily entail difficulties with the latter. For example, Baba (2002) showed that children who were highly capable of making calculations in the streets were unable to solve similar problems in a classroom mathematics test.

2.2.2 Ethnomathematics in Schools

While ethnomathematics can be defined as the drawing of connections between mathematical content and the culture of the learners, its curricular relevance goes beyond designing study programs that fit the local interests and customs of the culture that studies them. Such a focus might limit the curriculum only to mathematics that the students find relevant or interesting, that they see as connected to their cultural role or vocation (D'Ambrosio 2000). The purpose of education should rather be to provide students with mathematical content and skills that allow them to successfully command modern mathematics, and an ethno-mathematic approach can be a means to achieving this end (Davison 1989).

Various approaches have been suggested for diversifying mathematics education, like Ladson-Billings and Tate's critical race theory (Ladson-Billings and Tate 1995; Tate 1997), which has been effective in improving the achievements of minorities in mathematics. Other studies have shown that students who learn mathematics in an ethnomathematical program do better on standard mathematics tests (Lipka 2002). As Adam et al. (2003) pointed out, integrating mathematical principles and methods from the learners' culture into the formal academic curriculum can help them to draw upon their own mathematical experience in order to better understand and apply mathematical principles.

Drawing on the learners' general mathematical knowledge can help present conventional mathematics more accessibly, and give learners a greater appreciation of its practical value. Davidson (1989) also noted that harnessing cultural values as a means of conveying mathematical content helps to emphasize the relevance of mathematics to the learners' lives, which in turn makes the lesson more interesting and enjoyable.

Everyday applications also spark students' curiosity and motivation to work towards finding a solution to the problem (D'Ambrosio 1987). Other studies have shown that students who are exposed to other mathematical cultures and reflect on them discover that they know more than may be indicated by formal mathematical assessment, a discovery that boosts their self-esteem and motivation to learn (Powell and Frankenstein 1997).

2.2.3 The Bedouins in the Negev

The word *Bedouin* originates in the Arabic word for *desert*, and means *desert dweller*. The Israeli Bedouins are part of the Arab population of Mandatory Palestine that remained in the State of Israel after its foundation in 1948. They are the descendants of a semi-nomadic tribal people who, for hundreds of years, dwelt and roamed in the desert areas of the Middle East and North Africa, including the Negev desert.

While the Bedouins who live in the Negev are Muslims, they are a distinct sub-culture due to their close ties to the desert landscape and the lifestyle that evolved here. The Bedouins of the Negev now constitute an ethnic minority within the State of Israel, geographically and culturally distinct from other Israeli Arabs and Jews (Levinson and Abu-Saad 2004). Their traditional lifestyle was structured around seasonal migration with their herds, with women, children and elders left behind to tend the specific familial territory and men returning to their designated homes periodically in accordance with the seasons.

However, for many Bedouins, this traditional lifestyle has undergone a fundamental change, moving from a nomadic existence based on herding to a far more sedentary one. Such a change has been necessitated by the close proximity and cultural influence (both in Israel and in surrounding countries) of other sedentary populations with vastly different lifestyles, and was further expedited by the resultant sharp decrease in land left available for the Bedouins' use, as areas they had been accustomed to live on were reallocated by the state for other uses (Manor-Rosner et al. 2013). These changes in lifestyle have, in turn, begun to generate additional changes in the Bedouin social structure, customs and cultural values, as the day to day experience of the younger generation drifts further away from that of their elders and ancestors.

Despite this gradual move towards Israel's modern cultural mainstream, the Bedouin population of the Negev is still quite distinct from the rest of the country's inhabitants in several ways. First, it is still strongly characterized by a distinct set of social and cultural norms that are deeply rooted in long-held Bedouin tradition. These include, for example, adhering to a certain style of dress and maintaining a variety of traditional views and customs regarding marriage, family unity and the respective roles and status of men and women. Second, the Bedouin population of Israel remains distinct in that both its socioeconomic status and its level of education tend to be amongst the lowest in the country (Ministry of Education 2013).

Many older Bedouins suffer from financial hardship, and are either unemployed or working in low paying jobs. Most of the older adult population has very little formal education. The younger generation often has difficulty in school, and those who graduate high school are often unable to meet the academic criteria for acceptance into institutes of higher education. As a result, they often enter the unskilled job market directly out of high school, or move straight into the ranks of the unemployed. In recent years, extensive efforts have been made (both by top-down government funded initiatives and by grassroots movements from within the Bedouin community itself) to find ways of breaking this cycle of hardship through action on a variety of fronts, including employment, infrastructure, welfare and education.

2.3 Methodology

The main goal of this study was to uncover ethnomathematical elements in the Negev Bedouin community, to develop teaching units that integrate these elements into the school curriculum, and to assess the affective and cognitive impact of those units on the Bedouin students who study them.

2.3.1 Research Questions

The study consisted of the following five stages:

Stage 1: Identifying various ethnomathematical elements in the Negev Bedouin community through a series of interviews and conversations with village elders.
Stage 2: Analyzing these elements and organizing them into formal mathematical categories in preparation for use in an integrated teaching unit.
Stage 3: Creating teaching units that combine the ethnomathematical elements from stages 1 and 2 with elements from the standard national mathematics curriculum.
Stage 4: Implementation: teaching the integrated units to Bedouin junior high school students.
Stage 5: Analysis: assessing the impact of the integrated units by comparing the achievements and attitudes of the students who learned them to those of a control group.

It is important to note at this point that this paper presents only part of the study's full findings. The data from Stage 1 yielded material for three teaching units, but due to the limited scope of this paper, it will present only the results pertaining to the first of these: the ethnomathematical unit designed to teach measures of length and weight.[1] Specifically, this paper will address the following questions:

1. What ethnomathematical elements and authentic mathematical strategies can we identify in the Bedouin community that pertains to units of measurement?
2. How, and to what extent, does the implementation of an ethnomathematical teaching unit influence students':

[1]The two additional units were designed for and implemented with students in the 8th grade. They addressed 2D geometry, as reflected in embroidery, and 3D geometry, as reflected in the traditional tents.

(a) Cognitive aspects, such as achievements, solution strategies?
(b) Affective aspects, such as self-confidence, attitude towards mathematics, attitude towards school and society?

2.3.2 *Study Population*

This study included two separate study populations, one for the gathering of the ethnomathematical data, and another for the implementation of the teaching unit.

2.3.2.1 Population for Stage 1: Identifying Ethnomathematical Elements

The first stage of the study was conducted with the general population of one of the largest and most prominent tribes among the Bedouins in the Negev Desert, which will be referred to in this paper only as the *tribe*. Specifically, it focused on 35 of the tribal elders: sheiks, teachers, religious teachers, and other members of the older generation, who are well acquainted with the customs of the Bedouins in the Southern Region. A widespread search was conducted in this community to uncover the system of concepts and tools used in these Bedouins' daily life to address issues relating to measures for length and weight. The data was collected through videotaped personal interviews due to the elder generation in the tribe's inability to read or write.

Several points are worth noting about this particular population. First, videotaping is not common among the Bedouins for cultural reasons, particularly amongst the older generation, so some of the people interviewed for this study refused to be videotaped. Second, the traditional measurement units that were the focus of this study's interest are not part of the tribe's daily life, and are relatively marginal in comparison to other things. This means that, while the phenomenon is still in use as a preservation of their cultural character, many of the younger generation are unfamiliar with these concepts, even though they are from their own society. Thirdly, our survey of the tribespeople showed that their society is changing as it undergoes a process of modernization. These Bedouins are transitioning from their traditional existence as nomads, who continually move from place to place in search of water and pastures, and becoming modern, sedentary Bedouins, who settle permanently in a single territory. This change is impacting their social structure, their cultural and economic situation, and the patterns according to which they live their lives.

2.3.2.2 Population for the Implementation of the Weights and Measures Study Unit

The population for this part of the study consisted of two groups of 7th grade students, a control group and an experimental group. All of the students were from the same tribe, and attended one of two local junior high schools. Each group was made up of students from two different classes (i.e. one class from each school in every group) but this division is not reflected in the data analysis, which combines the classes into a control group and an experimental group.

The *control group* consisted of 70 students (34 boys and 36 girls). The students in this group learned about weights and measures according to the standard national curriculum provided by the Ministry of Education. In other words, they were not introduced to the ethnomathematical study unit.

The *experimental group* consisted of 75 students (38 boys and 37 girls). The students in this group learned about weights and measures according to the ethnomathematical study unit that was developed for this study. This unit took up about 30 h.

2.3.3 Data Collection and Analysis

Data for the first stage, which addresses question one, was gathered through video or audiotaped personal interviews and conversations with adults: tribesmen, elders, sheiks, sons of sheiks and women. The interviews were semi-structured, so as to allow the interviewees freedom to take the conversation in unexpected directions and offer information unlooked for by the interviewer.

The interviews were structured around questions like: What units of measurement have you known? What does the name of each of these units mean? Why is it called that? What is the origin of the name? Are there other aspects to this unit? Who mainly uses it, men or women? Does it have a particular cultural connection? Does it say something about the person who uses it? Do you know if it is different in any way in other tribes? Are there any units that indicate a direction? Do you know any like that? Are there units that refer to different things, other than weight or length?

Interview analysis was qualitative. The material was looked at/listened to again and again, with the aid of two Arab speaking mathematics teachers and one linguistics teacher. Traditional measures of weight and length were first extracted separately by each individual analyzer, and then refined through common discussion until a consensus was reached about the measures, their literal meaning, and their equivalent in universal, standard measures.

Data for research question two, regarding the new program's influence, was gathered using "*tailor-made*," anonymous questionnaires, administered pre and post intervention. These questionnaires were divided into 22 questions, some of which addressed students' motivation and some of which addressed their self-esteem. The students were asked to answer each question with a rating on a scale of 1 (not at all) to 5 (very much).

The questions addressed a range of topics, ranging from their attitudes toward school ("To what extent would you say you value going to school?"), toward mathematics ("To what extent do you like and value studying mathematics?"; "How important are mathematics in your life?"), and to the relationship between mathematics and their own cultural identity ("To what extent do you think Arabs have contributed to mathematics?"; "Do you think Bedouins have the proper tools to learn mathematics?").

Descriptive statistical analysis was used to determine each group's average questionnaire score, and *t*-tests were used for comparative analysis of the two groups. Moreover, to determine the reliability of the data from the questionnaires, we ran a Cronbach's alpha internal reliability test. Results were: Pre-experiment motivation: $\alpha = 0.796$, Post-experiment motivation: $\alpha = 0.860$, Pre-experiment self-perception: $\alpha = 0.777$, Post-experiment self-perception: $\alpha = 0.945$. (Note: though the reliability of the tests from different periods was slightly different, it was still relatively high overall).

To further track the process and progress of the new program, the teachers recorded personal interviews with students and maintained a teachers' journal, in which they documented what was done and said in class. The interviews were conducted with approximately 60 students from the experimental group. They included questions like: "How would you summarize your experience over this process that we went through?"; "What units do you remember?"; "Which of those would you use?"; "Did this change, this integration that we made, affect you or your achievements? How and to what extent? In a good way or a bad way?"; "Would you like to learn more topics in the same way?" However, the interviews were semi-structured, so as to allow the students the freedom to take the interview in new and unexpected directions rather than simply answering the specific questions they were asked.

Data regarding the students' achievements were obtained from tests conducted by the students' schools both before and after the intervention. The tests were designed by the school's mathematics coordinator, in cooperation with the teachers, with each participating school generating one uniform test for all the students in each grade. For the purpose of comparison, the tests were based entirely on the formal school curriculum, with no mention of the ethnomathematical elements. The topic was *units of measurement* and the questions focused on conversion from one unit to another, and on the efficient use of the proper units in the proper context.

Examples of conversion questions: (1) If two of a rectangle's sides are one meter long, and the other two are 75 cm long, what is the circumference of the rectangle? (Note: the students were not limited in the units with which to describe the circumference.) (2) The circumference of an Isosceles triangle is 2 m and 30 cm. The length of the base is 70 cm. What is the length of the two sides?

Examples of efficient use questions: (1) To measure the height of your desk in class, would you use meters? Centimeters? Kilometers? (2) Iman measured the length of the school football field and the result was: 500 km? 500 m? 500 cm? Which one makes sense to you?

2.4 Results

Results are divided into three parts. The first part describes the units for measuring length and weight that were gathered by the interviews in the first stage of the study. The second part provides examples from the integrated teaching unit that was created, showing how the ethnomathematical elements were combined with the standard mathematics curriculum. Finally, the third part describes the results of that teaching unit's implementation.

2.4.1 Units of Length and Weight Described by the Elders of the Tribe

The interviews and conversations with the elders of the Tribe yielded a variety of units of measurement. The most prominent of these are presented below, divided into two categories: units of length and units of weight.

2.4.1.1 Units of Length

The most prominent units of length used by the elders are:

(a) Arm ذراع **Read: Dera'a**

(Fig. 2.1)

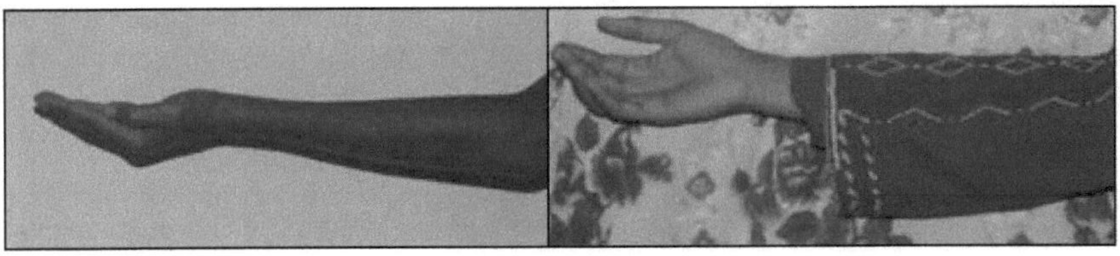

Fig. 2.1 Dera'a (arm). *Source* Personal file

One of the most basic units of measurement among the Bedouins, going back even before the time of Mohammad, is the distance from the elbow to the tip of the fingers. However, there are those who say that the *dera'a* is 24 fingers, or six palms, placed side by side.

It is important to note that there are many forms of dera'a, and that its meaning depends on where the user lives. One elderly woman claimed she was sure that the term referred to six hairs from the back of a horse, and that it can also be measured as 144 grains of wheat placed side by side. Islamic scholars claim that in Islam's golden age the dera'a was only 49 cm, but today it averages approximately 60. This unit has historically been used primarily for measuring length and height. Some Arab nations, like Yemen, still use it, so its use to this day is still considered acceptable and even honorable.

Additional concepts related to this measure are:

- Fist القبضة Read: Alk'bda Approx. length: 8 cm
- Finger الاصبع Read: Ala'sba Approx. length: 5 cm

(b) Stick Throwing Distance مقرط العصا **Read: M'krat ala'sa**

This term is one of the most common amongst the Bedouins, especially amongst the older generation. To understand this concept, it is important to clarify that most Bedouins make their living by herding sheep, goats, camels or other animals. The man in charge of the herd, not necessarily a shepherd, would generally hold a stick with which to lead or direct the flock. The length of this stick was (approx.) no less than 80 cm and no more than 150 cm.

This measure can be said to refer to a certain estimated distance. While the term *m'krat ala'sa* literally refers to the distance to which the shepherd can throw his stick, the measurement *m'krat ala'sa* refers to how far the herd can walk in a day before nightfall, or how far a man can walk in a day. *M'krat ala'sa* primarily marks the direction of the walk; some claim that it ranges between 3 and 7 km in the direction indicated by a finger.

Others claim that it refers to a specific distance between two bodies located at a reasonable distance from each other. One of the sheik's sons, perhaps the most knowledgeable, told us that *m'krat ala'sa* was approximately 50 m, and that this was the space around a given tent that no stranger is allowed to enter without permission. Even if someone enters that space in flight from someone else, he cannot be touched without the owner's permission.

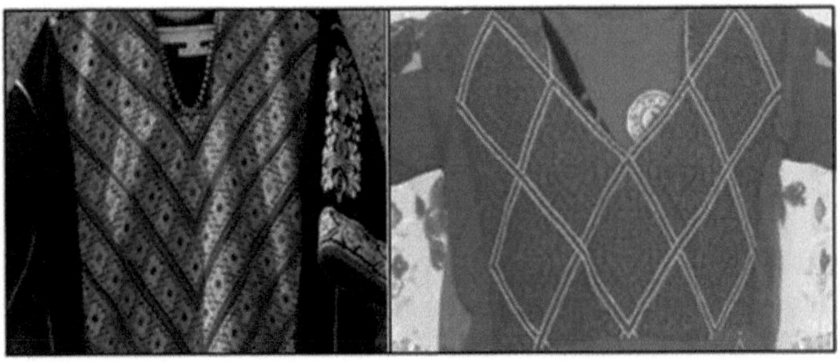

Fig. 2.2 Alendasa: front of a woman's dress. *Source* Personal file

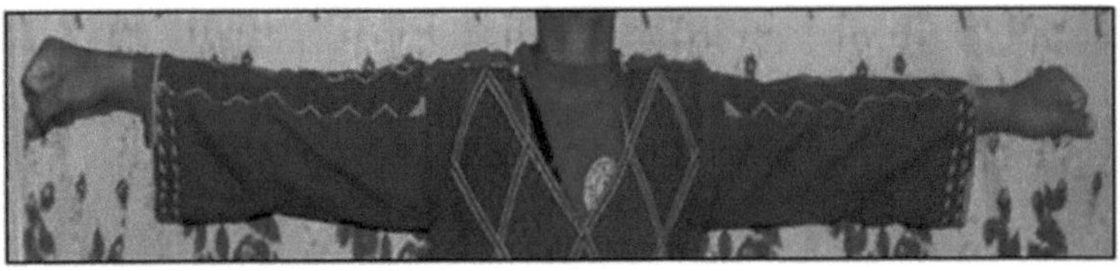

Fig. 2.3 Ba'a (arm-span). *Source* Personal file

Additional concepts related to this measure are:

- Stone throwing distance (sling) مقرط الحجر Read: M'krat alh'gar Means: Equal to: 1-3 km
- Stone throwing distance (sling) مقرط الدمس Read: M'krat aldms Means: Equal to: 3-7 km.

(c) Front of woman's dress الاندازة **Read: Alendasa** (Fig. 2.2)

This is a measurement used by a Bedouin woman making herself a dress, and is equal to approx. 65 cm. Some say that this is for all intents and purposes a dera'a, but is used only for measuring women's fabrics. It is worth noting that this term was used *only* by women. 65 cm. is the breadth of a woman's chest, with an addition of both sides under the arms, making it the sum of the woman's chest and her sides, taking into account that the garment is not meant to be tight.

(d) Arm-span باع **Read: Ba'a** (Fig. 2.3)

The literal meaning of this concept is: the distance between the tips of the fingers when each hand is open in its own direction. This is one of the oldest and most famous measures. When we asked how many centimeters it was, we were surprised by the number of answers: One *ba'a* equals 4 *Adr'aa*, or 24 fingers, or two meters.

(e) Height of a man قامة **Read: Kamah**

This is one of the more common measures, used for measuring height (especially humans) and for measuring depths. It is meant particularly for measuring animals, camels and humans. When asked, some claimed that it was about 170 cm, or that a *kamah* was equal to four dera'a.

(f) Foot **قدم** **Read: K'dm**

This is a basic unit of measurement among Bedouins that is also common today. Some of the interviewees claimed that a foot is equal to half a dera'a and others said it was about 24 cm. This unit is especially common in measuring plots or large tracts of land, which are measured by walking heel to toe.

(g) Stride **خطوة** **Read: H'toh'**

This term refers to the length of one stride of a man of average height. It is used to measure intermediate distances, not large ones, such as the race tracks of camels and horses, or the distance between tents. It is equal to about 80–120 cm, or the equivalent of three k'dm.

(h) Horse-run **شوط** **Read: Shoot**

Literal meaning: the distance a horse rider can cover at a run in one burst without stopping. This is one of the more common measures today, and was designed for measuring particularly long distances. When we asked how far it was, we were told that it was the distance between the town of Lod and the town of Ramle, approximately 18 km.

(i) Horse-leg rope **عقل حصان** **Read: A'kl alhesan**

Literal meaning: The string with which you tie a horse's legs so he does not move or run away. The term refers to the distance between the two feet of a standing horse (between 15 and 35 cm.). *A'kl alhesan* is an entirely Bedouin use, indicating that something is close by.

Units of Length: Summary

It is important to remember that there are many more measuring concepts than those noted here, that these are just the ones we deemed most important. An overview of these concepts yields several conclusions:

(a) All of the concepts are taken from the Bedouins' daily life. In place of more exact mathematical tools for measuring length and weight, the Bedouins drew on tools and concepts that reflect their own lives. Though the concepts they use are varied, they can be roughly divided into two major categories, those based upon the bodies of animals (*A'kl alhesan*) and those based upon the bodies of humans (*Dera'a*).
(b) Exactness in these units of measurement is not always important, but in many of them there is a great deal of significance to the direction, as in the case of concepts like *m'krat ala'sa* and *shoot*, concepts that refer to relative distance rather than exact distance in km.

Additional concepts worth noting:

1. Stars in the sky (on a clear night), meaning *infinite* نجوم السماء ليلة هلالها
   In the desert, particularly when the moon is new and the night is especially dark, the number of visible stars is immense, uncountable. This term therefore refers to the concept of "infinity."
2. Noonday shadow, meaning *zero* ظل شمس الظهرية
   In the middle of the day, when the sun is at its zenith, nothing in the desert has a shadow. This term therefore refers to "none" in the context of distance.
3. The measurement *finger* (اصبع) is equal to 5 cm.

2.4.1.2 Bedouin Units of Weight

Before presenting the concepts for units of weight, it is worth noting that these units are more accurate than the units of length, and there is more general agreement about what they are worth.

(a) Read: Retel رطل

This is the Bedouins' most basic unit of weight, and it is used to this day. This term has no literal reference, and no meaning other than as a unit for weighing. Retel is used by the Bedouins in several ways, particularly in commerce. It is the equivalent of 3 kg, and while there are many kinds of retel, depending on the country or area where the speaker lives, there is no change in how much it is worth.

(b) Read: Wakeh الاوقية , وقية

This is the most basic Bedouin unit of weight, and it is still used in many tribes today. Some claim that there are four wakeh in a kilogram, so that it is worth 250 g. One interviewee claimed that it was worth one twelfth of a retel, and if the retel is worth 3 kg, then 1/12 of that also comes to 250 g.

(c) Bucket دلو **Read: D'alo**

This is a unit of weight used in barter, or to weigh water or wheat. It refers to the amount of wheat that fits in a bucket, though it was designed for weighing water.

(d) Read: Gentar قنطار

An ancient unit of weight that older adults use up to this day. It was designed for measuring heavy weights, and is equal to one ton, or 1000 kg. This unit of measurement was common before Islam, and is even found in an old parable that says, الوقاية خير من قنطار علاج (*taking care to avoid disease is better than a gentar of medicine*).

(e) Handful كف **Read: K'af**

A unit of weight that measures the amount of flour that fits in the palm of the hand. This unit is not accurate, but it gives some indication of the amount of material. The k'af translates into approximately 30 g. An accompanying or alternative concept is دبت ايدك (*Dbt aedk*, which means: *what fills your hand with flour*).

(f) Drinking Vessel كوز Read: Kooz

Kooz is the name of a traditional clay vessel with a spout, from which one can drink water directly. This is a unit of weight that has gradually faded away, especially as the Bedouins stopped using the kooz. But it is not yet entirely gone. There are people, like the father of one of the authors, who still believe that the kooz keeps water better than a refrigerator by keeping it cool but not too cold. As a unit of weight, it was meant for the sale of herbal medicines. Every kooz is worth 1 kg (Fig. 2.4).

(g) Vessel for Carrying Water or Milk قربة Read: Kerbh

The kerbh is a vessel made of goatskin for keeping milk in the tent or cooling water. This unit of measurement was used mainly for the sale of milk or its products, though some claimed that it was also a unit for weighing water when it was brought from the well for drinking, especially if more than one kerbh was brought up. One kerbh is worth 30 kg (Fig. 2.5).

Read: Seaan صاع/صعن (Fig. 2.6)

Some claimed that seaan is a vessel that the king drinks from, but others claim it is a vessel for making cultured goat's milk. The seaan is a unit of measurement for expensive and valuable things, and is important for weighing wheat. This unit is

Fig. 2.4 Kooz—a traditional drinking vessel worth 1 kg. *Source* Personal file

Fig. 2.5 Kerbh—a vessel for liquids, worth 30 kg. *Source* Personal file

Fig. 2.6 Seaan—a 2 kg vessel for weighing wheat, both traditional (made of entire goatskin) and modern (made of iron). *Source* Personal file

mentioned in the Koran, and is still in use today. It is equal to 2 kg. Some interviewees claimed that one seaan was worth 8 wakeh. Each *wakeh* is 250 g, so 8 do indeed come to 2 kg.

Units of Weight: summary

Like the units for measuring length, the units for measuring weight are also connected to the Bedouins' daily lives, and are also alternatives to using exact mathematical measuring tools. The tools the Bedouins use for measuring also serve their daily needs, meaning that the same tool serves an immediate practical purpose (holding water), and serves as a unit for weighing when necessary.

2.4.2 The Integrated Teaching Unit

The units of length and weight gleaned from the interviews with the Bedouin tribal elders were used in the construction of an ethnomathematical teaching unit for Bedouin 7th graders. The unit was designed to acquaint the students with standard units for measuring length and weight (centimeters, kilometers, grams, and kilograms), as well as with units of measurement from the students' own culture. The traditional Bedouin units were designed not to replace but to *support* the standard units, by making the material clearer and more accessible to the students.

The students were required to learn about, apply and solve problems using both the standard and the traditional units of measurement, and to conduct comparisons between the two. The unit consisted of four sections, which were divided into two categories, standard measurement units and traditional ones. The first two sections included exercises that introduced the students to the universal mathematical tools for measuring length and weight, respectively. The third and fourth sections included exercises for measuring length and weight that introduced students to the traditional values and tools that are used in their own culture and society.

The learning functioned so that students learned the standard units and the traditional units simultaneously. In the universal-tools section *measuring length*, the students learned units of measurement like kilometers, meters and centimeters. As they did so, they also covered the same topic in the traditional section, learning units of length like *ba'a*, *dera'a*, and *kamah*. Similarly, the students learned universal units of weight like kilograms, grams and tons, whilst learning the same topic in the traditional measures section with units like *wakeh*, *retel*, and *gentar*.

The following are several examples of exercises from the teaching unit. The first two require students to measure various objects using a traditional unit for measuring length or weight. The third is an exercise in converting one traditional unit to another, while the fourth explicitly connects the traditional cultural units to the standard units, asking the students to conduct measurements using both.

Example 1: Units for Measuring Length
In this exercise, the students are asked to use their own "dera'a" to measure the length of the bodies in the exercise (Fig. 2.7).

Example 2: Units for Measuring Weight
In this exercise, the students are asked to measure the weight of the bodies in the picture using the traditional tool *retel* (Fig. 2.8).

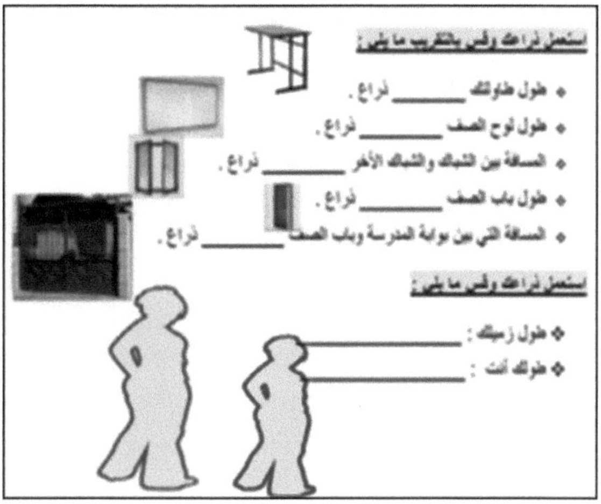

Fig. 2.7 Conducting measurements using *dera'a*. *Source* Personal file

Fig. 2.8 Using *retel* to calculate objects' weight. *Source* Personal file

القياس بالقامة	القياس بالباع	القياس بالذراع	المقاس	الرقم
		8	طول باب صفك	1
5			طول اللوحة الخارجية لمدرستك .	2
	8		[illegible]	3
		80	طول المسافة التي تقع أمام الصف	4
	6		[illegible]	5
4			[illegible]	6
		12	[illegible]	7

Fig. 2.9 Exercise for using and comparing several traditional units of length. *Source* Personal file

الوزن بالكيلو غرام	الوزن بالوقية	الموزون	الرقم
	16	وزن بطيخة كبيرة	1
60		وزن مربي الصف	2
1000		وزن سيارة متوسطة	3
	200	وزن كيس قمح	4
	40	وزن خروف رضيع	5
3000		وزن بالة قش	6
	3	وزن حاسوب شخصي	7
1/4		وزن حفنة ذهب	8

Fig. 2.10 Exercise for comparing traditional and standard weights. *Source* Personal file

Example 3: Using Multiple Units for Measuring Length
In this exercise, the students are asked to measure the lengths of the various objects noted in the table using three traditional tools, first the dera'a, then the ba'a, and then the kamah (Fig. 2.9).

Example 4: Traditional and Standard Units for Measuring Weight
Here, the students are required to learn the weights of the things in the table, first using the traditional tool "wakeh," and then in kilograms. The goal of the exercise is to fill in the missing weights in the table and thus compare the traditional tool to the universal one (Fig. 2.10).

2.4.3 Results of the Implementation of the Teaching Unit

This section begins with the results of the statistical analysis of the students' motivation and self-perception questionnaires. To determine the significance of the

Table 2.1 Pre and post motivation and self-perception for control and test group (*$p < 0.001$)

Group	Variable	Avg. (SD) before unit	Avg. (SD) after unit	Avg. (SD) t-test statistical value	No. of participants
Test group	Motivation	3.44 (0.31)	4.16 (0.31)	13.17*	75
	Self-perception	3.41 (0.41)	4.43 (0.51)	11.41*	75
Control group	Motivation	3.56 (0.78)	3.54 (0.79)	−0.72	70
	Self-perception	2.09 (0.44)	1.94 (0.46)	−4.19	70

Source Personal file

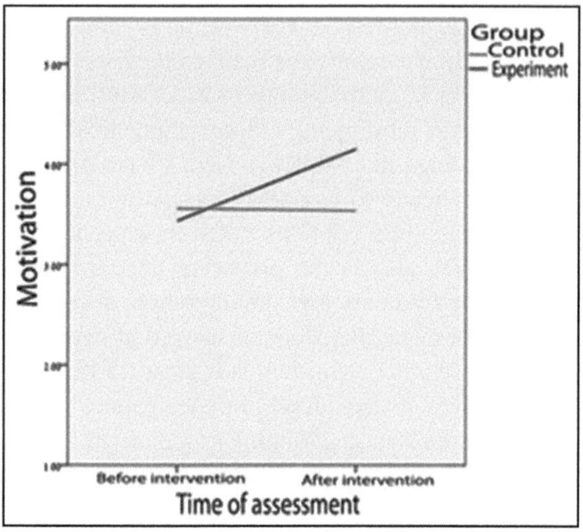

Fig. 2.11 Trajectory of student motivation in control versus experiment group. *Source* Personal file

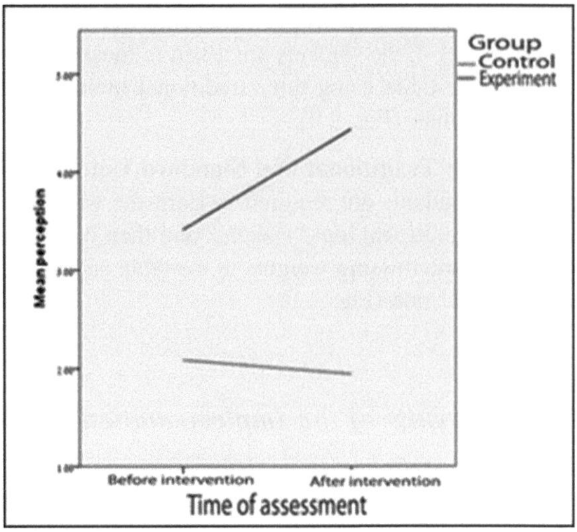

Fig. 2.12 Trajectory of student self-perception in control versus experiment group. *Source* Personal file

difference in the two variables before and after the study unit, we performed a *t*-test for dependent variances on both, as shown in Table 2.1, and in Figs. 2.11 and 2.12.

The data shows that in the test group's statistical test, which included 75 observations (per variable), there were significant differences for both variables between the pre and the post test. Both motivation and self-perception were significantly higher after the unit than before it. In the control group's test, which included 75 observations, no significant differences were found between the two tests, with the values actually going down between the pre and post-tests.

The quantitative data from the questionnaires were supported and expanded upon by data from the teachers' journals and the student interviews. For example, the rise in motivation reflected in the statistical analysis was also evident in the students' preparedness and class participation, which changed markedly as the teaching unit progressed. Early in the program, the students participated very little. Perhaps they were wary, or perhaps they were just accustomed to not cooperating with teachers or participating in class. But, as the integrated program progressed and we began introducing ethnomathematical concepts, they became enthusiastic, telling us things like "this is a term I have only ever heard at home".

Students began to compete to see who would remember the most units, and to try and master the relations between units and the conversion from one unit to another. They also began preparing for participation in the next lesson by interrogating their parents and grandmothers about upcoming topics. By the time the unit was complete, the students arrived at every lesson prepared, and at the end of the lesson we always finished late, since 45 min were "not enough". As one student told us, "I was always afraid of mathematics, I didn't like it, wasn't interested. Now I'm very interested and when I'm interested I learn more".

The journals and interviews also clearly showed that the teaching unit had impacted the students' perception of the connection between their home life and their life at school. Moreover, they showed that this change had altered the students' perception of mathematics, as well as their perception of their families, their heritage and themselves. They suddenly began connecting home and "grandma" with school, and their surprise at this connection highlighted the fact that until that point they had not seen the two as connected in any way. Until then these had been two different worlds, in which parents did not ask students what they did in school and children did not ask their parents or grandparents for help with their schoolwork.

Suddenly, the students found that their parents and grandparents had knowledge that could help them prepare for class. They were surprised to find that the people at home could be knowledgeable about something related to school. One student responded very positively to the unit, but added that "the only thing that makes me mad is that I didn't know that something like conversion is a subject they know at home". Having found out that her family can be an authority on school matters after all, another student commented, "I'll check of my grandma knows any physics, because I'm still having trouble there".

Many students reported taking a teacher's answer to a question home and asking their grandmother to check it, and vice-versa. One student approached an aunt for help, and found that she was able not just to confirm her new knowledge, but to expand upon it: "I came prepared today. I sat with my aunt and she had a lot of units. She explained a different method to me, but it came to the same result. It's fun to know that there is more than one method of solving the same question". As this student's comment shows, the integrated unit not only taught her something new about her aunt, it also showed her a new and *fun* aspect of mathematics, namely that a single problem can be solved in more than one way.

Bringing material from the integrated teaching unit home helped students bridge the gap between home and school by showing them that the former world could be a useful and relevant part of the latter. It also seems to have bridged the gap in the *other* direction by showing the people at home the relevance of what the children are doing at school. As one student told us, "For the first time in my life I ask my grandma and she tells me that my teacher is smart". All in all, it was evident that the unit had brought the students closer to their culture and heritage, and their surprise at its value and relevance highlighted how disconnected from it they had felt before.

The connection that the unit forged between school and home also created a positive surge in the students' self-perception, especially among the girls, who suddenly saw the connection and the application of mathematical concepts to their home environment, amongst their mothers and grandmothers. (Note: in Bedouin society girls spend much more time around the house than boys, who are free to move around beyond the living area. As a result, they have closer relationships with their mothers and grandmothers than the boys do.)

For example, one student whose achievements had been mediocre told us, "until now I gave up on mathematics and didn't make an effort. If I had known that my mother knows so much math, I wouldn't have given up". Knowing that their mothers and grandmothers could do mathematics made mathematics seem more accessible and more *doable*. One student said, "listen teacher, I always thought math was hard, but when grandma did a correct conversion from Dera'a to Ba'a, and my grandma can't even write her own name, I was surprised. Math is easier than Arabic!!" This realization also gave the students a feeling of *ownership* over mathematics, which suddenly became close and familiar rather than remote and abstract: "I felt that math belonged to me and to grandma and wasn't just X and Y."

The impact of the integrated teaching unit resonated beyond the boundaries of the topic of units of measurement, and beyond the bounds of the group of students that were chosen to participate in the experiment. First of all, many of the participating students that we interviewed expressed the wish that the rest of the mathematics curriculum would also integrate concepts from their culture or from their daily lives, rather than just teaching from the textbook. Some were also eager to connect their heritage even further to their school experience, not just in mathematics but in other fields as well. Secondly, as word of the teaching unit's success began to spread, more than half of the students in the control group approached us and asked if they could join the experiment, because they had heard from the participants that they were *learning mathematics with fun* and that the fun was coming from their own homes and culture.

We later found that in questions that required conversion from one unit to another, the students from the integrated experimental group were helping students from the control group with their standard questions. The resonance of the program was such that eventually the principals of both participating schools approached us directly, after speaking with the students' homeroom teachers—about implementing it in other classrooms as well.

Unfortunately, our ability to gather specific data regarding improvements in individual students' grades was impeded by the schools' ethical restrictions against releasing personal student information. The principals and the teachers in both schools were extremely cooperative, particularly after seeing the experiment's impact on the students and the community, but because they were not authorized by their supervisors to provide us with detailed information, they were able to offer only a more general indication of overall trends.

According to the principals, there were no significant differences between the control and experiment groups immediately after the intervention. On the one hand, this was reassuring, since the achievements of the experimental group were not *diminished* by the fact that they had to learn more material than the control group within the same period of time. On the other hand, it was somewhat disappointing, since the researchers had hoped that the achievements of the experimental group would be noticeably improved.

At a later time, another test was administered only amongst the experimental group, asking similar questions, but requiring them to use the ethnomathematical units of measurement rather than the standard ones. In these tests the group performed very well, but there is no basis for comparing these scores with those of the control group.

It is worth noting that a second intervention, set to take place in the next year, *has* been granted formal access to the students' specific grades. This will allow us to perform pre and post tests for the intervention, as well as an additional test at a later time to determine the intervention's long-term effects. At this point the results show significant changes only in the students' attitudes and motivation, but we can surmise that, over time, these may have a positive impact on the students' achievements as well.

2.5 Discussion

The study presented here arose from the need to find a way to help underprivileged students from the Bedouin tribes improve their mathematical achievements, motivation and self-confidence. In light of the limited success of the various other approaches that have been tried, we experimented with the integration of mathematical elements from the Bedouins' own cultural background into the standard school mathematics curriculum. To do this, we first had to discover what sorts of mathematical elements have traditionally been used in the Bedouin community, and determine how these could be associated and interwoven with elements from the standard mathematics curriculum.

Our conversations and interviews with the elders of the Bedouin tribe yielded a wealth of mathematical concepts (of which only those related to measuring length and weight are presented here). Like other studies that have examined the mathematical cultures of tribal peoples (Agarkar and Amit 2016; Amit and Abu Qouder 2016), we found that the lack of access to 'standard' units of measurement like centimeters or kilograms was in no way a hindrance to these tribespeople's ability to conduct sophisticated calculations of distance, weight and proportion using a wide range of units based on elements from their daily lives, like the length of a human arm or the size of a goatskin bag.

Systematically collecting, recording and categorizing these culturally specific units of measurement is important, not least because doing so will help preserve the aging remnants of a culture that—in light of the recent marked changes to the Bedouins' way of life—may soon disappear. But perhaps even more important for our purposes is the fact that the mathematical elements we found can be usefully compared to—and incorporated in—a standard mathematics curriculum.

As Orey and Rosa (2007) pointed out, the question of how the study of the mathematical ideas and practices of different sociocultural groups is realized in the classroom remains problematic, because many of the ethnomathematical investigations that identify ethonomathematical forms of mathematics "do not continue to develop the pedagogical actions" that make practical use of the forms they have found (p. 64). It was therefore significant to our purposes not only to collect mathematical concepts from the tribe, but to find a way to implement them in a classroom setting.

Like Lipka et al. (2013), who examined the mathematics embedded in the everyday activities of Yup'ik elders and found that they relate directly to ratios and proportional thinking and form a coherent and generative set of concepts, which incorporate geometry, fractions, ratios, and proportional reasoning, we found that the mathematics used by the elders of the tribe were more than a mere curiosity for historians and cultural anthropologists to explore. On the contrary, they were valid educational tools that could be used to teach students mathematical concepts and methods that are relevant to the requirements of a standard curriculum as well.

This latter point was one we found we had to emphasize in our attempt to "sell" the idea of the ethnomathematical teaching unit to the parents of the students and the principals of the participating schools. Orey and Rosa (2007) note that one of the challenges facing the introduction of ethnomathematical programs in schools is that "many teachers are not trusted or allowed to work away from required texts and curriculum, and therefore lack the support and cooperation required to make significant changes to the content that they teach" (pp. 68–69). Indeed, our proposed changes were initially met with suspicion, and with the worry that the ethnomathematical elements would come at the expense of the *real* mathematics the students needed to pass their exams.

These concerns are reflective of one of the common critiques of ethnomathematics, namely that spending time on alternative mathematical content will set students back rather than helping them to progress. According to this critique, introducing ethnomathematical ideas in school can function as "a factor for exclusion" because it is *formal mathematics* that gives students "access to a privileged world" (Pais 2011, p. 213). Teaching ethnomathematics therefore runs the risk that while "the students from the 'dominant culture' continue to learn the academic mathematics that allows them to compete in a more and more *mathematized* world, students from other cultures will only learn a local and rudimentary knowledge that scarcely contributes to their emancipation" (Pais 2011, p. 213).

We addressed this concern by emphasizing that the ethnomathematical elements were in no way designed to *replace* the standard mathematics curriculum. Instead, the specific cultural content was designed to *supplement and support* the standard curriculum, making the material more accessible and relevant to the students. In this, our teaching unit was similar to those being developed by Lipka, based on the assumption that "the promise of culturally based education" is not to replace mainstream education, but "to establish the basic conditions under which teaching and learning should prosper" (Kisker et al. 2012, p. 76). The implication of this assumption is that, for students who come from a cultural background that differs from the one that is dominant in school, these basic conditions are not being met by the standard curriculum alone.

Over twenty years ago, Alan Bishop pointed out that once we free ourselves of the *myth* that mathematics is a *culture-free* form of knowledge, we suddenly become "starkly aware of the fact that many young people in the world are experiencing a dissonance between the cultural tradition represented outside school (for example in their home or their community) and that represented inside the school" (Bishop 1994, p. 16). He added that such dissonance can occur not just in rural societies, but in societies where 'westernization' has happened rapidly, as is the case with the Bedouin population, which is in the midst of a process of rapid westernization. This dissonance was strongly evident in the qualitative data from our study, which suggested that the students were accustomed to thinking of school and home as entirely separate worlds.

Thus, for instance, the students were surprised to suddenly be discussing concepts that they *had only ever learned at home* in school, and that they could suddenly bring their schoolwork home and ask their families for help. The danger of such dissonance is that it can create an *either-or* situation, in which students feel that they must choose "between the culture of their natal community and the culture of schooling, and that getting an education will require them to leave their own culture behind" (Kisker et al. 2012, p. 76). One goal of ethnomathematics is therefore to mitigate this opposition and blur the sharp division between school and home.

One way that ethnomathematics does this is by showing that knowledge of mathematics can be acquired and used *outside* of school, and that it can therefore be a part of a student's home culture even if their culture does not have a history of formal schooling (Rosa and Orey 2011). The students in our study, for example, were surprised to discover that their illiterate grandmothers, who could not even write their own name, were nevertheless capable of mathematical calculation and conversion. By presenting students with the wealth of mathematical concepts and uses that pervade their own culture, ethnomathematics "turns perceived conceptual poverty into conceptual richness" (Orey and Rosa 2007, p. 66) by showing them that mathematics is not the sole province of the world of school, but also a strong presence in the world of home.

Helping students see the mathematical richness of their home environment is important in part because of the social value that is ascribed to the knowledge of mathematics today. As Pais (2011) points out, mathematics is commonly perceived as "one of the biggest achievements of humankind (…) the main pillar of our technological society, and an indispensable tool to becoming an active participant in a more and more mathematized world" (p. 217). This, he adds, imbues mathematics and the people who know it with a measure of *prestige*, and the result is that mathematics empowers people not because of any particular knowledge or competence that it gives them, but because the knowing itself makes people more socially valuable.

If students perceive mathematics as prestigious and socially valuable, the question of whether or not it is present in their culture can significantly impact their perception of that culture, and of themselves. In other words, seeing their culture as wholly separate from one of humanity's biggest achievements may cause them to feel that they and/or their family are somehow inferior, and that maybe they are not worthy or capable of learning mathematics.

On the other hand, knowing that mathematics is an integral part of their cultural background, and that other people in their family are knowledgeable and competent in mathematics, can give students the confidence that they need to persist. This was stated explicitly by at least one student in our study, who told us that if she had known that her mother knows so much mathematics, she would not have given up on it so easily herself. In this sense, ethnomathematics can serve as a means of increasing students' self-confidence and reinforcing cultural dignity and self-respect (Rosa 2000).

According to Rosa and Orey (2011), by seeking a cultural congruence rather than dissonance with the home environment, an ethnomathematics curriculum serves as an indication of "teachers' respect for the cultural experiences of their students" (p. 33). Our results suggested that this gesture of respect also prompted an equally important gesture of respect from the students' families. One of the challenges that arise from the dissonance between traditional Bedouin culture (in which formal schooling was not a prominent feature) and the *Western* culture prevalent in the state of Israel is that Bedouin students' families do not always see the value and relevance of formal schooling to their children's lives. Integrating cultural elements into the mathematics curriculum encouraged the students' family members to become actively involved in their school work, and even to declare, for the first time, that *the teacher is smart*. While students whose families do not see their studies as relevant may well find it easier to drop out and go to work, students whose families are involved and invested in their schooling may find it easier to persevere.

All in all, the results of our study were very encouraging. For the participating students, discovering that mathematics can be found all around them, particularly in the desert, was a thrilling experience. Moreover, they discovered that it was the older members of their tribe, those who do not drive cars or use cellular phones, who are in possession of all this mathematical knowledge. The study unit led the students to ask their elders about the mathematics of their culture and helped raise that generation somewhat in the estimation of their descendants. Nevertheless, it is important to point out that the scope of this study remains very limited.

While it did show that it is indeed possible to use ethnomathematics to effect positive change in students' motivation to learn and their perception of themselves and their culture, there is still a long way to go. The program needs to be expanded to include additional topics and additional student populations, and its effects, both cognitive and affective, must be assessed both in the short and the long term. Much work still remains to be done, but, to paraphrase an old Bedouin proverb, the march of a thousand miles begins with one small step.

References

Adam, S., Alangui, W., & Barton, B. (2003). A comment on Rowland & Carson: Where would formal, academic mathematics stand in a curriculum information by ethnomathematics? A critical review. *Educational Studies in Mathematics, 52,* 327–335.

Agarkar, S. C., & Amit, M. (2016). Understanding the units of length measurement used by tribal people in India. Paper presented in the 13th International Congress of Mathematics Education (ICME 13). Hamburg, Germany: University of Hamburg.

Amit, M., & Abu Qouder, F. (2016). Bedouin ethnomathematics: How integrating cultural elements into mathematics classrooms impacts motivation, self-esteem and achievement. Paper presented in the 13th International Congress of Mathematics Education (ICME 13). Hamburg, Germany: University of Hamburg.

Amit, M., Fried, M. N., & Abu-Naja, M. (2008). The mathematics club for excellent students as common ground for Bedouin and other Israeli youth. *International Perspectives on Social Justice in Mathematics Education, 1,* 105.

Anderson, S. E. (1990). World-math curriculum: Fighting Eurocentrism in mathematics. *Journal of Negro Education, 59*(3), 348–359.

Ascher, M. (1991). Symmetric strip decoration. In M. Ascher (Ed.), *Ethnomathematics: A multicultural ideas* (pp. 156–179). Pacific Grove, CA: Brooks/Cole, Publishing Company.

Ascher, M. (2000). Ethnomathematics for the geometry curriculum. *MAA NOTES*, 59–63.

Ascher, M., & Ascher, R. (1986). Ethnomathematics. *History of Science, 24,* 125–144.

Baba, T. (2002). Significance of ethnomathematical research: Towards international cooperation with the developing countries. In T. Nakahara & M. Koyama (Eds.), *Proceedings of the Conference of the International Group for the Psychology of Mathematics Education* (Vol. 2, pp. 1–11). Hiroshima, Japan: PME.

Bishop, A. J. (1988). *Mathematical enculturation: A cultural perspective on mathematics education.* Dordrecht, The Netherlands: Kluwer Academic Publishers.

Bishop, A. J. (1994). Cultural conflicts in mathematics education: Developing a research agenda. *For the Learning of Mathematics, 14*(2), 15–18.

D'Ambrosio, U. (1985). Ethnomathematics and its place in the history and pedagogy of mathematics. *For the Learning of Mathematics, 5*(1), 44–48.

D'Ambrosio, U. (1987). Reflections on ethnomathematics. *Newsletter: International study Group on Ethnomathemathematics, 3*(1), 3–5.

D'Ambrosio, U. (1994). Cultural farming of mathematics teaching and learning. In R. Biehler, R. W, Scholz, R. Straber, & B. Winklelmann. (Eds.), *Didactics of mathematics as a scientific discipline* (pp. 433–455). Dordrecht, The Netherlands: Kluwer Academic Publishers.

D'Ambrosio, U. (2000). Etnomatemática e modelagem [Ethnomathematics and modeling]. In M. C. Domite (Ed.), *Anais do Primeiro Congresso Brasileiro de Etnomatemática* (p. 142). São Paulo, SP, Brazil: FE-USP.

D'Ambrosio, U. (2002). Ethnomathemathematics: An overview. In M. de Monteiro (Ed.), *Proceedings of Second International Conference on Ethnomathemathematic (ICEm2)* (pp. 3–5). CD Room. Ouro Preto, MG, Brazil Lyrium Comunicação Ltda.

Davison, M. (1989). An ethnomathematics approach to teaching language minority students. In J. Reyhner (Ed.), *Proceedings of the Annual International Native American Language Issues Institute* (pp. 2–7). Billings, MO: ERIC.

Gerdes, P. (1990). On mathematical elements in the Tchokwe sona tradition. *For The Learning of Mathematics, 10*(1), 31–34.

Gilmer, G. (1990). An ethnomathematical approach to curriculum development. *Newsletter: International Study Group on Ethnomathematics, 5*(1), 4–6.

Keitel, C., Damero, P., Bishop, A., & Gerdes, P. (1989). Mathematics, education, and society. *Science and Technology Education*. Paris, France: UNESCO.

Kisker, E. E., Lipka, J., Adams, B. L., Rickard, A., Andrew-Ihrke, D., Yanez, E. E., et al. (2012). The potential of a culturally based supplemental mathematics curriculum to improve the mathematics performance of Alaska Native and other students. *Journal for Research in Mathematics Education, 43*(1), 75–113.

Ladson-Billings, G., & Tate, W. F. (1995). Toward a critical race theory of education. *Teachers College Record, 97*(1), 47–68.

Levinson, E., & Abu-Saad, A. (2004). *Statistical yearbook of the Negev Bedouin.* Beer Sheva: Negev Center for Regional Development and the Center for Bedouin Studies and Development, Ben Gurion University of the Negev.

Lipka, J. (2002). *Schooling for self-determination: Research on the effects of including native language and culture in the schools.* Charleston, WV: Clearinghouse on Rural Education and Small Schools.

Lipka, J., Wong, M., & Andrew-Ihrke, D. (2013). Alaska native indigenous knowledge: Opportunities for learning mathematics. *Mathematics Education Research Journal, 25*(1), 129.

Manor-Rosner, Y., Rofè, Y., & Abu-Rabia-Queder, S. (2013). The unrecognized Bedouin villages–internal spatial order as a basis for development. The State of the Educational System in the Bedouin Sector. Jerusalem, Israel: Ministry of Education.

Orey, D., & Rosa, M. (2007). Cultural assertions and challenges towards pedagogical action of an ethnomathematics program. *For the Learning of Mathematics, 27*(1), 10–16.

Pais, A. (2011). Criticisms and contradictions of ethnomathematics. *Educational Studies in Mathematics, 76*(2), 209–230.

Powell, A. B., & Frankenstein, M. (1997). *Ethnomathematics: Challenging Eurocentrism in mathematics education*. Albany, New York, NY: State University of New York Press.

Presmeg, N. C. (1998). Ethnomathematics in teacher education. *Journal of Mathematics Teacher Education, 1*(3), 317–339.

Rosa, M. (2000). From reality to mathematical modeling: A proposal for using ethnomathematical knowledge. College of Education. Unpublished master thesis. Sacramento, CA: California State University, Sacramento (CSUS).

Rosa, M., & Orey, D. C. (2011). Ethnomathematics: The cultural aspects of mathematics. *Revista Latinoamericana de Etnomatemática, 4*(2), 32–54.

Tate, W. F. (1997). Critical race theory and education: History, theory, and implications. *Review of Research in Education, 22*, 195–247.

Chapter 3
Listening to the Voices of the Knowledge Holders: The Role of Language in Ethnomathematical Research

Mogege Mosimege

Abstract Language plays a central role in ethnomathematical research. It allows the researcher to interact with the knowledge holders and research participants at a level and in a context where they can express themselves in a language that they are more comfortable with. The use of indigenous languages in research empowers knowledge holders to be free, expressive and more engaging and willing to share more of their knowledge. It is essential that the researcher familiarises himself with the language of the knowledge holder to derive more benefit and gain access to forms of expressions associated with the aspect of investigation. In cases where there is no knowledge of the language, the researcher needs to explore ways in which this limitation can be reduced for the enhancement of the interaction and collection of data.

Keywords Language · Ethnomathematical research · Culture · Cultural village · Mathematics education

3.1 Ethnomathematical Research

Ethnomathematical research and focus in mathematics education is traced back to the seminal work of Ubiratan D'Ambrosio, followed closely by the work of Paulus Gerdes. Both have contributed extensively to the definitions of ethnomathematics and to the conceptual development of this area in mathematics education. Their definitions and ideas have subsequently been embraced, extended, and critiqued by other mathematics educators working in this area. In one of the earlier references to ethnomathematics D'Ambrosio (1985) argues that cultural groups are important in ethnomathematical discussions.

D'Ambrosio (1985) then goes on to define the cultural groups as national tribal societies, labour groups, children of a certain age bracket in the following way: "we

M. Mosimege (✉)
Human Sciences Research Council, Pretoria, South Africa
e-mail: mmosimege@hsrc.ac.za

M. Rosa et al. (eds.), *Ethnomathematics and its Diverse Approaches for Mathematics Education*, ICME-13 Monographs,
DOI 10.1007/978-3-319-59220-6_3

will call ethnomathematics the mathematics which is practised among identifiable cultural groups, such as national-tribal societies, labour groups, children of a certain age bracket, professional classes and so on" (p. 45).

Gerdes (1994) defines ethnomathematics as "the field of research that tries to study mathematics (or mathematical ideas) in its (their) relationship to the whole of cultural and social life" (p. 20). Gerdes (1996, 1997) goes on to indicate that, as a research field, ethnomathematics may be defined as the cultural anthropology of mathematics and mathematical education.

Although the focus of this chapter is on ethnomathematics and specifically on the role that language plays in ethnomathematical research, the chapter fits broadly in the sociocultural aspects of mathematics education and, in this context, it refers, albeit briefly, on another, but closely related area of critical mathematics education (François and Stathopoulou 2012).

This chapter first looks at the importance of culture in ethnomathematical studies. The major focus and emphasis is on the role of language in ethnomathematical studies, so the discussion then looks at language as a component of culture and how it is used in mathematics education. In order to give context to the challenges of language, the chapter then looks briefly at the languages in South Africa, bringing to the fore the complexity of a number of official languages and how this affects developments in both basic and higher education sectors.

Three episodes from ethnomathematical research in South Africa are included in the chapter for illustration of language challenges: the first episode illustrates the interaction between an ethnomathematics researcher and knowledge holders in the production of an artefact, the second episode illustrates the use of language during an indigenous game whereas the third episode follows along the lines of the first in which elderly knowledge holders are interviewed on beadwork.

3.2 Culture as a Central Component in Ethnomathematical Studies

Mathematics educators who work specifically with culture as it relates to mathematics or mathematical activities define culture in different ways. This may be viewed as an attempt to locate the idea of culture within a particular form of understanding and in the context of that writer.

3.2.1 *Definition of Culture in Mathematics Education*

For Mellin-Olsen (1985) culture is "a concept which considers the relationships between the formation of a society, the creation and construction of the individual in this society and the way he himself lives according to this" (p. 105).

Shared meanings imply a number of common aspects within members of a particular society, whereas relationships are an indication of the closeness of various aspects within that society. Shared meanings also imply some common understanding in relation to a group of people who engage in a common activity. The common activities could include, among others, engaging in socio-cultural activities like the playing of games which are cultural specific; activities like problem solving which are not necessarily sociocultural, but of interest to the group of people engaging in the activity.

The shared meanings give the same sense of shared experiences as given in the definition of culture by Begg et al. (1996) in which they define culture or sub-culture as a set or subset of people who have a set of shared experiences. Although shared meanings and shared experiences imply a form of closeness between the members of a society being referred to, the meanings suggest common understanding whereas experiences imply something that members of a given group have gone through together, but do not necessarily have the same understanding of or give the same interpretation to.

This definition given by Begg et al. (1996) may be used to explain the experiences that members of a society may have about a game. Although each member may have the same experiences about the game, at the time when the game is played or referred to the same people may not have the same knowledge.

In his earlier work *Evolution of Mathematical Concepts* (EMC), Wilder (1981) defined "culture as a collection of customs, rituals, beliefs, tools, mores, and so on, called cultural elements, possessed by a group of people who are related by some associative factor (or factors) such as common membership in a primitive tribe, geographical contiguity or common occupation" (p. 7).

In reflecting on this definition, Wilder (1981) says that he should have included language among the cultural elements, especially since it forms the cement which binds together the world view of a people. He also identifies some cultural elements, which include language. This cultural element had not been included initially.

All the cultural elements are important, but it is of interest that these cultural elements also include beliefs and language, two of the three important features discussed by Amir and Williams (1994). They do not necessarily exclude experiences as Wilder (1981) indicates that the list is not exhaustive. Language is seen as the basis of communication in a culture, which shows how a culture is constituted.

3.2.2 Language as a Main Feature of Culture

Much of the writing on ethnomathematics uses language as a determining feature of culture and as an important component of mathematical activity (Barton 1996). Emphasising the importance of language in culture and the relations between the two, Prins (1995) expresses the "view that there seems to be no language without culture and no culture without language" (p. 94). She continues with the view that the language of a people reflects their culture and since cultures differ, they do not use language in the same way.

Fitouri (1983) is more explicit about the relations between language and culture. Developing his argument from the research with young bilingual Tunisian children, Fitouri (1983) categorically states that language and culture are inextricably linked. He continues to state that the acquisition of a second language for a child is not a mere mechanical skill but a psychological understanding and emotional acceptance of the culture to which that language is the key.

3.3 Language in Ethnomathematical Research

Language plays an important role in ethnomathematical studies. Mosimege (2012) identified a number of methodological issues that needs to be addressed when ethnomathematical studies are undertaken. One of the methodological issues that needs to be reflected upon is language. In this respect Mosimege (2012) asks:

- How does language feature in ethnomathematical studies?
- Is it important to share a language with the cultural group?
- What happens if the researcher cannot speak the language of another cultural group? Does it mean that the researcher cannot continue with the research?
- Is it impossible for the researcher to continue with the research because of language problems?

Most of the studies reported at the International Conferences on Ethnomathematics (ICEM) (from ICEM 1, in Granada, Spain, in 1998 to ICEM 5, in Maputo, Mozambique, in 2014) have reflected on the language that has been used throughout the ethnomathematical research activities. These studies have largely reported on the language used during the interaction with knowledge holders and elders in communities who prefer to communicate in their home languages. Although most of these studies have not focussed on language as a central component of the discussion, the use of language has continuously emerged as important for purposes of communication with various community members.

Stathopoulou and Kalabasis (2006) discuss about power relations that come from the use of language, not just in research conducted outside the classroom, but also about the impact that they have on the understanding of learners in mathematics classrooms. They also argue that research that is not conducted in English or recorded in English is largely unseen by the international community. In the context of the argument by Stathopoulou and Kalabasis (2006), how much do the issues of language affect interviews conducted in ethnomathematical studies and how do translations from indigenous language to English and other languages affect the extent of the knowledge provided in ethnomathematical settings?

The language of interaction in ethnomathematical research is very important and central in the advancement of such research. Mosimege (2012) concludes that it is "given that most elderly members of the various communities are more comfortable to express themselves in their own indigenous and local languages" (p. 70–71). Thus, the challenge for researchers is to either find a way to understand the indigenous language spoken by the research participants (interviewees) in order to facilitate access to their knowledge, or to find a way to translate from the indigenous languages to the language in which the research results are being communicated.

This, therefore, means that the inability to speak the language of the knowledge holders limits access to the kind of information related to the exploration of the artifact or indigenous practice. However, this limitation can be reduced through various ways and approaches and a deeper interest and some form of basic understanding of the language of use in ethnomathematical research.

3.4 Languages in South Africa

South Africa has 11 official languages and a number of unofficial languages. According to the 2011 census (Statistics South Africa 2012), isiZulu is the mother tongue for 22.7% of the South African population, followed by isiXhosa at 16%, then Afrikaans at 13.5%. The next two languages are English at 9.6% and Setswana at 8%. The other official languages range from 7.6% (Sesotho) to 2.1% (IsiNdebele). Although English is ranked fifth as a home language, it is used as the language of business, politics, and media.

IsiZulu, isiXhosa, siSwati and isiNdebele are collectively referred to as Nguni languages and have many similarities in syntax and grammar. Setswana, Sesotho sa Lebowa (referred to as Sepedi in the South African Constitution) and Sesotho are collectively known as the Sesotho languages as they also have a lot in common in terms of syntax and grammar.

This means that a person whose home language is isiZulu would be very familiar with other Nguni languages and finds it easier to communicate in such languages. The same applies to any of the Sesotho languages. A person whose mother tongue is Setswana would find it relatively easy to communicate and follow other Sesotho languages.

Language has been and continues to generate many tensions and debates in South Africa. The linguistic tensions between European languages (English and Afrikaans) and African (referred to as indigenous) languages have manifested themselves at various levels and in a number of ways. The most memorable and demonstrated tension is the Soweto riots of 1976. In fact the riots started in Soweto but also spread to other parts of the country.

The basis of the riots was the enforcement by the government of the day to use Afrikaans as the language of instruction in schools. The black youth revolted against this and resulted in one of the major uprisings in South Africa.

3.5 Episodes Illustrating the Use of Language in Ethnomathematical Research in South Africa

The three episodes reported below are drawn from ethnomathematical studies that have been conducted in South Africa. The first and the third are part of an ongoing study on *Indigenous Mathematical Knowledge at South African Cultural Villages* and the second is part of a study on *Indigenous Games and related Mathematical Knowledge*.

Reports related to these studies (and other episodes not referred to in this article) have been presented at various conferences and published elsewhere, however, the emphasis has been on mathematical analysis of indigenous activities and mathematisation processes. The emphasis in the context of this chapter is on language and how it impacts on ethnomathematical research.

3.5.1 Episode One: The Use of Language During an Interview on How to Make 'Sesiu'

The excerpt reported below is part of the interview that was conducted with Mr X, one of the members at the Basotho Cultural Village in the Free State Province, South Africa (Mosimege 2017). He is actually the most important member in the village as he plays the role of the Chief of the village. The excerpt explores how a container made of grass (Sesiu) is constructed. It also reveals the extent of the use of mathematical concepts in this particular activity as done by the male inhabitants of the Basotho Cultural Village.

The study was conducted by two researchers, named R1 and R2. M represents Mr. X. The interview was conducted in the Sesotho language which is the main language used at the village. Using this language also enabled Mr. X to express himself freely without going through the difficulties of trying to cope with a foreign language which is not spoken daily at the village nor as part of the home language of the Basotho nation.

R1: Ooh Ntate, o ka qala joale o re bolelle hore ha ke qala ke etsa jwang, re ke re kgone ho ho latella. O hle o re rute joale Ntate [Father, you can start now, tell us how we begin and what we do, so that we can follow you. You may proceed to teach us].

M: Ke tla le ruta hee hore le ke le tsebe, le be le ye ho ruta ba bang. Joale ha ke qala sesiu sena, ke qala ka ho nka hlotswana se se kana, joale ke se finye lefuto. Lefuto lena ke la ho etsa hore ho na le ntho e ke tla hlaba mona e tlabe e se e tla tshwara. E tlabe e be e se e tshwara hobane ke tla nka thapo e se e kene mono. Lehlabo leno ke hlaba ka lona. [I am now teaching you so that you can know, and then you can go and teach others. When I start 'sesiu', I start by taking this amount and make a knot. I make this knot so that when I use something to put through, it will hold because I will take a string. This needle I use to thread through].

R2: Ntate, o re le nka joang bo bo kana. He ke a bona o bo bala, fela wena o tseba hore bo lokile. O tseba jwang hore bo lokile? O tseba jwang hore O tsee selekanyo se se kae? O tseba jwang hore o tshwanetse ho tsaa ngatana tse lesome kgotsa tse masome a mabedi, hoba tse nne, ho ba tse kae? [Chief, you say you take so much grass. I notice that you have not counted it. How do you know which amount to take? How do you know which amount to take, whether it be ten grasses, or twenty, or four, or how many?].

M: Ooh, ha ke etse joalo. Ke tsipa fela sehlopa. Ha ke batla ho etsa lefitole leholo ke tsipa haholo fela. Ha ke batla ho…[I do not do that. I merely take a certain amount. When I want to make a big knot, I take a big amount. When I want…].

R2: So, ho tsipa hoono, ha se hore o a dibala. [So taking an amount, it does not mean you count them].

M: Ha ke di bale. [I don't count them].

R2: Ha o dibale. O di lebella ka mahlo, wa fetsa ka hore tse di ka etsa sena, tse di ka etsa sena. [You don't count them. You just estimate by mere looking at them and decide that this amount will give this and this].

M: Eke. Ha ke batla ho etsa ho ho holo, ke tla sheba ka mahlo fela hore ha e le mona ke nkile ha kana, bo tla etsa boholo bo bo kana. Ha ke nkile ha kana, bo tla etsa bonnyane bo bo kana. [Yes. When I want to make a big thing. I just use my eyes to estimate any size I want to make. If I have taken this much, it will give me this small amount].

R1: Joale ha le nke le kgathatseha hore ha le sa di bale, joang bona ba lona bo tla fela ka pele, pele le qeta sesiu sena? (…) Ke dumela hore le tshwanetse le di bale.[Now don't you get worried that when you don't count, the grass will get finished before you complete the 'sesiu'? (…) I believe that you should count them].

M: Hee. Ha re di bale hoo hang hobane le joang bona ha re bo reke, re bo hela fela le he bo ka fela. Re ntse re ya. Ebile o tseba hore o tlile ho etsa sesiu se se boholo bo bokae se se tla nkang joang bo bo kae. [No, we don't count at all. After all, even this grass we don't buy, we merely pick it up].

3.5.1.1 Mathematical Concepts Related to *Sesiu*

In this excerpt from the interview with Mr. X, a number of mathematical concepts are referred to or used throughout the interview. Prominent among these is the *estimation* about how much grass is needed for which step in the making of this container. Figure 3.1 shows Basotho cultural village male workers showing the *sesiu*.

Mr. X explains the accuracy of his estimation from the experience which he gained from many years of working with the grass. He does not go on to *count* the number of grasses that he uses for the different sizes of the container, although this does not mean that he does not know how to count.

Another part of this interview which is not reflected in the excerpt above is the *size of the container* (*sesiu*). The size is determined by what the container will be used for, so that the sizes of the containers vary. So, the amount of grass used for the different sizes of the containers is estimated.

Fig. 3.1 Basotho cultural village male workers showing the *sesiu*. *Source* Mosimege (2017)

3.5.2 *Episode Two: The Use of Language in a String Figure Game*

The excerpt reported below was conducted in a mathematics classroom in a school in the Limpopo Province, South Africa. The learners who knew the Malepa Game (String Figure Games) were given an opportunity to give demonstrations to their fellow learners on any Gate that they had the knowledge of. The gate is a final configuration that can be made using a string comprising of triangles and quadrilaterals. Gate Two would therefore comprise of a figure which has two quadrilaterals and four triangles.

Each learner who gave such a demonstration (of the Gate) was given an opportunity to use the language that they were comfortable in and the learner giving the demonstration below decided to use the Sesotho sa Lebowa language which was understood by all the learners.

(1) Tseang wa boseven le tsentsheng mo. Le ka go gongwe diang. [Take the seventh and put it in here. Do the same on the other one] {The presenter already had the string hooked on the thumbs and the small fingers, so she started with the very next step. One of the learners asked her to wait a bit before she continues so that they also hook the string on the four fingers. The learner has not started by numbering the fingers, and by the seventh she refers to the pointing finger on the right hand side. When she says that they must also

do on the other side she points to the side by the pointing finger of the left hand}.

(2) Ntshang wa bo five. [Remove the fifth] {she removes the thumbs at the same time}.

(3) Le thieng ka fatshe. A kere le dirile so, e buseng ka mo, le e gogeng so, e be so. Haaa. Goga wa mo fatshe. Ka mo go o monnyane. [Let it pass underneath. You have done like this, turn it back this side, pull it, it must be like this. No. Pull the one underneath. On the small one.] {By passing underneath she is referring to the thumbs. As she starts to hook the string with thumbs the learners murmur to indicate that they are not following the demonstration very well. At this point the demonstrator reverses the thumbs and starts to explain what they have already done up that point. She then starts the step again. One of the learners then asks her to wait a little bit, and she does. Then she looks at what that learner has done and comments about her method, commenting about what she seems to be doing wrong, particularly as it relates to the string underneath. She illustrates to this learner without starting from the beginning nor moving nearer her for assistance but helps her still standing at the front}.

(4) Go e ya bo seven gogang e enngwe, e, ye, e, e, e be so. [On the seventh, pull the other one, yes, yes, yes, yes, must be like this] {She first spends a few seconds looking at her string and then talks about the seventh. Then a male learner asks her whether it is the seventh she is referring to. She points to the string on the seventh finger to be removed with thumbs, then she uses the thumbs to pull the string from the seventh, without specifying the use of thumbs. Then she keeps saying yes - on about three occasions - as the learners check with her whether they are doing the correct thing}.

(5) Ntshang o monnyane. [Remove the small one] {Although she has actually referred to one small finger she removes both small fingers}.

(6) Le goge e, e be so. Mo e tshwanetse le e dire so, e tshopagane so. E tshopagane so. [Pull it, it must be like this. Here you must do it like this; it must be entangled like this. Entangled like this] {By pulling the demonstrator refers to using small fingers, although does not refer to them explicitly. Another female learner calls her by her name - Sophy - and once more asks her to wait}.

(7) Ntshang so, e, e megolo, e, ntsha e megolo, le sale ka one le seven. [Remove like this, yes, the big one, yes, remove the big one, you must remain with one and seven] {The learners ask her if she is referring to the big ones, and whether she means both big fingers, and she says yes on both occasions. She refers to fingers one and seven without clearly showing them but just looking at the right hand, and the learners do not ask her what she means by one and seven, giving an indication that they either see exactly which fingers she is referring to or that they know what she means by one and seven}.

(8) Tatang ka mo so, le tla dia? Le ka go gongwe dia ka mouwe [Make a twist here like this, are you doing it? Do the same the other side] {She starts to make an anti-clockwise twist with the fourth finger on the right hand side. By doing it on the other side she is not referring to the left hand but to the small finger of

the right hand side, symmetry between the fingers and not necessarily the hands as has been the case thus far. Before she asks them whether they will do it, she looks at them and seems to be getting a feeling that they are either not sure what to do or they are finding it difficult to do it. Unfortunately the video camera did not get focussed on the learners to verify their activities at this point. The presenter laughs a little about the difficulties experienced by the learners in this step}.

(9) Le ka mo left diang. [Also on the left hand side do the same] {She then makes similar twists on the left hand side, although not exactly similar as the twists on the left hand side are actually clockwise twists i.e. the learner does not make a distinction between anticlockwise and clockwise twists, but still knows that the same activity must be done on the right hand side. However, this time she does not mention the twist to be done on the small finger but rather follows that doing on the left hand means doing it for both fingers. A female learner then asks 'and then'? This suggests that she had made this step and was now eager to see what will happen next}.

(10) Tsentsha mo, go e mennyane. E, e be so. [Put it in here, in the small ones. Yes. Must be like this] {In this step the thumbs are used to pull the string on top of the small fingers, however no reference to thumbs but to small fingers, not even indicating which string on the small fingers as there are two. Another learner asks whether the figure must look like the one this learner had made, and the demonstrator says yes. However it is interesting that she does not say that it must be like this learner's figure but refers to her's (the demonstrators) i.e. responds to the learner's question by referring to the model being used upfront}.

(11) Tsea o wa boseven, o e tsentshe mo. Le ka mo go o. [Take the seventh one, and put it here. And also on this one] {The demonstrator starts with the string on the left hand, whereas the twists at step number 8 were started on the right hand. This has important implications as it means that you do not always have to start with the right or the left hand all the time, but in many instance you can start with any hand, and don't have to keep to what you started with throughout the activity all the time. This is also different from the how the hands are used in step 12 below}.

(12) Le ntshe so, le ka mo lentshe. [Remove it like this, also this side remove it]. {The demonstrator first removes the string at the back of the right hand thumb, and then follows with the left hand thumb. At times, especially for beginners, this step is best done through the use of a mouth so that you don't lose the other string that must remain on the thumb. This illustrates how adept the demonstrator is about making the gates and the manipulations involved}.

(13) Tsentsha mo, mo, e, wa boseven [Put it here, here, yes, on the seventh finger] {Female learner asks where, and the demonstrator points again at the triangles underneath the thumbs without mentioning the geometric figure involved}.

(14) Ke ka moka lentshe e mennyane. [Then remove the small ones].

Fig. 3.2 Demonstration of a string figure gate 6 by a learner. *Source* Personal file

(15) Le e goge. [Pull them]. {Pulling here means turning the Gate away from you to face the learners}.
(16) The demonstrator then shows the learners how String Figure Gate 6 looks like (Fig. 3.2). {This is done without an accompanying explanation}.

The researcher then asks the learners if they had managed to do it and finds out that only one learner had managed to make this Gate.

3.5.2.1 Mathematical Concepts in String Figure Games

There are mathematical concepts that can be identified in string figure games.

Variety of Geometric Shapes and Figures

In any string figure gate the following geometric figures may be identified: angles, triangles, quadrilaterals (particularly rectangles and squares). The number of the geometric figures increases as the number of gates increases. The same can be seen from the illustration in Fig. 3.2.

Patterns, Relations, and Functions

An analysis of the String Figure Gates reveals a variety of relationships between the triangles and quadrilaterals, quadrilaterals and intersecting points, and the generalisations that result from these relationships. The generalisations between the different geometric figures are:

- triangles and quadrilaterals: $y = 2x + 2$.
- quadrilaterals and intersecting points: $y = 3x + 1$.
- quadrilaterals and the number of spaces (spaces are given by the combination of triangles and quadrilaterals): $y = 3x + 2$.

Symmetry

Working with gates, a variety of symmetry types may be identified. Most of the gates show bilateral symmetry, particularly even numbered gates. Some of the gates exhibit reflection symmetry, for example Gate 2, while others have rotational symmetry, for example Gate 1. It is also possible to show radial symmetry; translational symmetry; and anti-symmetry.

After making a specific gate, disentangling the string along a specific line of symmetry also ensures that the string does not get entangled. An interesting classroom activity could be to investigate which types of symmetry are associated with which String Gates.

3.5.3 *Episode Three: The Use of Language During an Interview on Beadwork*

The following excerpt reports on the interview conducted with two Ndebele ladies at the Lesedi Cultural Village, Ms. Y and Ms. Z. The Lesedi Cultural Village is located in the Gauteng Province, South Africa. In the excerpt, the questions focus on how the ladies engage in beadwork activities and explore related mathematical concepts. It focuses specifically on how they use beads to engrave names of people on beadwork activities, but also refers to other beadwork activities.

In the excerpt, R refers to the researcher and SS to Ms. Z and LM to Ms. Y. The ladies speak Isindebele, however, they also understand the Sesotho languages like Setswana and Sesotho, as a result the interviewer used both Isindebele and Sesotho as the latter is mostly understood by the researcher in comparison to Isindebele.

R: Le rutilwe ke mang ho sebetsa ka dibeads? [Who taught you to work with beads?].

SS: Si fundiswe uGogo. [We were taught by our grandmothers].

R: Ni fundiswe nini? Le rutilwe leng ho etsa dibeads? [When were you taught to work with beads?].

LM: Si fundiswe sise bancane. Si ne minyaka e 10. [We were taught when we were very young. We were 10 years old].

R: Nkgono yo a le rutileng, ene o ne a rutwa ke mang? [Who taught the grandmother who taught you to work with beads?].

LM: U fundiswe ngo mama wakhe. [She was taught by her mother].

R: So ho raya hore hangata Gogo o ruta ngwana, ngwana yo ha a setse a hodile o ruta bana ba hae. Jwalo jwalo. [So it means that many a times grandmothers teach their children, and when these children have grown up they also teach their children, and it continues like that].

LM: Njalo njalo [It continues like that].

R: Ho raya hore ha ho hlokahale hore le ye sekolong ho ithuta ho sebetsaka dibeads? [It means you do not need to go to school to learn to work with beads].

SS: A siyanga a skoleni [We have never attended school].

R: Jwale le entse tsohle tse di leng mona [So have you done all beadwork items that are here?] (Researcher asks pointing to all the items displayed around the ladies).

SS: Si enze konke, na nga se stolo. [We have done all the beadwork things here, including all the things in the store] (Sophie says this pointing to the back where the store is and the bead artefacts are sold).

R: Joale mona o etsang? [Now what are you doing here?] (Researcher asks what Lenah is doing, pointing to the artefact she is working on).

LM: Ngi enza igama le Manager wethu. [I am working on a name tag of our Manager].

R: Le etsa joang hore ho be straight? [How do you ensure that this part of the ornament you are making is straight?] (The researcher points to the straight part of the ornament in which LM is writing the name of the Manager at Lesedi Cultural Village. The Manager's name is Xolani).

LM: Indaba ise nhloko. O ya yazi [The matter is here in the head] (Linah says this pointing to her head. Later on, she further explains that they take two beads at a time). O I stopa kabini ngale [You take two beads that side].

SS: Si khetha umqamu o linganayo. [You choose beads of the same size]. (This is an additional explanation from Sophie about how the straight lines are made and maintained. Sophie then continues to explain how various shapes are made, for instance indicating that when you want to start at the centre of any artefact, you start with a big bead to indicate the centre).

R: Le tseba joang hore mona ke bead e kgolo, mona ke e nnyane? [How do you know that here you put a large bead and here you put a small bead?].

LM: Si bona nga mehlo. [We can see with our eyes].

R: Manje, ni bona ka njane ukuthi ni fake esingakhi? [Now, how do you see that you Must put so many beads at a particular point?].

SS: Si ya zi bala. [We count them].

R: Kanti ni yazi kanjani ukubala? Ni the a niyanga esikolweni. [How do you count them? You told me you have not gone to school].

SS: Si ya zibala. [we count them]

R: Hai. Ni zi bala ka njani? Le mpoleletse hore ha le a ya sekolong. [No. How do you count them? You told me that you have anot attended school].

SS: Si ya zi bala. Sithi Kunye, Bili, Thato, Kune, Hlano. Ku hla ngapha ngi ya jika Ngi bheke le [We count them. We say one, two, three, four, five. Then we make a turn to move in the other direction].

R: So, kutsho ukuthi ufuna u ku yenzani. If o batla ho etsa ntho e e riling, o a bala then o jike [So it depends on what you want to do. If you want a specific artefact, you count and then make a turn].

SS & LM: Yebo [yes] (The two ladies respond at the same time. The interview then continues to explore other artefacts and how they were made and what they mean in Ndebele culture).

3.5.3.1 Mathematical Concepts in Beadwork

In this excerpt of the interview with the two ladies, it emerges that they are using a variety of mathematical concepts that are part of the artefacts they are making. Firstly, they refer to *straightness of lines* in making some of the artefacts. Figure 3.3 shows AmaNdebele ladies in their traditional attires at Lesedi Cultural Village, in Johanesburg.

One of the ladies attributes this to their *sense of estimation* and actually mentions that they just watch and get a sense that the line is a straight as they need it to be. She suggests that this is based on the experience they have gained in using this skill many times. However, Sophie further indicates that the straightness is also maintained through the size of beads that are used. In fact, she mentions that *turns (angles)* are made using different sizes of beads. Even though they have not attended school, they clearly demonstrate that they know how to count by *counting*

Fig. 3.3 AmaNdebele ladies in their traditional attires at Lesedi Cultural village, Johanesburg. *Source* Personal file

Fig. 3.4 Beadwork artefacts at Lesedi Cultural village. *Source* Personal file

from one to five. Figure 3.4 shows the Beadwork artefacts at Lesedi Cultural Village.

This counting is crucial in their activities as it determines the *patterns and shapes* they make. In the context of indigenous knowledge which is passed from generation to generation, the ladies indicate that the knowledge of working with beads is generally passed from mother to daughter, and such skills can actually be taught to others, and in this case the young are nurtured into these activities, ensuring that the skills do not die but are kept alive for the benefit of the greater society, in this case, of the Ndebele people.

3.6 Language Issues from the Three Episodes

It can be noted from all the three episodes that the knowledge holders were more at ease and expressed themselves freely in the language that they are familiar and comfortable with. This can be seen in the interaction between the researcher and the knowledge holders.

For instance, in Episode One, the knowledge holders use the following language to illustrate to the researcher the extent of their knowledge:

> M: Eke. Ha ke batla ho etsa ho ho holo, ke tla sheba ka mahlo fela hore ha e le mona ke nkile ha kana, bo tla etsa boholo bo bo kana. Ha ke nkile ha kana, bo tla etsa bonnyane bo bo kana. [Yes. When I want to make a big thing. I just use my eyes to estimate any size I want to make. If I have taken this much, it will give me this small amount].

This response by the knowledge holder does not only illustrate free expression and use of language but a demonstration of deeper understanding of the process of language to express depth of knowledge in making this artefact.

The second episode is long due to the fact that it is based on the instruction given by the learner on how to make String Figure Gate Six. Even for the young knowledge holder (student), it is interesting to note that she is more confortable to be using her home language to give the demonstration.

At the beginning of the demonstration the researcher gave the learners an opportunity to use a language of their choice and the learner opted to use her home language. It can be deduced from this illustration that indigenous games are able to be conducted and demonstrated easier in the setting and environment in which they are usually played. This can be seen from the choice of the language by the learner.

3.7 The Implications of the Language and Its Use in Ethnomathematical Research

The three episodes used above illustrate the use of the different languages in ethnomathematical research in the South African context. As indicated earlier, South Africa has 11 official languages and it is expected that these languages will receive equal status in use in official communication. It is also expected that when research is conducted, it will take into account the status of such languages for the purpose of interaction, especially with the elders and knowledge holders in indigenous communities.

In fact acknowledgement and use of these languages is likely to enrich the interactions between researchers and those who hold the knowledge. Even though this situation of official languages may not be exactly the same in other countries, the use of language (in the official and unofficial sense) is equally important.

The episodes (at least one of them) have also shown that the lack of knowledge and basic understanding of the language limits accessibility to the depth of knowledge that is held by the knowledge holders and also limits a deeper interaction between the researcher and the knowledge holders. This is in line with the argument made by Alexander (2005) in the power of language and the language of power.

Alexander (2005) says that "Language is the main instrument of communication at the disposal of human beings; consequently, the specific language (s) in which the production processes take place become(s) the language(s) of power. To put it differently, if one does not command the language(s) of production, one is automatically excluded and disempowered" (p. 2).

In line with Alexander's argument, a researcher is disempowered when he cannot communicate in the language of production i.e. the language of the knowledge holder in this case. He cannot access the deeper meaning, trend of thought, historical meanings and their interpretations as they are held and understood by the knowledge holder. In this case the language of production is not skewed in favour of the researcher but in favour of the knowledge holder.

This chapter does not argue that it is not possible to conduct research when one is not knowledgeable in a specific language spoken by the holders of the knowledge even though this will surely affect the level of interaction between the researcher and the knowledge holder. It argues for recognition of this limitation and emphasizes efforts that need to be undertaken by those involved in ethnomathematical and other forms of ethnographic research to ensure that the results obtained are as representatives of the views of the knowledge holders as possible

References

Alexander, N. (2005). *Language, class and power in post-apartheid South Africa.* Lecture Presented at the Harold Wolpe Memorial Trust Open Dialogue. Iziko Museum, Cape Town, South Africa: University of Cape Town.

Amir, G., & Williams, J. (1994). The influence of children's culture on their probabilistic thinking. In J. P. Ponte & J. F. Matos (Eds.), *Proceedings of the 18th International Conference for the Psychology of Mathematics Education* (pp. 24–31). Lisbon, Portugal: PME.

Barton, B. (1996). Making sense of ethnomathematics: Ethnomathematics is making sense. *Educational Studies in Mathematics, 31*(1), 201–233.

Begg, A., Bakalevu, S., Edwards, R., Koloto, A., Sharma, S. (1996). *Mathematics and culture in Oceania.* Paper Presented to Working Group 21. ICME 8. Sevilla, Spain: Campus Universitário de Reina Mercedes.

D'Ambrosio, U. (1985). Ethnomathematics and its place in the history and pedagogy of mathematics. *For the Learning of Mathematics, 5*(1), 44–48.

Gerdes, P. (1994). Reflections on ethnomathematics. *For the Learning of Mathematics, 14*(2), 19–22.

Gerdes, P. (1996). Ethnomathematics and mathematics education. In A. Bishop, K. Clements, C. Keitel, J. Kilpatrick, & C. Laborde (Eds.), *International Handbook of Mathematics Education* (pp. 909–943). Dodrecht, The Netherlands: Kluwer Academic Publishers.

Fitouri, A. (1983). Working with young bilingual children. *Early Child Development and Care, 10*(4), 283–292.

Francois, K., & Stathopoulou, C. (2012). In-between critical mathematics education and ethnomathematics: A philosophical reflection and an empirical case of a Romany students' group mathematics education. *Journal for Critical Education Policy Studies, 10*(1), 234–247.

Mellin-Olsen, S. (1985). *Mathematics and culture: A seminar report.* Bergen, Norway: Caspar Forlag.

Mosimege, M. (2012). Methodological challenges in doing ethnomathematical research. *International Journal of African Renaissance Studies, 7*(2), 59–78.

Mosimege, M. (2017). Research protocols and ethical considerations in indigenous knowledge systems. In P. Ngulube (Ed.), *Handbook of research on theoretical perspectives on indigenous knowledge systems in developing countries.* IGI Global: Hershey, PA.

Prins, E. D. (1995). *The influence of readability of examination questions on achievement in senior secondary school mathematics: A study on verbal problems with special reference to second language readers. Unpublished Doctoral Dissertation.* Stellenbosch, South Africa: University of Stellenbosch.

Stathopoulou, C., & Kalabasis, F. (2006). Language and culture in mathematics education: Reflections on observing a Romany class in a Greek school. *Educational Studies in Mathematics, 64*(2), 231–238.

Statistics South Africa. (2012). *Census 2011.* Pretoria, South Africa: STATS SA.

Wilder, R. L. (1981). *Mathematics as a cultural system.* Oxford, NY: Pergamon Press.

Chapter 4
Techniques and Learning Processes of Craftswomen in Brazil

Maria Cecilia Fantinato and José Ricardo e Souza Mafra

Abstract This paper brings partial results on an Ethnomathematics research grounded on ethnographic roots, aimed at studying the techniques, processes and devices applied on curved surface registry called *cuias*, with a group of eight craftswomen, members of a craftswomen regional association, in Aritapera region, rural area of Santarém, Pará, Brazil. This paper intends to analyse informal learning processes developed during *cuias* crafting production and ornamenting activities. Fieldwork includes observation, free interviews, photographing and recording. Results indicate that the reflected iconography on these *cuias* shows multiple techniques and identified cognitive strategies, with wide-ranging metallic instruments and organic materials usage. The strategies are organised and multifaceted, related to each craftswoman's singular unity, and, simultaneously, to a collective unit capable of applying dynamism in constant relationships with nature. The dialogue between empirical evidence with the literature on informal learning processes allowed us to understand some important aspects of knowledge production/transmission of these craftswomen's community of practice, as well as their vicissitudes due to socio-economical and temporal factors. This paper aims at contributing to further studies on ethnographic researches in Ethnomathematics, as well as its unfolding in educational experiences.

Keywords Ethnomathematics · Ethnographic studies · Knowledge techniques · Informal learning processes · Craftswomen

M.C. Fantinato (✉)
Federal Fluminense University (UFF), Niterói, Brazil
e-mail: mcfantinato@gmail.com

J.R.S. Mafra
Universidade Federal do Oeste do Pará (UFOPA), Santarém, Brazil
e-mail: jose.mafra@ufopa.edu.br

M. Rosa et al. (eds.), *Ethnomathematics and its Diverse Approaches for Mathematics Education*, ICME-13 Monographs,
DOI 10.1007/978-3-319-59220-6_4

4.1 Introduction

This study, based on Ethnomathematics perspective and with sociological and anthropological theoretical contributions, is linked to a post-doctoral research carried out by the second author and supervised by the first author, during the year of 2015 (Mafra and Fantinato 2016). This research had as main objective the development of a study on techniques, processes and tools involved in the preparation of ornament patterns on the curved surfaces of vegetable gourds, called *cuias*, by a group of craftswomen who belong to a riverbank community dwellers of the north region of Brazil. The mathematical aspects involved in the techniques of ornamentation of the gourds, as well as the craftswomen learning processes, constituted our interest of research.

The western portion of Pará, one of the states located within the northern region of Brazil, is a diverse and singular area. Among its characteristics, it presents an important riverbank population, very particular in terms of cultural dynamics. There, handcrafting is the permanent and singular element in their lives and takes part in the local culture and in the natural dynamics of those who live there. Particularly, for a group of eight craftswomen who live throughout five neighbourhoods comprised within Aritapera[1] an area within the municipality of Santarém. These women express their daily lives, traditions, social relationships and the core of their existence in regional *cuias*[2] carving.

By means of a natural and singular representative knowledge and practice of their identities, they frequently get together to develop instrumental actions related to the productions of these *cuias*. They are organized as an association, *Santarem Riverside Craftswomen Association* (ASARISAN), started in 2003 and that joins dwellers of the area. Its main interest lies on the dissemination of natural handcrafting know-how as well as on the economic and cultural empowerment of *cuias* as a product of traditional knowledge. Through an anthropological, social and instrumental point of view, this research aims at learning more and deeper about these craftswomen's processes, motivations, techniques and tools used while developing carving registries on the surfaces of these regional *cuias*.

An additional purpose of this article lies on going back to empirical indications of this Ethnomathematics research, grounded on theoretical frameworks based on informal learning processes. We intend to do so by analysing existing processes and knowledge involved in the making of *cuias*. Conceptual tools from informal

---

[1]These areas are: Enseada do Aritapera, Aritapera Center, Carapanatuba, Cabeça d'Onça, and Surubim-Açu.

[2]This gourd (our *cuia*) is the fruit of a tree called *cuieira*, which is a traditional tree in the Amazon area. It blossoms different shapes, sizes, dimensions, such as roundish, ovallish or even slightly flattish. It is applied, specially, to drink *tacacá*, which is a typical Amazonian juice/soup, from indigenous origin. However, as time went by, many other artefacts have been produced applying the same raw material.

education studies were also considered, in dialogue with craftswomen's cultural practices, focusing on informal learning processes and knowledge transmission.

4.2 Ethomathematics and Ethnographic Studies

Throughout the last years, plenty of conducted researches and studies (D'Ambrosio 1993; Vergani 2009; Rosa et al. 2016) show the transdisciplinary nature related to the studies under Ethnomathematics perspective. The relationship between these studies and cultural anthropology, education, sociology and philosophy, indicate a meaningful broadness in a theoretical and methodological framework, capable of grounding researches and studies under the sociocultural bases.

A descriptive and thoughtful analysis of such opposing forces in a certain study context can deliver substantial meaningful convergence, validating discussions that involve indicators found at various knowledge areas, such as mathematics, history, anthropology and education. These indicators are unmistakably identified in Vergani's considerations, when she states that:

> Counting, localization, measurement, arrangement, play, description are activities which involve theoretical and practical perspectives besides critical expression. Ethnomathematics does not only fit anthropology, cognitive psychology, verbal language and aesthetics or playful expression. Its epistemological approach connects it to history, to general welfare and social justice. Its pedagogical approach gives room to common sense, social changes challenges and technological development (Vergani 2000, p. 37).[3]

Such considerations aim at the merging of a holistic conception of knowledge in the way D'Ambrosio (1985, 1993, 1997) suggests, that is, knowledge creation, its areas of acceptance, values and rules to be transmitted and spread widely. As such, they offer a connection and interaction with the natural environment, in such way that theoretical assumptions in Ethnomathematics perspective are present and alive from socially mentioned practices standpoints.

This knowledge can be linked to the development of symbolic codes and to the movement of such codes in their social lives according to the importance of its information and communication circle to the development of several activities. This improvement depends on the social environment stability and on its interrelationship with the surroundings, where such characteristics are produced and organized, consequently providing an "understanding/reading of the world from the same movement that (re)shapes such symbolic codes" (Vergani 2003, p. 134).

Related social cultural practices studies and investigations, as well as their connections with studies perspectives perceived by Ethnomathematics have been providing meaningful contributions in the last decades. At first, Ethnomathematics studies were closely related to ethnography and approached identified cultural

---

[3]This quote and the followings along this paper have all undergone free translation.

groups' know-how: "these groups could be based on their ethnicity, on their professional occupation, age, and even other aspects" (Palhares 2008, p. 14).

With the growth and improvement of this area of research, Ethnomathematics concept, developed by the leading Brazilian theoretician Ubiratan D'Ambrosio, was both re-evaluated and improved. D'Ambrosio's Ethnomathematics Programme (1993) replaced the idea of an ethnic mathematics, referring to the study of these different groups' know-how and their inherent cultural dynamics, by adding cognitive, philosophical, historical, sociological, political and, naturally, educational aspects. As a result, different sociocultural group studies under ethnographic framework emerged, prevailing for many years, even though this has been changing lately.

When analysing the graphs that classify presented studies in the four previous Brazilian Ethnomathematics congresses (Fantinato 2013), one can notice how representative the thematic axis *Mathematical education in different cultural contexts stood*, mainly in 2004 and 2008 congresses, matching *Ethnomathematics and pedagogical practices* axis, which concentrated more papers in 2000 and in 2012. Therefore, although these studies were once part of Ethnomathematics area, the ethnographic studies have been giving room to studies aimed at ethnomathematical ideas on pedagogical applications. From eight Ethnomathematics articles published by BOLEMA journal between 2006 and 2010, Costa (2012) found five of them on educational themes, and only one of them approached mathematical ideas present in different social groups' practices.

Even though we agree with Pais (2011), when he says that "a significant part of ethnomathematics research has educational aims" (Pais 2011, p. 210), we believe in the ethnographic studies liveliness in Ethnomathematics, especially when under transdisciplinary perspectives empirical data analysis (Vergani 2009). According to Clareto (2009), Ethnomathematics allows knowledge to be approached as an invention, as inventiveness and not as recognition of something that was already there. We propose, then, Ethnomathematics as "occidental mathematical re-identification" in other sociocultural settings, moving us towards originality, creating "possibilities to reach out for other senses, opening up other perspectives towards cognition and learning" (Clareto 2009, p. 130). Dasen (2004) says that:

> All societies, either from the North or the South, have their peculiar ways of transmitting their culture from generation to generation, away from formal education represented by schools [...] good knowledge about informal education can help schools adapt these cultural contexts they are immersed into the environment realities they are placed in, in much better ways (Dasen 2004, p. 23).

In this study, concepts on informal education studies (Chamoux 1978; Dasen 2004; De Vargas 2009; Greenfield 1999; Lave and Wenger 1993) are brought into light in dialogue with an Ethnomathematics research on *cuias*' craftswomen cultural practices.

4.3 Contextualizing the Research

This investigation was developed in riverbank communities in Aritapera region along the Amazon River, in the state of Pará, in the north of Brazil, three hours away from Santarém by boat. Lowlands such as Aritapera go through lots of natural events due to severe geographic changes in times of flood/draught along the Amazon River, the first half of the year, the region undergoes the flooded season, while during the second half they strive through the dry season. Riverbank dwellers live their lives based on seasonal influence in order to maintain their livelihood; fishing, agriculture, cattle farming, household caring. These *cuias* crafting production is an antique traditional cultural practice developed by the women in this area, especially during the flooding season (Maduro 2013).

The *cuias* harvesting—collected from a tree called *cuieira Crescentia cujete*, as well as the following phases of their production, such as cutting, scraping, dying and crisscrossing—benefits from natural resources. By using natural resources at each stage of production (fish scales for sanding and an organic purple pigment for dyeing) this activity has no impact on the local ecological balance or on the forest's regeneration.

There is a variation on the devices and techniques applied on the carvings. Every craftswoman seems to dispose of a wide range of creative possibilities (D'Ambrosio 1997), regarding both the size and decoration of the *cuias*, depending on their different uses. There are basically two kinds of ornament patterns: the floral or the *tapajonic* (Fig. 4.1).

The floral ornaments, based on the mixing of indigenous techniques and European rococo, are very traditional, and were learned through informal processes with grandmothers and mothers. Symmetrical patterns inspired on indigenous culture of the region of Tapajós River, named *tapajonic*, which were first

Fig. 4.1 Handcrafted *cuias* with *tapajonic* (in the *front*) and floral (at the *back*) patterns. *Source* Photo by the authors, with ASARISAN's authorization

introduced by a group of anthropologists working in the community, however, slowly became part of craftswomen's repertoire of patterns. The different patterns drawn on the *cuias* demonstrate a rich modelling diversification, both at an individual and craftswomen's group level, passed on throughout the years by the influence of their cultural heritage.

The *cuia* assigned for a floral carving has to be totally smooth and the craftswoman performs her work by freehand on the roundish surface. As for the *cuias* with *tapajonic* patterns markings, characterized by geometric carvings of indigenous origins, the first step consists of carving two parallel circumferences on the widest part of the bases which has the half-scooped sphere shape. The drawing with geometric patterns is then started within the delimited half-scooped sphere space between the circumferences.

The *cuias* are applied to many different usages in riverside people daily routines, such as drinking, water or food storing, bathing, water removal from canoes etc. Although commercial activities involving such cultural artefacts have been happening since the 18th century (Costa 2013), they have deeply increased within urban contexts in the Amazon Region, when such *cuias*:

> (…) settled an acquired visibility as a mandatory vessel for the drinking of *tacacá*, regional cuisine dish which became widely known in the touristic circuits, as much as turning into a point of becoming an eloquent national representative identity symbol from Pará (Carvalho 2011a, p. 25).

In 2002, by means of *Cuias de Santarém* project, aiming at supporting the production and commercialization of such objects, fruits of this *cuieira* tree, such as bowls, cups, fruit bowls, *maracas*,[4] besides the *cuias* themselves—, the Folklore and Popular Culture National Centre (CNFCP/IPHAN) carried out researches, photographic and audiovisual documentation, courses and workshops, *cuias* sales and exposition, always with these craftswomen's intense participation in the decision making. Later, CNFPC became a partner of Santarém Riverbank Craftswomen Association (ASARISAN), founded in 2003, rather than a direct enforcer in the communities. Among the first changes that arose from such support was the printing of an almanac presenting ornament *cuias* standard patterns based on Brazilian museums' research.

Such work has been useful for encouraging these *cuias* decoration practices, promoting a process of memory rebuilding and diffusion of a rich iconographic gamut. At the time when *Cuias* de *Santarém* Project was implemented, severe carving ornament depreciation was into process as the market was truly interested in plain or poorly decorated *cuias*, which meant low sales prices (Carvalho 2011a, p. 14).

In 2003, in Aritapera community, the creation of ASARISAN brought a new work concept for these craftswomen, as part of the individual activity was replaced by group work, performed within the core of each community. Production increased

[4]Maraca is an indigenous musical instrument, consisting of two hollow containers filled with beans or small stones. They are shaken to provide rhythm.

and these *cuias*' ornament started getting more and more refined as usages have also diversified. ASARISAN has also brought both ornament patterns changes and pieces grading to attend national market and cultural exhibitions. The craftswomen, for example, started using measurement instruments to classify the *cuias* according to four sizes, as when they were to be commercialized in sets solely for ornamental purposes.

The establishment of such Association also had its consequences in the familiar subsistence tasks' distribution. After the creation of ASARISAN, men alone have been almost entirely responsible for many of the other activities such as fishing, agriculture and cattle farming. Women kept household responsibilities and their children's upbringing, alternating with the *cuia*s' production, which gets a lot more intense in flooding season. This circumstance has also contributed to strengthen female identity along these riverbank communities.

In this research, we followed a group of eight ASARISAN affiliated women. The study was developed through fieldwork and consists of a detailed description of the making of *cuias*. It involves a detailed description of labour group activities, applied resources, instrumental techniques and know-how intrinsic characteristics of this group. Its central focus lies on the process study, motivations, techniques and instruments applied on the curvy surfaces, named *cuia*s in the region.

The investigation lies on qualitative nature (Bogdan and Biklen 2003) with ethnographic fieldwork development, carried out between 2014 and 2015. Data collection methods used included field notes, recordings, open interviews, photographs and video recordings. This research intends to depict a "thick description" (Geertz 1973) of the cultural context in which this activity is practiced, to develop a clearer understanding of the cultural dynamics implied, as well as the search for the sense of understanding on peculiar cultural dynamics and its complexities. The investigation took ground on the immersion of the researcher in the environment where the studied object lies as the main theoretical principle (Bogdan and Biklen 2003), designating the observer-researcher as the main instrument able to both collect and analyse the data.

As part of this research program, we also carried out an analysis of informal learning processes (Dasen 2004) involved in the making of *cuias*. We attempted to use some conceptual tools drawn from informal educational studies with these craftswomen, focusing on informal learning processes and knowledge transmission, as related to the subjects' cultural practices.

4.4 Techniques and Processes at Handcrafting Know-How: Mathematical Aspects

The material and immaterial order mechanisms applied in Aritapera craftswomen's work activities show instrumental actions, both in obtaining the raw material and in crafting the final product, that is, the ready or *scratched cuias*. Their shaping and their related products stages portray conceptual and concrete element knowledge from the visual perception of the one performing the action. Logic involved in the tasks performed moves towards a constructed and woven relationship with the physical environment in which they work. For instance, stylized flowers, fruits and animals are part of the ornamental repertoire.

The handling aspects involved in the act of shaping performed by these craftswomen, according to our observations, record procedures very much alike those performed in academic calculations, such as counting and application of proportionality principles. Some craftswomen perform calculations by estimate, that is, "mind calculations". Based on one of the craftswomen's speech: Avanilda: "we do it like that, we perform the calculation on our minds and get hands to work". Supposedly they have a very clear notion of the models and shapes to be performed as they have visualizing them on their minds for a long time.

Other cognitive activities which are mathematized[5] are related to the construction of bigger and or smaller pieces according to the dimensions of the storage bulk of the gourd. Such activities are performed according to proportional relations between the size of the biggest and smallest pieces to be made within an established set. Statements such as Lélia's, another craftswoman "we look at the model and reproduce it", or even according to Silvane, craftswoman: "depending on the size we want (…) we do it on our minds", to indicate the existence of a wide flexibility in terms of handcrafting activity cognitive variation.

Empiric knowledge, built throughout the years in the making of these *cuias*, offers us distinct elucidations to the detected episode. According to Silvane and Lélia: "to follow the same carving, we follow what we know, (…) we look at the catalogue or model and reproduce it".

During these *cuias* making processes, the existence of certain regularity in the dimensions of the produced pieces it may be noticed. The item's dimensions, as well the type of use it intended to, will serve as basis for these craftswomen when setting a suggested price. However, as it does not consist of a mandatory standardized procedure, pieces of similar size/characteristics may be found at different prices. According to the craftswomen, the prices are set during the moment of selling and are dependent on the mediation between them and the buyer or middleperson.

There is a variation on the devices and techniques applied on the carvings. As it seems, the craftswomen group imaginary and creative organization supply a wide range of possibilities (D'Ambrosio 1997) in the cravings and markings, related to different aims to the produced *cuias*. From the observed carvings, the talent towards different models, reflected in the imaginary artwork and patterns, shows rich modelling diversification and growth, identified in a personal unit (the craftswoman unique characteristics) and the collective unit (the craftswomen group),

[5]The use of the *mathematized* concept refers to the understanding, explaining, knowing, etc., referring to or daily needs. It suggests capabilities of classification, inference, problematization, and ability to relate things, among others, very similar to the characteristics observed throughout our empiric studies.

acknowledged by their gathered cultural heritage throughout the years, as Vergani (2003, 2009) states.

It seems there is an unavoidable absorption capability, offered by the interaction among distinct cultural dynamics, notorious in practice group communities living in the area. The endorsement made by Angeli, a craftswoman, who described one of the procedures applied in the reference markings: "Sometimes we calculate. Sometimes it doesn't work (Showing the way they perform the calculation). Half and half, naked eye". In other words, the imagination and creation/elaboration capability overlaps the predefined needs and aims, mainly screening related elements with attempt/error procedures as one of the core elements to the making of the final products (Lave and Wenger 1993).

While following up the work of the craftswomen, we identified along the production the development of many aspects, related to a kind of mathematical logic present in instrumental activities identified in the work with the *cuias*. Such aspects allowed us to make considerations about aspects of the manipulative act of modeling. In our observations, procedures not so distant from those used in calculations of the academy were identified, as described in the following.

4.4.1 The Use of the Compass in the Preparation of the Sections

Throughout our observations, we had the opportunity to observe one of the craftswomen perform the carvings on the *cuias* by means of a compass (cf Fig. 4.2). Lenil's handling of such instrument reveals unique skills in the records making and

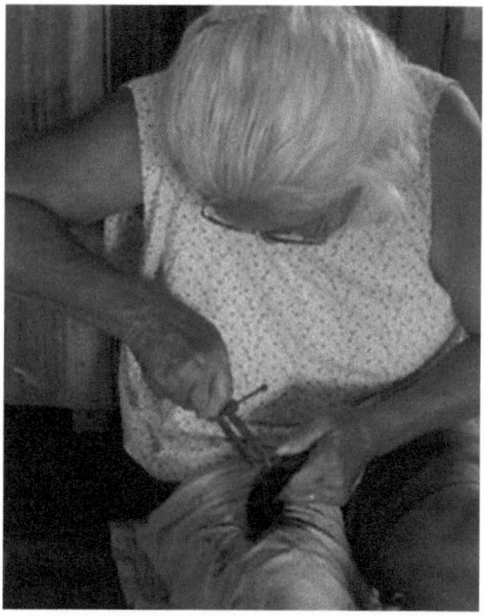

Fig. 4.2 Craftswoman handling the compass. *Source* Photo by the authors, with ASARISAN's authorization

in the layout, as it seems she is the only one to handle such tool in her instrumental activities.

The craftswoman Lenil, while performing records with the use of the compass, follows a few simple steps of instrumental organization in the initial markings, which are described as follows:

(1) First of all, a starting point is defined, being such reference required for the following actions. The geometrical place assigned as starting point is the center of the curved surface.
(2) The preparation of lines and thin traces in the form of circles is performed using a kind of visual metric. The center of the compass provides—using the point located at the center of the curved surface—successive openings, thus distinct circles are obtained, all of concentric rays, as indicated in Fig. 4.3.
(3) After the initial circles markings, a set of smaller circles and circular auxiliary sectors is drawn, for more specific ornament settings, originating, thus, the floral patterns evidenced in the sections, as shown in Fig. 4.4.

The tracing developed during the process of incised records points to a successive division of circular sectors, such as those indicated in Fig. 4.5.

However, the distribution of such segments along the surface, as it seems, is determined on the basis of successive approximations between the lines initially

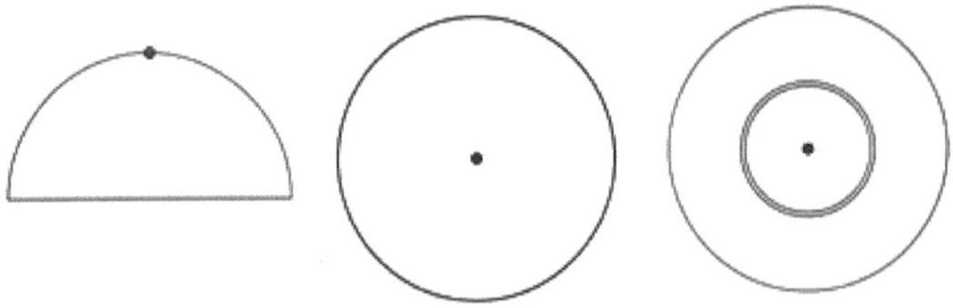

Fig. 4.3 Initial steps, with the use of the compass. *Source* Own authorship assisted by software

Fig. 4.4 Incisions made in a *cuia* using the compass. *Source* Photo by the authors, with ASARISAN's authorization

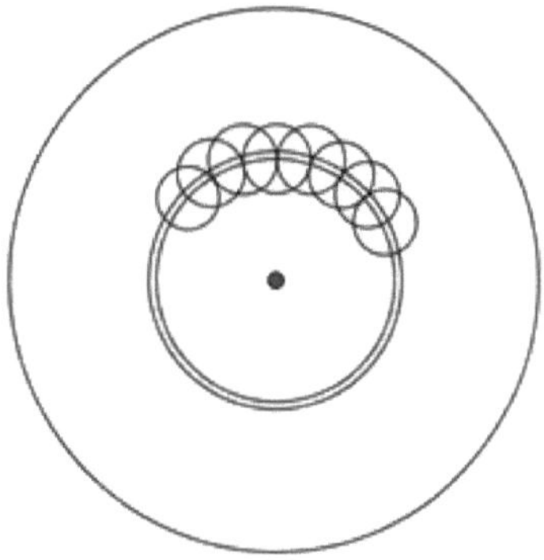

Fig. 4.5 Initial markings of floral patterns by using the compass. *Source* Own authorship assisted by software

drawn. Intersection of lines provides the creation of reference points, used for the preparation of the following sections.

Clearly there are evidences of a kind of qualitative mathematics, and also of the use of mental calculation for the composition of *eye* estimation, seeking to locate references of divisions and distributions of lines.

(4) It seems that, as the craftswoman uses the compass, employing a personal system of reference and qualitative measures the circle acts as the generator for the formats created to set a series of incised records.

The compass instrument itself is used to set the initial floral arrangement development. This first design is used as basis for the following steps—of scraping and scuffing, which will result on the final floral pattern. Considering such context, we infer that the craftswoman has gained experience through her connection with the instrumental activities developed in her working context. That is, by employing her compass handling skills, obtained from her instrumental activities know-how and from her personal life story, she transfers her values and essence to the pieces, bringing out a meaning which adds beliefs, emotions and feelings.

White (1988) states that the individual acquires a culture by means of inherited customs' learning. This is what is noticed when observing the craftswoman, as she learned to appreciate nature and reproduce/reflect herself, in a way it would portray knowledge and understanding produced through processes and controls developed in the making of *cuias*. When questioned about her procedures and techniques applied to the development of *cuias*' carvings, Lenil, the craftswoman, gives us evidences on how knowledge organization and transmission are performed when producing those *cuias*, making ornamental patterns with a compass.

> Lenil: each one of us here in our area has a particular type of (…) (referring to the *cuia* making). Lélia (another craftswoman) knows every little thing when she sees the mark (the scuff on the *cuia*) (…). When it's my turn to do it, it's different. She hasn`t seen me doing this type.
>
> Researchers: Are you an expert, Mrs?
>
> Lenil: Does it look the same (the other *cuia*)?

Researchers: No way!

Lenil: With my mother, in those days we already had card holders… (talking about other types of pieces) back then we had to use the compass to mark.

Researchers: When do you use the compass, Mrs? Or don't you use it anymore?

Lenil: I still use it, always. I still do. I'm going to make this one here (performing the layout with the compass).

Researchers: Do you always use the compass this much, Mrs?

Lenil: I always use. They always ask me to make a fern tree, or any other thing, I am always coming up with something new.

Researchers: But, do you decide on the layout to be applied on it yourself?

Lenil: *Yeah*. It can look really nice if we remove this little white here. Do you *wanna* see it looking different? When my husband buys big *cuias,* I get it and cover it with wide circles, and then I just do it.

Researchers: Do you choose to use the compass on those which are truly round?

Lenil: It draws and nothing gets lost (making lines with the compass) (04/08/2015).

The records made with the compass showed an iconography reflected in the *cuias*, which points out to motivations especially related to social representations evidenced in the community dynamics. Although the activity developed with the compass is considered a singular instrumental action, we identified how much social interactions exist in its environment, contributing to a flow of permanent and diverse knowledge exchange.

4.4.2 Two Possible Forms of Mathematical Distribution for the **Tapajonic** *Patterns in* **Cuias**

Many patterns are used as forms of records during the process of marking the *cuias*. The establishment of sequences is one of the most common mathematical ideas that can be found in different cultural contexts.

While following up the work with the group, we were able to identify some strategies of markings employed in the *tapajonic* patterns. Two possible aspects are highlighted in the work of craftswomen. Léia does it in one way (equally spaced markings, all at once), while Silvane performs the markings one by one (at the end she makes approximations for the last marks, in such a way that they remain at equivalent distance).

The spaced markings, developed by craftswoman Léia, are performed in a uniform manner, according to the following steps:

(1) The *tapajonic* patterns, for the most part, are produced on the lower surface of the *cuia*, delimited by two parallel traces and with no defined metric distance (see Fig. 4.6).

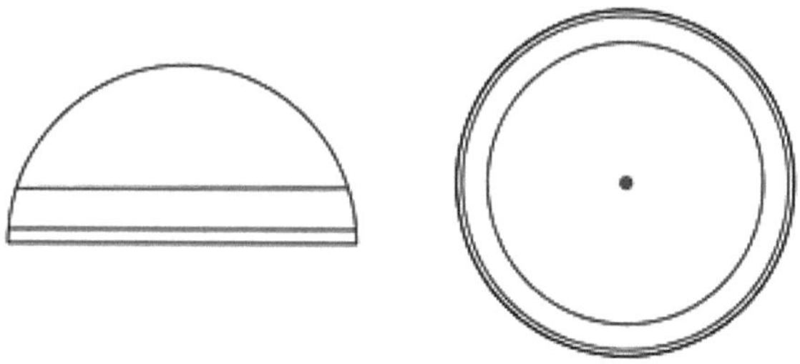

Fig. 4.6 Initial markings of *tapajonic* patterns making. *Source* Own authorship assisted by software

(2) The internal records are drawn within the lower and upper limits of the rectilinear line, in a multiplicity of possible patterns and formats. Among these formats, triangles, squares, lozenges or other polygons are used to compose the drawings.
(3) The spaced records have a uniform incision field, generated from the intervals between the markings, by choosing an initial segment, perpendicular to the two parallel tracings, on the edge of the *cuia*. There is a diversified repertoire, in relation to these standards. Some examples of patterns are presented in Fig. 4.7, drawn out from Carvalho (2011b, pp. 184–185).
(4) The marking of the lower and upper limits is done only once, by distributing the reference points throughout the field of recordings, so that these benchmarks might be used for the geometric construction of the illustrations. The craftswoman constructs the spaced segments visually, without the use of standardized metrics. It seems that her metric is conceived from the visual and global scaling of the *cuia* to be worked on, as illustrated in Fig. 4.8.

The other strategy, adopted by craftswoman Silvane, shows a style of marking according to a technique that we have chosen to denominate *marking by approximation*, as described in the following steps:

Silvane's steps 1 and 2 are identical of the strategy adopted by Léia.

(3) Silvane initially marks the interval spaces, in which the illustration will be recorded and works one segmented space at a time.
(4) In order to *close* the distribution of segments within the *cuia*, Silvane uses a sort of body metric (her thumb), drawing up estimates of approximation between the distances of the last markings, as to minimize any distortion in terms of the distance between the segments, as illustrated in Fig. 4.9.

The procedures applied in the making of the carvings show us a strong sensorial aspect, by means of our basic senses, mainly the sight, which shows estimate identification possibilities and measurements using the naked-eye (Shockey 2002). Additionally, it demonstrate intersected elements such as intuition, understanding and apprehension of available aspects within a certain context, elements which are meaningful to the construction of local knowledge, produced according to the needs and intrinsic limits of the natural environment.

Fig. 4.7 Some *tapajonic* patterns used in the registry of sections in edges. *Source* Carvalho (2011b)

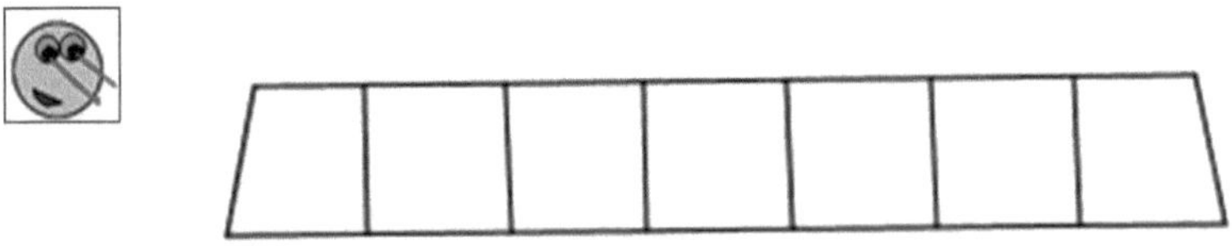

Fig. 4.8 Illustration of spaced segments, in the edge of *cuias*. *Source* Own authorship assisted by software, combined with image from https://gartic.com.br

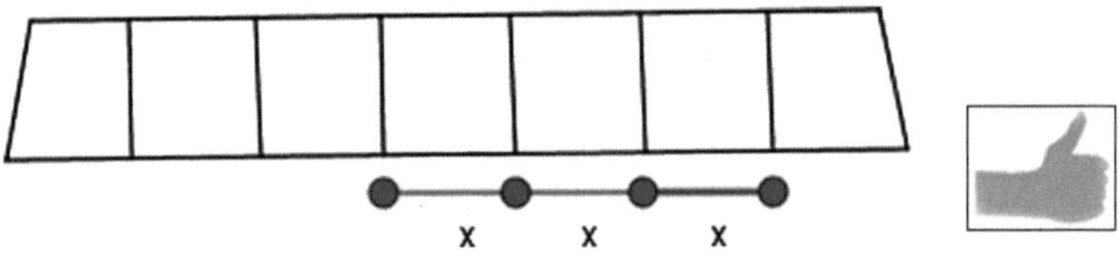

Fig. 4.9 Strategy adopted by Silvane in the final distribution of markings. *Source* Own authorship assisted by software, combined with image from http://bose-lady.net

The activities identified in the ornamental *cuias* making phases reflect multiplicative procedures, considering the essential elements in the pieces making process. On the other hand, the peculiar character perceived in each work is undeniable. The singular touch of personal elements, such as shapes and portrayed representations on curved surfaces, suggests elements related to *the skills* which each one of them owns in the development of their instrumental activities.

Such characteristics, we believe to be grounded on aspects such as the collective point of view, which coincides with the anthropologic concept of human being, capable of producing, and also with the knowledge disseminating perspective, considering day to day labour needs.

The records, revealed along the work performed with the compass, show an iconography reflected on the *cuias* which leads to motivations specially related to social representation highlighted in the community dynamics. Although the activity performed with the compass is considered a singular unit in terms of instrumental action, we noticed how much this environment of existing and probable social interactions pervade a flow of permanent changes of diverse knowledge.

Diversified surface patterns are used during carving recording outlining process. Patterns setting is one of the most recurrent mathematical ideas in different cultural contexts. The usage of models and compasses establishes a strategy and labour production enhancement technique. Therefore, mathematical ideas present in the carving records conception on *tapajonic cuias* reflect work dynamics related to the structural organization of the social context it belongs to, considering local processes, organization and aesthetics.

The (holistic) harmony of such principles allows us to show the movement developed by the related knowledges with Ethnomathematics. The way the craftswoman displays the images reflected on the *cuias*' surface is strictly connected to these principles as emotions are still attached to the craftswoman throughout the carving records creative process, as perceived during the ethnographic research.

As such, we understand identified Ethnomathematics ideas reflect the manifestation of multiple emotions by the craftswoman. The emotion to afford a professional activity, to explain depictions when developing *cuias*' scribbling process, which makes us think about a rich array of possible outcomes, especially in terms of its application in educational environments.

Such line of thought demonstrates how much the cultural heritage printed in the dwellers' knowledge is exposed to diverse ponderings, among which we point out ideas and identified mathematical strategies in their relation with other knowledge mechanisms, interlocked, likely to be maximized for example, in educational environment placed around Aritapera communities.

4.5 Informal Learning Processes Among Aritapera Cuias' Craftswomen

Production phases involve a very diverse set of knowledge that could be analysed from an Ethnomathematics perspective. In this topic, the focus will be on informal learning processes and knowledge transmission among Aritapera's craftswomen.

These *cuias* production and ornament ownership process can be defined as Aritapera general technical knowledge, however private (Chamoux 1978) to the women's group. Although some men can eventually help in some phases, as in the fruit harvesting, it is understood as a typically female activity, passed from generation to generation: "it has been running through our forefathers, it is already historical. I learned with my mother, with my grandfather and there it goes" (Lélia, 03/8/15). Since young age, girls observe the daily activity of the women of the group in their process of *cuias* production.

We can define Aritapera craftswomen group as a community of practice (Wenger 1998), as they share the same knowledge and practices repertoire, gathering in order to search for common objectives, and as their collective activity creates an identity feeling for the group. Another relevant concept applied in this study that provides an important tool to approach learning constituted as a social practice refers to *legitimated peripheral participation* (LPP) (Lave and Wenger 1993), which analyses "forms of adhesion and of identity construction, embodying place, acquired practice organization, development, reproduction and social transformation cycles in a community of practice" (Fantinato and De Vargas 2006, p. 3). By LPP we mean:

> (…) to draw attention to the point that learners inevitably participate in communities of practitioners and that the mastery of knowledge and skill requires newcomers to move toward full participation in the sociocultural practices of a community. 'Legitimate peripheral participation' provides a way to speak about the relations between newcomers and old-timers, and about activities, identities, artefacts, and communities of knowledge and practice (Lave and Wenger 1993, p. 29).

In the craftswomen's collective context, the knowledge transmission process related to these *cuias*' process happens by means of LPP. The apprentices, joining the craftswomen group, start performing carving and ornamental activities gradually, directed by the craftswomen through a well-defined progressive teaching/transmission process. The apprentices start off by learning the types of pattern which are considered the easiest.

In Carapanatuba community, Marinalva does not scratch. On the other hand, Silvane can only draw *tapajonic* carvings while Lélia is very good at drawing floral patterns. Therefore, knowledges related to this outlining can be considered as particular technical knowledge (Chamoux 1978), which depend on some specific skills. Angeli, from Enseada community, says that not all of them are skilful in graphics, and "she (Lenil) was the only one skilful with the compass" (05/8/2015). This lady chooses the *cuias* with round basis to apply this outlining tool.

The little girls and the apprentices learn through *impregnation*, process to which Chamoux (1978, p. 63) supposes two conditions:

> Firstly, it is based on a corporal training common to every single member of the village group: gestures, postures, material perception ways, language […] This training is connected to what is generally called group culture. Secondly, it implies in observation repetition of different techniques and gestures experimentation.

However, the same writer points out that in case one of these two conditions is not fulfilled, impregnation does not happen by itself. So, it requires a master to allow the conveyance of such know-how (Chamoux 1978). In the craftswomen's community practice context, this role is performed by the mother, the grandmother or even a more experienced person, such as a friend.

Throughout the fieldwork, we managed to observe an interaction situation between two craftswomen, Lélia and Marinalva, in which Lélia played the role of the master (cf Fig. 4.10). Marinalva was working with a large knife, building the support base for a fruit bowl, adjusting the piece gradually into a spherical shape. Then, Lélia took the piece Marinalva was producing and carried out a test placing a fruit bowl in the hollow semispheric shape on the piece Marinalva had been

Fig. 4.10 Two craftswomen interacting during *cuias* production. *Source* Photo by the authors, with ASARISAN's authorization

working on in order to check the balance of the piece. Lélia showed the set to Marinalva, who was attentively watching her friend and master, and pointed where the piece needed a few adjustments.

After the demonstration, Lélia handed Marinalva both pieces. Then, Marinalva applied the large knife in the areas in need of modifications, while looking at her friend/master. Right after verbal confirmation, she resumed to her scraping activity with the large knife. Minutes later, Lélia asked Marinalva to hand her the piece which had just been fixed and carried out the same balance test, saying:

Lélia: Look Marinalva, look.

Marinalva: Yes.

Lélia: There is just this one more little thing here to fix, but reduce it from here, can you see?

Then Marinalva gets hold of the large knife to sharpen another knife which she was working with and remarks:

Or we break it or we fix it (04/08/15).

And Marinalva carried on with her scraping work with the large knife.

The last step in the production of *cuias* consists of outlining, when the women make use of small knives to carve the black surfaces of these *cuias*, already dyed with *cumatê*[6] dye. The floral ornaments which represent a mixed traditional crafting, "based on the mixing of indigenous techniques and European rococo, were preserved and were transmitted in the craftswomen's collective memory" (Costa 2013, p. 41). Therefore, the floral pattern is learned through informal processes with grandmothers and mothers.

In contrast to the floral patterns, the *tapajonics* follow a different learning process as they were not used on the *cuias* ornament from Aritapera before the creation of ASARISAN and CNFCP/IPHAN enterprises. As previously mentioned, such

[6]*Cumatê* is a purple natural pigment produced out of an Amazon tree bark.

projects resulted in the printing of a catalogue of patterns from indigenous origin, as well as on the conduction of courses and workshops to encourage its spreading in the community. Raimunda, one of the craftswomen, confirmed these facts as she talked to the researchers:

> Raimunda: These ones here, these carvings here, we started working with them, when we got together, after getting everybody together. Then, there was this young man and then we managed to do it, thanks God. We managed to create the association, then we improved these carvings, more and more.
>
> Researchers: Did you use to make these *tapajonic* patterns carvings before the course?
>
> Raimunda: Only floral. We weren't taught all kinds of patterns at the course. There were few. Few graphisms. From that moment, we started creating, imagining, thinking how to do it. Sometimes, we picture a drawing on our minds and figure out how to do it (05/08/2015).

Raimunda and Lenil say this motivational work "awoke the creation of many kinds of pieces" and that it represented a cultural identity recovery. From *tapajonic* models learned in the workshop, the oldest craftswomen started creating plenty of different variations on such patterns. Little by little, they started sharing this knowledge with the youngest craftswomen. Angeli stated, pointing to two older ladies of her group "they both took part in a workshop, somebody came to present there. I did not take part in it, but I learned with her" (05/08/15).

Consequently, we found variations of *cuias*' decoration knowledge acquirement and transference in Aritapera craftswomen community. Raimunda said she picked *tapajonic* patterns in a course, therefore she learned in a formal situation. However, she transferred all that knowledge to a friend of hers, Angeli, informally.

We agree with Greenfield (1999) when she says that "as cultures change over time, the very processes of cultural learning and cultural transmission also change" (p. 57). However, the distinctions between formal and informal education, depending on the context they refer to, hold an array of shades, they are not dichotomous. De Vargas (2009) grounded on Greenfield and Lave (1979), mentions the existence of an informal/formal *continuum*:

> Greenfield and Lave stand their characterization grounds on the dichotomy which can be, in many cases, over generalized, in such way that it becomes more relevant to have a continuum as a reference, a grading among several levels and not merely an opposition (De Vargas 2009, p. 195).

This craftswomen's knowledge towards the outlining on these *cuias* is, therefore, dynamic. The *cuias* making art traditional knowledge and floral decorations started interacting with knowledge brought from the outside and learned by more formal learning means. In turn, this new knowledge related to these *tapajonic* carvings was incorporated by these women in the exercise of their practice, resulting in the generation of even newer knowledge, including that associated with this type of pattern.

We could deduce so when observing these women carve the *cuias,* as they create varied patterns without following any printed templates. Lélia even admitted she knew the catalogue with the *tapajonic* patterns, but she would rather follow, as she says, the ones she had on her mind. Raimunda's words summarize this creative process: "we start doing it, start creating, it is imagination flow" (05/08/15).

This dynamic aspect can also be observed towards the characteristics of a specific technical knowledge (Chamoux 1978). This represents a good example of gender-specific knowledge, that is, in this case, knowledge transmitted strictly between women. However, Angeli told us one of her sons is learning to carve as well. As it is a prominently female activity, this fact called our attention. We suppose it is the direct result of a respect process of these craftswomen activity in the familiar context, as after the creation of the Association, they started making an outstanding difference in the family income.

4.6 Craftswomen's Techniques and Informal Learning Processes: Issues for Ethnomathematics

A wide range of aspects are still to be investigated. What is behind this knowledge? What is the meaning and importance of these instrumental activities developed in the *cuia* handcrafting for these communities in Aritapera? How social representations, reflected on the round surface carvings, become meaningful to the craftswomen group? These questions will find room when theoretical constructions in further investigations are viable and its focus rely on open questions inquiries.

Based on our observations, nearly the whole group runs or develops actions together or close to identical action, related to the established phases through the construction process of their activities. However, these arguments do not grant actual occurrence of a single reproducibility towards the pieces' decoration as we suppose each craftswoman acknowledges distinctive perceptive behaviour, established by their own peculiar nature.

It shows us, regardless the broadness of the context, that the cognitive construction elaboration is to be expected, diverging only on how they are performed, through constituent strategy of each specificity. Knowledge nature and work motivation established throughout *cuias*' preparation phases can set a dynamic aspect to it, from the universal point of view, so elements' organization which are available and accepted as necessary to the construction and to the mobility of this knowledge must be possible.

Material and immaterial access elements in our context, including biological factors, deep rooted costumes and rooted cultural speech, allow us to identify contexts' specificities, based on the activities construction and operational procedures from nature itself. Thus, we acknowledge a man not excluded from society, as his prime stimuli or *prime science* as an apprehension pivotal element, key to his cognitive knowledge development process.

Therefore, we have chosen to highlight the artefacts' making as an internal order factor, an internal stimulus and it also allows whoever is the raising kind of *cognitive mobility* perpetuity, important to transcend the present moment in search of a model or a different representation of what had been previously perceived, and highly connected to the produced models, set and rooted on the craftswomen's mind. Thus, this perpetuity lies merely on the inexistence of atypical shapes in variability terms and never made by the craftswomen.

The craftswomen have been making these very much similar records on *cuias*' surfaces for a long time. When ASARISAN started, a new and more meaningful amplitude emerges and embraces even unimaginable possibilities. For example, the request for new shapes, demanding improved skills from them, such as the making of atypical layouts, and the elaboration of a catalogue with diversified dimensions, shapes and iconography.

The incorporation of new layouts and shapes demanded a new organization in terms of possible settings, however it is noteworthy how the craftswomen refused to neglect their heritage and traditional layouts and incorporated their own cultural background on their work. Permanent change perceptions, in terms of instrumental procedures, seem not to jeopardize their own set of practices. In our specific study, the craftswomen's group provides important data to exemplify what Almeida (2001) calls *science* of tradition, which may take into consideration sensitive, subjective relations and *un-rational* as meaningful importance in existing proceeding systems understanding in sociocultural context.

Several questions are yet to be investigated. What lies beneath this knowledge? When it comes to labour activities in pottery, what is the real meaning and importance of these communities in Aritapera? When and where do these beliefs become meaningful to the craftswomen's group? All these research questions are going to find significant answers if we establish a systematic mapping considering the meaning each of them hold to the life to which they belong. So they become viable by setting proper theoretical constructions in further investigations, concentrating on open ended questions.

When liaising with this group while in our study, we have noticed the existence of heterogeneity in terms of procedures and attitudes by their members. This is quite ordinary in any context where diverse and universe co-exist. In our study, we have tried to emphasize the study of applied procedures in the *cuias*' carvings record making, along with the broad understanding of the guiding and structural elements in theses labour activities.

These studies on non-formal learning processes help us understand production/transference knowledge aspects of practices communities, such as the craftswomen, producer of *cuias*, in Aritapera, as well as their own transformation. Over time, social and environmental factors, the cultural dynamics of the encounters, as well as worldwide economic globalization have influence on practices once considered as traditional. Greenfield (1999) stated that in traditional societies which are in changing processes, knowledge transfer, which was primarily performed by scaffold guidance shifts to being done through creation processes and trial-and-error experimentation.

As the craftswomen in Aritapera do not live isolated, interacting with people from big urban centres, they fit their productions to the needs of these centres, constantly re-elaborating their knowledge. We can discuss a formal/informal *continuum* in the production knowledge transfer during these *cuias*' production and ornament. Knowledge related to *tapajonic* patterns outlining was introduced into the community by means of formal processes, but were transmitted from the craftswomen to the youngest ones informally.

We wonder if the ornament practice has not been learned through impregnation and by means of LPP, even though there had been a learning situation promoted by a more formal structure afterwards. This idea is corroborated by the fact that, the new aspects (the *tapajonic* patterns) were transmitted to the youngest craftswomen informally, as it used to be.

This matter raises questionings towards Ethnomathematics researches, mainly those attempting to articulate school and sociocultural knowledge for some groups. Most of the times these researches are mainly carried out in schooling contexts, unaware of the deep complexity existing in communities of practice (Wenger 1998) where they are taken from and very superficially adapted to the curricula by means of mathematical knowledge examples.

One cannot forget learning goals at schools are basically propaedeutic, very different from survival and transcendence goals which (D'Ambrosio 2006) summons knowledge production in domestic and professional lives contexts. Shall this *bridge* become more difficult at school due to the assumed artificiality of the context? Close attention to traditional knowledge learning and transfer processes in ethnographic studies in Ethnomathematics can truly contribute to deepen our debates.

It is important to avoid in Ethnomathematics researches an ethnocentric and legitimist look, which intends to notice merely manifestations of what can be understood as mathematics, and which ends up treating sociocultural group knowledge as exotic. The transdisciplinary and holistic view is central to perceive the quantitative and spatial multiple relations within social, cultural, economic, geographical and historical contexts. We also think the knowledge and cultural practices which these students share in their original groups, when incorporated in the school curriculum, can contribute to the permanence and renewal of these practices and traditional knowledge within their original contexts.

The range of artistic possibilities and technical, strategic and material investigation applied, lead our thoughts towards future pedagogical purposes related to Ethnomathematics truly, specially framed for teachers training courses. Learning environments placed around riverbank communities, close to Santarém, Pará, can take advantage of the identified principles in the making processes of these *cuias*, in such way they can be brought inside the classroom.

This guidance aims at partnering with many other mathematical instrumentation mechanisms, in such way that knowledge gathered from these craftswomen's work may be turned into pedagogical action organizations, so as to improve skills and competences related to measurement, counting aspects and geometrical topology. The conceptual characteristics of such mechanisms can be settled in these environments, along with other subject areas, suggesting a possibility of interdisciplinary integration among knowledge areas capable of showing students a wide range of possibilities to handle daily circumstances.

4.7 Some Possible Methodological Developments Towards the Teaching of Mathematics

The topological notion of such sizing settings in curved spaces and surfaces makes us wonder about sizing, outlining and drawing referential construction's capacity on these types of shapes. Once again, we refer to the setting and making of a *qualitative* mathematical inference, in which agreed references on surfaces allow setting both guiding points alongside structural points applied in the making of such records.

These performed action expressions show elements connected to knowledge of distinct mathematical nature, in a highly-articulated manner, in which mathematical knowledge reflect transdisciplinary and holistic knowledge perceptions. Multiple relations are set due to the strategic action context applied to symbolic and spatial elements which blend with social and cultural contexts and these craftswomen's experience. These interactions allow some elements' prognoses and constructive action factors related to Ethnomathematics goals in terms of educational framework elements.

This investigation, under Ethnomathematics program principles, allows us to foreknow elements of influence over a series of aspects within mathematics teaching reality, mainly when considering the aspect related to life-like reality in which the student is in and mathematics knowledge acquisition. Such approximation can be discussed in terms of interlocution, discussion and understanding of distinct solving procedural methods of similar activity, either formal or informal. This analysis perspective lies on the assumption that it should be possible to understand the mathematics processes interconnected with meaning, with an existing reason which should clearly promote the ability of gradual abstraction and reasoning ability, creativity and cognitive process development.

We might mention, for future researches, the systematic study of patterns and metrics applied by the craftswomen and their similarities in terms of applied strategies in educational settings, by students. Mental calculation elements provided by these craftswomen, through the development of their activities suggest strong connections to cognitive activities like the ones we face under didactic teaching environment and therefore, we suggest further studies to be sensitive to such specificity.

Another possible example for this situation derives from geometry taught at school—the same that is strictly connected to Euclidian Geometry, that is, Geometry, which allows us to appoint answers when we propose them through constructions with rulers and compasses. When we witness, these works performed by these craftswomen and the perfection with which the carvings are done, by means of strategic constructions and cognitive elements, we understand it to be very close to geometric principles susceptible to regular school learning.

When analyzing the different activities developed by the craftswomen, outsiders, such as students, may understand that such strategies represent techniques and processes conceived in order to solve daily need situations. From the point of view of school curricula, the endorsement of this knowledge as learning and categorical elements placed in educational environment becomes indispensable, in such way that balance between these dialogical areas is likely to be accomplished.

A proposal which can be made to contribute towards formulating a response to the questions placed previously consists of *immersing* parts of the educational community—students, teachers, parents, directors, legal representative etc.—directly into the context of these craftswomen's work, aiming at identifying, locating and understanding social cultural practices which include these women's and their own community life dynamics. However, pedagogical procedures compatible to local reality are necessary.

It is understood as sort of an intervention which can evolve to an action research, aiming at shifting the existing temporary scenario into a long term one. D'Ambrosio (2006) says that "An educator's intervention aims at refining practices and thoughts and instruments for criticism. This improvement does not happen as an imposition, but as an option" (p. 81).

Such actions, which are subjected to the recording of these singularities, have specific time and specific space yet to be figured out. Therefore, this knowledge allows technological organization (to the purest concept of the term) towards the purpose of providing these craftswomen's basic needs. From a knowledge philosophy standpoint, this could result in an improvement in school curricula, so as to establish a dialogue between the multiplicity of thinking and acting ways of this specific group and syllabuses taught in the classroom.

By that, we mean not only subjects such as Mathematics, but any others somehow connected to this process, as History, Geography, Arts or any other area seeking a work based on an experienced reality. Such curricula could contribute to students' intellectual and human growth.

These instrumental activities developed by these craftswomen show us a new perspective in the shape of distinct implicit and singular characteristic knowledge, reflected through codes and some representative patterns set in performed carvings and records. Through these signs and social representations shared by Aritapera dwellers and acknowledged as elements of value and meaning, there is a permanent transcendent and dynamic flow connecting these river bank residents.

Knowledge and culture promotion must be an educational permanent practice in the classroom. Our perspective somehow is close to the fifth approach, presented by Adam et al. (2003), when they classify ethnomathematical researches that establish relations with education. We quote Pinxten and François (2011), who believe that Ethnomathematics must move on from being the *car repair* department of mathematics education, to become its main avenue:

> (…) what goes wrong in the design and sales department of (mathematical) cars will be mended and repaired by the culturally sensitive zealots of Ethnomathematics, we say in an oversimplified and challenging way. We suppose that since we consider Ethnomathematics to be the generic category, its educational part would be the main avenue of mathematics education (Pinxten and François 2011, p. 267).

As seen throughout the craftswomen's work, mathematics ideas come along and provide support to the development of their final work. Therefore, getting students closer to such aspects could help the process of promoting an awareness increase towards the topics to be taught in the classrooms. Thus, knowledge and appreciation of the culture of socio-cultural groups, such as those of the Aritapera craftswomen, should be a constant in the educational practice in the classroom.

References

Adam, S., Alangui, W., & Barton, B. (2003). A comment on: Rowlands and Carson "where would formal, academic mathematics stand in a curriculum informed by ethnomathematics? A critical review". *Educational Studies in Mathematics, 52*(3), 327–335.

Almeida, M. C. (2001) *Complexidade e cosmologias da tradição*. Belém, PA: EDUEPA; Natal, RN: UFRN/PPGCS.

Bogdan, R., & Biklen, S. (2003). *Qualitative research for education: An introduction to theories and methods*. Boston, MA: Allyn and Bacon.

Carvalho, L. G. (2011a). Artesanato e mudança social. In L. G. Carvalho (Ed.), *O artesanato de cuias em perspectiva: Santarém* (pp. 19–47). Rio de Janeiro, RJ: IPHAN.

Carvalho, L. G. (Ed.) (2011b). *O artesanato de cuias em perspectiva: Santarém*. Rio de Janeiro, RJ: IPHAN.

Chamoux, M. N. (1978). La transmission des savoir-faire: un objet pour l'eth-nologie des techniques? Techniques et culture. *Bulletin de l'Équipe de Recherche, 191*(3), 46–83.

Clareto, S. M. (2009). Conhecimento, inventividade e experiência: potências do pensamento etnomatemático. In M. C. C. B. Fantinato (Ed.), *Etnomatemática, novos desafios teóricos e pedagógicos* (pp. 125–134). Niterói, RJ: Editora da UFF.

Costa, W. N. G. (2012). Imagens da etnomatemática em periódicos brasileiros. *Unión Revista Iberoamericana de Educación Matemática, 32,* 165–180.

Costa, A. G. M. (2013). Iconografia atualizada das cuias do Aritapera. In A. M. S. Santos & L. G. Carvalho (Eds.), *Terra, água, mulheres & cuias: Aritapera, Santarém, Pará, Amazônia* (pp. 40–48). Santarém, PA: UFOPA.

Dasen, P. R. (2004). Education informelle et processus d'apprentissage. In P. R. Dasen & A. Akkari (Eds.), *Pédagogies et pédagogues du Sud* (pp. 23–52). Paris, France: L'Harmattan.

De Vargas, S. (2009). Estratégias não-escolares de ensino-aprendizagem e formação de professores da EJA. In M. C. C. B. Fantinato (Ed.), *Etnomatemática: novos desafios teóricos e pedagógicos* (pp. 193–201). Niterói, RJ: Editora da UFF.

D'Ambrosio, U. (1985). Ethnomathematics and its place in the history and pedagogy of mathematics. *For the Learning of Mathematics, 5*(1), 44–48.

D'Ambrosio, U. (1997). *Transdisciplinaridade*. São Paulo, SP: Palas Athenas.

D'Ambrosio, U. (1993). Etnomatemática: um programa. *A Educação Matemática em Revista, 1*(1), 5–11.

D'Ambrosio, U. (2006). *Ethnomathematics: Link between traditions and modernity*. Rotterdam, The Netherlands: Sense Publishers.

Fantinato, M. C. (2013). Balanço da produção acadêmica dos congressos brasileiros de Etnomatemática. *Unión Revista Iberoamericana de Educación Matemática, 33,* 147–161.

Fantinato, M. C., & De Vargas, S. M. (2006). Rural past marks in migrant young and adults mathematical knowledge. *Proceedings of the Third International Conference on Ethnomathematics*. Auckland, New Zealand. Retrieved from: https://www.math.auckland.ac.nz/Events/2006/ICEM-3/GivenPrez.html

Geertz, C. (1973). *The interpretation of cultures: Selected essays*. New York, NY: Basic Books.

Greenfield, P. (1999). Cultural change and human development. *New Directions for Child and Adolescent Development, 83,* 37–59.

Greenfield, P., & Lave, J. (1979). Aspects cognitifs de l'éducation non scolaire. *Recherche, Pédagogie et Culture, 44,* 16–35.

Lave, J., & Wenger, E. (1993). *Situated learning: Legitimate peripheral participation.* Cambridge, MA: Cambridge University Press.

Maduro, R. G. A. (2013). A cuia nossa de cada dia. In A. M. S. Santos & L. G. Carvalho (Eds.), *Terra, água, mulheres & cuias: Aritapera, Santarém, Pará, Amazônia* (pp. 32–38). Santarém, PA: UFOPA.

Mafra, J. R., & Fantinato, M. C. (2016). Artesãs de Aritapera/PA: técnicas e processos em uma perspectiva etnomatemática. *Revista Latinoamericana de Etnomatemática, 9*(2), 180–201.

Pais, A. (2011). Criticisms and contradictions of ethnomathematics. *Educational Studies in Mathematics, 76,* 209–230.

Palhares, P. (2008). A etnomatemática um desafio para nossos dias. In P. Palhares (Ed.), *Etnomatemática*: um olhar sobre a diversidade cultural e a aprendizagem matemática (pp. 11–21). V. N. de Famalicão, Portugal: Edições Húmus.

Pinxten, R., & François, K. (2011). Politics in an Indian canyon? Some thoughts on the implications of ethnomathematics. *Educational Studies in Mathematics, 78*(2), 261–273.

Rosa, M., D'Ambrosio, U., Shirley, L., Alangui, W., Palhares, P., & Gavarette, M. E. (2016). *Current and future perspectives of ethnomathematics as a program.* ICME-13 Topical Surveys. Hamburg, Germany: SpringerOpen.

Shockey, T. L. (2002). Etnomatemática de uma classe profissional: cirurgiões cardiovasculares. *BOLEMA, 17,* 1–19.

Vergani, T. (2000). *Educação Etnomatemática: o que é?.* Lisboa, Portugal: Pandora edições.

Vergani, T. (2003). Etnomatemática e contemporaneidade: (com) fluências. In C. A. Silva & I. A. Mendes (Eds.), *A surpresa do mundo: ensaios sobre cognição, cultura e educação* (pp. 127–140). Natal, RN: Editorial Flecha do Tempo.

Vergani, T. (2009). *A criatividade como destino: transdisciplinaridade, cultura e educação.* São Paulo, SP: Editora Livraria da Física.

Wenger, E. (1998). *Communities of practice: Learning, meaning and identity.* New York, NY: Cambridge University Press.

White, L. (1988). *La ciencia de la cultura: un estudio sobre el hombre y la civilización.* Barcelona, España: Círculo Universidad.

Part III
Pedagogical Action of Ethnomathematics: Classroom Applications

One of the objectives for an ethnomathematics program is learning to understand the student's own reality and create pedagogical action in a natural manner by using a cognitive focus and develop a cultural basis for the curriculum. The sociocultural context of mathematics should be emphasized in the field of education because it suggests that the study of mathematics, as it is traditionally practiced in Western academic contexts, does exhibit a cultural bias. Any given mathematical ideas, procedures, or practice are a product of a particular culture, and we are primarily concerned with the way in which mathematics is taught in schools in relation to the pedagogical action of the ethnomathematics program.

Chapter 5
Once Upon a Time… The Gypsy Boy Turned 15 While Still in the First Grade

Charoula Stathopoulou

I understand access and participation as Social Justice.
—D'Ambrosio (2012)

Abstract The need to develop a bottom up curriculum for Roma students (preschool education) in order to support their learning at school—language and mathematics—and with a view to contribute to their social inclusion through an ethnomathematical perspective led us to conduct fieldwork on the Roma students' community of origin. The ethnomathematical perspective supported the combination of a critical ethnographical fieldwork using critical communicative methodology (CCM) for exploring students' funds of knowledge as well as the parameters that affect Roma children's education. Poststructural ideas such as power/power relations contributed to understanding how inequalities are constructed through discursive practices, making the inclusion of Roma (and other marginalized groups) merely rhetorical. The pragmatological material, discourse, discursive practices, practices, representations etc.; derived from the community informed both our practices/our interventions in the kindergarten, and our future actions in the community aiming to respond in social justice issues, important for both Roma and non Roma communities.

Keywords Ethnomathematics · Roma children education · Critical ethnography · Sociopolitical turn · Poststructural

---

Project: "Education of Roma children in the Epirus, Ionian Islands, Thessaly and Western Greece", 2010–2015 (E.U. Lifelong Learning, action code: 304263).

C. Stathopoulou (✉)
University of Thessaly, Volos, Greece
e-mail: hastath@uth.gr

M. Rosa et al. (eds.), *Ethnomathematics and its Diverse Approaches for Mathematics Education*, ICME-13 Monographs,
DOI 10.1007/978-3-319-59220-6_5

5.1 Introduction

Five years ago, because of my ethnomathematician's background, I was invited to contribute to a longitudinal project on Roma children's education during the period of 2011–15, addressing all educational levels. Together with a linguistic colleague, the project concerned preschool education, meeting the design and implementation of educational practices that would allow for students' involvement in classroom communication, and for the broadening of their communicative repertoire through the use of Greek language and mathematical literacy practices.

(Mathematics) education and particularly mathematics learning in classroom contexts, is not a simple task for such populations, depending on cultural, social and political aspects that are strongly involved. Aware of this, our decisions were supported theoretically by research on the sociocultural and mostly sociopolitical field, including in these ethnomathematics that recognize the complexity of (mathematics) education. We began with a critique of approaches that ignore crucial dimensions of learners and educators, and of real-life situations that involve or/and affect, explicitly or implicitly, the procedures of Mathematics education (Gutiérrez 2013; Pais and Valero 2012; Mesquita et al. 2014).

Despite the long-term recognition in mathematics education of the need to develop models of mathematics education that address the diversity of populations that they serve (Brown 2010), there persists a need to go beyond mere depiction of the cultural conflicts and their impacts toward explicit responses. We sought potential transformation of situations involving inequality and exclusion through processes of data collection and our critical ethnography on the Roma community through the theoretical lens of Critical Communicative Methodology (Gómez et al. 2011).

This chapter is organized as follows: first, we present: (a) briefly, the situation of Roma education in Greece together with the role of preschool education; this is followed by, (b) our theoretical and methodological perspectives; (c) our findings from the fieldwork in the community; (d) some interventions in Kindergarten classes with Roma students; and (e) a closing discussion.

5.2 Roma Children's Education: Preschool Education

Despite the rhetoric of the formal curriculum about equality and inclusion, Roma children face serious conflicts when they are required to participate in a formal school setting. For example they have to deal with boundary identities, bilingualism (Gana et al. 2016), and the lack of efforts to exploit their informal knowledge. What we find is a setting characterized by serious cultural and cognitive conflicts that makes their schooling process an acculturation process.[1]

[1]See Bishop (1988, 2002).

The *top-down* curriculum adopted by common teaching practices, together with the ignorance of the mathematics knowledge children acquire through their involvement in their family professional activities (Stathopoulou 2005), seem to further marginalize these children and their families, constructing them as occupying inferior positions in terms of both learning and cultural identities.

My main contribution to this project was my extensive prior experience working with Roman communities, an ethnomathematical study I previously conducted in order to explore how the cultural context of Roma children is related to mathematics teaching/learning (Stathopoulou 2005). I brought the findings from this previous research, along with my understanding of the Roma community and my personal relationships with several individuals in the community with whom I had worked in that previous study, to the specific focus on Roma children's preschool education, and to my official responsibility for the mathematics education components of the curriculum.

The research *on the spot* was conducted both in a first grade class of Roma students[2] and in the community of origin of these students. This particular community, it should be noted, more or less reflects the greater Roma community, and anyway is similar with the community of reference in this paper[3] as it is documented by the fact that people of the two communities were relatives. The fieldwork pointed out how their schooling process and particularly mathematics learning is affected directly or indirectly by the cultural peculiarities of the community.

The main characteristics related to Roma children's education in general, and particularly mathematics learning, were (Stathopoulou and Kalabasis 2007):

- *Semi-nomadic way of life*. There are obvious consequences of this characteristic for formal schooling; for example, it often results in a delayed start to schooling, and creates an inconsistent attendance.
- *Socio-economic organization.* Businesses are usually organized within the framework of the family group. As a result, children are involved in their families' activities.
- *Language orality.* Roma have only oral language; the orality of language has as a consequence students to memorize lot of information, like a list of shopping or easiness to make mental calculations, while the negative aspect of this cultural element is the fact that Roma children have no previous experience of written texts.
- *A different perception of education.* Formal education is not an activity, yet, integrated into Roma culture in the way it is in the broaden society. However, despite it is appeared that Roma people dispute the need for formal education the reality is more complicated and many questions are emerged. Is this

[2]Despite, the official rhetoric at the school where I conducted the research there were two classes purely of Roma children, constituted by children from 7 to 12 years old.

[3]Our project included all Thessaly area, a big part of all Greece. The communities we were working with are characterizing by diversity about the broaden society, among them and inside them.

> education designed to include them? Is it adaptable to their needs? Have they the information of what is education really? Could help them in a direct way to respond to their problems' problems of surviving? If they need to travel all around the country how could they leave their children at home alone? So what it is attributed to the luck of interest why it is not luck of the state to be adaptable to their needs; needs for surviving?

An important conclusion through my research at this time has to do with the relative strength of the knowledge students acquire through their involvement in family activity. I therefore argue that it would be far more effective to bridge this informal knowledge with the formal school knowledge, instead of ignoring or disputing this knowledge, as often transpires in their school experience. The choice to bridge such knowledge in the case of my research, which, in this project, also involved teaching, resulted in a class with students' voices present –a classroom of participation (Stathopoulou 2005). Such instructional practice is not a common in a typical mathematics classroom where teachers follow textbooks and curriculum designed for the mainstream student.

My own experience in the Roma project is consistent with findings from other recent research. Papachristou (2014) focused on teachers' conceptions about Roma students' formal education. This study found that teachers: (a) do not recognize background knowledge of Roma students; (b) do not exploit Roma students' language orality during teaching processes; and (c) maintain stereotypical perceptions on the school process of Roma students and their potential. Additionally, recent reports from UNESCO (2014) highlight the fact of Roma students' inadequate attendance and high dropout rates in comparison with other sub-populations in their schools; despite recent improvements from the past, they are still high.

Among the other conclusions from my fieldwork at this time emerged the crucial role of Preschool education: the two children that had attended kindergarten programs were familiarized with the norms and practices of formal education, facilitating their adaptation to classroom culture and school learning processes. In addition, through several researches the role of preschool education is underlined, since preschool experiences contribute to building necessary knowledge and skills for the successful transition to, and subsequent attendance in, elementary school. Preschool education as an educational stage, appears to be especially crucial for students with minority cultural or even language origin, for whom it provides supports in overcoming numerous learning obstacles related to their sociocultural background; preschool education thus contributes conditions that ensure equal educational opportunities for all (Becker and Tremel 2011; Gana et al. 2016).

Although discussion about the role of preschool education in ensuring equal educational opportunities for children from immigrant families, ethnic minorities, and socially marginalized groups is not new, it has recently increased more rapidly due to interest in PISA findings and their potential correlations with possible school failure that children from the above-mentioned groups face. Research has emphasized the importance of the timing for addressing knowledge deficits, specifically deficits in the language of instruction (Fuchs-Rechlin and Bergmann 2014), and connections to the creation of conditions that ensure equal educational opportunities for all (Gogolin 2009; Becker and Tremel 2011): the duration of school attendance and the quality of educational practices are included as main parameters that improve learning outcomes (Hasselhorn and Kuger 2014).

Despite the fact that the literature highlights the positive role of preschool education, and the gradual accrual of insights regarding qualitative characteristics of pedagogies that ensure benefits of school attendance for students with multicultural and lingual diversity, Roma students do not yet appear to benefit from such research findings. A recent comparative study (UNESCO 2014) on the educational situation of Roma children in European countries states that the number of students attending compulsory education in European countries is still too low. Further research maintains that Roma children arrive at school without adequate preparation, and with little understanding in the majority language.

Top-down educational policies addressed to the mainstream students, combined with the ways that teachers interpret and materialize those policies in contexts including Roma students, appear to significantly account for diminished school attendance of Roma children. Relevant considerations have been described with regard to the Greek educational reality, as well. The Greek educational system, like most educational systems in European countries (Govaris 2005), is not yet in a position to effectively respond to a school reality characterized by linguistic and cultural diversity. The applied pedagogies formulate a field of unequal distribution of opportunities for recognizing and exploiting the learning resources included in the linguistic and cultural capital of a diverse student body.

In fact, Greek school practices tend to be guided by an assimilationist ideology that seem to ignore or understate fundamental characteristics of children's cultural identities. "Their" culture is usually assessed as insubstantial and worthless, and most teachers presume that non-Greek home languages do not contribute, or even stand as an obstacle, to their school performance; teachers are furthermore unlikely to use the home language as a resource. In such an educational context, Roma children's erratic school attendance and their dropout rate, which is among the highest in the country, could be strongly linked to the silence, marginalization and underestimation of their world that Roma children experience in classrooms (Noula et al. 2015).

5.3 Conceptual/Theoretical Perspectives

For a long time, research in mathematics education was based on psychological approaches that appear insufficient, since they could not respond effectively in mathematics learning mostly for students out of the main stream. In recent decades, the realization of the inadequacy of psychological perspectives, the failure of applying programs of modern mathematics all around the world and the consequences of globalization such as displacement of people (migrants, refugees, etc.), and the formation of multicultural societies with diverse educational needs moved international research to broaden its view, including approaches that explore social, cultural, and political dimensions as important for interpreting and responding to the above situation in mathematics education.

5.3.1 Sociocultural and Sociopolitical Turn in Mathematics Education

In 2000, Lerman inserted the term 'sociocultural turn' to describe this trend in research. In his work he includes a corpus of studies that challenge previous perceptions about mathematics knowledge and mathematics learning, and which document the term he inserted. He included Jean Lave's work (Cognition in Practice 1988) as an example of challenging cognitivism and transfer theory in mathematics learning. The research of Carraher (1988) on School and Street mathematics is another example that challenged the role of the context in problem solving. The book of Bishop (1988) entitled *Mathematics Enculturation: A Cultural Perspective of Mathematics Education* was mentioned by Lerman for its cross-cultural view of mathematical practices, and the critical exploration of the assumed universality of mathematical activities, showing how these depend on cultural context.

Two other contributions of this time that are of great interest for us is this of Valerie Walkerdine and of Ubiratan D'Ambrosio. Walkerdine, in The Mastery of Reason (1988) located through a Foucauldian analysis meanings in practice, and wrote about the construction of identities as discursive practices, bringing to the scene issues of power/power relations. And, finally, the ethnomathematical approach, a term introduced by D'Ambrosio at the 5th ICME in Adelaide, (D'Ambrosio 1985), was included by Lerman (2000) as a new direction of research "that played a large part in creating an environment that was receptive to the social turn" (p. 9). Also, Lerman, in this work, makes two additional important observations: He furthermore speaks about the influence of Vygotsky' work explicitly or implicitly in the research at this time; and he discusses the influence of other scientific fields: Anthropology, sociology, and cultural phycology.

Although the political dimension is detected in the research reviewed by Lerman, a few years later, Gutiérrez (2013) used the term *sociopolitical turn* in mathematics education to mark the movement beyond a sociocultural view toward the exploration of sociopolitical concepts and theories, highlighting identity and power, by researchers that focus on anti-racism and social justice issues (Gutiérrez 2013). She questioned about the late development of the research fields that led to this turn (sociopolitical turn):

> Ethnomathematics, which seeks to decenter Western mathematics and highlight the mathematical practices of people throughout the world, was created in the 1980s; critical and social justice mathematics has flourished just in the last 2 decades; critical race theory, LatCrit theory, and science and technology studies only gained momentum in the mid-1990s, and post-structuralism and postmodernism have been embraced in mathematics education only recently (Gutiérrez 2013, p. 43).

Gutiérrez supports acknowledging the contribution of sociocultural perspectives to challenging notions of *learning* and *participation*, and makes clear that adopting sociopolitical perspectives is a challenge to rethink terms such as *mathematics, who is good in mathematics, the role of resistance in relation to dominant circles*, and *quality teachers*. The process of deconstruction is particularly useful to expose current practices/knowledge/categories as socially constructed in a particular point in history. This approach opens up new possibilities, new views on learners and educators, and new arrangements within/beyond school upon which we can act. Highlighting gains from a sociopolitical posture, she includes as very important: moving *Beyond Essentialization and Victimization, Challenges of Common Notions, of Teacher Quality of Racial Hierarchy and of (School) Mathematics*.

Gutiérrez (2013) emphasizes in her work the important contribution of post-structuralism to the sociopolitical turn, noticing that post-structuralism offers additional theoretical tools for those who have adopted a sociopolitical stance. In this framework mathematics education, learners, teachers, and researchers are considered both results and producers of discourses. Discourse is not considered as individual, static, or referring only to language but involves other symbolic expressions, objects, and communities (Moschkovich 2007). Because discourses are inherently social, political, historical, and connected with the construction of meaning, these approaches share much with those ways of thinking about mathematics education that are connected to a concern with culture, considering culture not a stabile entity.

Meaning, reasoning, knowledge, action, learning, and so on, are products of discourses and discursive practices, constantly renegotiated in social and cultural contexts, finding their meaning in the outcomes of actions and interactions moment by moment (Appelbaum 2008; Walshaw 2007). In other words, meanings that people make of themselves and their world are constantly being created in and through interactions with others, in larger social and political contexts, with discourses that are themselves renewed and modified through these experiences and events (Appelbaum and Stathopoulou 2016). In the Foucauldian approach, knowledge is an effect of a primarily linguistic discursive formation, that is, a set of fundamental rules that define the discursive space in which the subject exists (Freitas and Walshaw 2016). Very often discourses do not represent the reality but construct it.

Gutiérrez (2013) uses as an example of reality's construction the achievement gap in U.S. mathematics education that it is presented and been conceived as an absolute truth. The importance of understanding discourses in this way is that they produce *truths*. In her framework, she also questions the notion of *success* that is largely driven by discourses of achievement and proficiency on standardized exams and tangible outcomes that can be measured in some way. In this context, a poststructuralist view is against singular meanings and challenges truths such that concepts like *success*, *proficiency*, *achievement gap*, and even *mathematics*. The way these notions are constructed are in line with what we think of as habits of successful learners or practitioners (a form of internal surveillance) (Foucault 1977).

Our definitions of success rarely include self-actualization, that is, the idea that we should be allowed to become better people by our own definitions, not just those prescribed by schooling. That is partly why discussions of identity and power are so important because the goals we have for students may be disconnected from the ways in which they see themselves now or in the future. And, yet, even in constructing and privileging certain truths over other possible ones, discourses are malleable, subject to outright rejection or (re)inscription (Butler 1999).

That is, teachers who have adopted a sociopolitical stance may decide not to judge their success only on whether they close the achievement gap (Gutiérrez 2009), but also look for ways in which students are being creative and imaginative when doing mathematics or for when students see a more positive relationship between themselves, mathematics, and their futures. One difference in the way discourse is interpreted by Foucault is that unlike other theories that imply an overarching metanarrative, where people are oppressed by the narrative, post-structuralism ascribes more agencies to individuals in recreating or shifting meanings of the discourse.

In a recent paper Stinson and Bullock (2015) discuss the complexity of mathematics education and the need of including multiple conceptual and methodological approaches in order to face this complexity. They suggest a critical postmodern methodology by exploring, hypothetically, the different and somewhat discomforting possibilities for data collection, analysis, and representation when research is framed with/in critical postmodern theory. As they claim, mathematics education community should encourage expanding the frontiers of science by supporting not only those who look toward science to answer concrete questions but also to those who look toward science to generate different questions that might produce different knowledge and produce knowledge differently: "In the end, we believe that the mathematics education research community should embrace chaos as opportunity and as evidence of a vibrant of a vibrant and vital field" (Stinson and Bullock 2015, p. 17).

5.3.2 Ethnomathematics in a Broaden Conceptual Landscape

The community of ethnomathematicians is both heavily responsible for promoting the social turn, and why not sociopolitical, and has remained as keepers of the flame, so-to-speak, maintaining a vanguard and marginal status that continues to search for its purposes and for forms of community building consonant with its aims (Stathopoulou and Appelbaum 2016). D'Ambrosio (1985) introduces ethnomathematics as the "Way different cultural groups mathematize (count, measure, associate, classify and draw conclusions). This is done using practices; knowledge, dialects and codes vary from culture to culture" (p. 45). In the 90s, Appelbaum (1995) called for a creolized interculture characterized by the poetry of Aimé

Césaire and the emerging discourses of Anthropology as cultural critique (Marcus and Fischer 1986).

Although the ethnomathematical approach originally emerged as a response to issues of mathematics/mathematics education and inequalities in particular contexts (i.e. non-Western countries), we consider it as a dynamic field of knowledge and action built around the notion of culture. Ethnomathematics considers other aspects of life and their connections to mathematics/mathematics education; therefore, as an approach, it can inform mathematics education in Western areas and respond to issues of social justice. In this perspective, in order to better understand parameters that are connected to our understanding of how inequalities are constructed and prevent the mathematics education of group like Roma, we can borrow tools from post-structuralism/postmodernity. In this section, ethnomathematics is also discussed together with ideas of multiculturalism and diversity education.

Since the notion of culture is crucial in an ethnomathematical framework, it is needed to be clarified that, here, it is conceived as a complicated analytical category: "*an historically transmitted pattern of meanings embodied in symbols, a system of inherited conceptions expressed in symbolic forms by means of which men communicate, perpetuate, and develop their knowledge about and their attitudes toward life*" (Geertz 1973a, p. 89) and those "*webs of significance people themselves spin*" (Geertz 1973b, p. 5). Although culture is considered as a way of understanding the others and as recourse, at the same time, it can be an obstacle since it is a way to homogenize people who may belong to the same community in order to cover the problems and inequalities by pretending that they are cultural differences. This is done mostly in regards to marginalized groups.

Considering the problems that confront in (mathematics) education these kinds of groups, identified as just cultural different, we ignore other dimensions: culture neither is produced in a social vacuum nor is independent of policies, both in macro and/or micro level, and power/power relations are crucial for their identity construction. So, in order to deal with these complicated frameworks we adopt moreover the idea that "*power relations forms a dense network or tissue that crosses the mechanisms and institutions not located precisely on them we exploit poststructural ideas*" (Foucault 1978, p. 120).

In the past several decades, researchers have approached ethnomathematics through a variety of points of view: as a research activity (Gerdes 1994); as a subject of study (D'Ambrosio 1985); as a way of behavior (Zaslavsky 1994); as a form of expression (Borba 1990); as a language of communication (Borba 1990) with the notion of culture that permeate all of them. Unfortunately, this early phrasing of ethnomathematics, consistent with the anthropological understanding of culture as defining difference, was not yet influenced by the discourses of cultural critique, nor by the post-colonial concepts of creolized intercultures, and instead re-established indigenous mathematical and pedagogical traditions (those not included in the standardized, normative, Colonialist curriculum) as inferior and less sophisticated than those set by developed nations as universal.

Critics of ethnomathematics sometimes misidentify the program as merely a field of research and dismiss ethnomathematics as political correctness gone too far. It is certainly challenging for many scholars to confront the realities of mathematics, as well as educational institutions, as the arms of a political and ideological posture. Nevertheless, ethnomathematics requires this (Stathopoulou and Appelbaum 2016). As D'Ambrosio (2007) asked, "If proposing a pedagogical practice which aims at eliminating truculence, arrogance, intolerance, discrimination, inequity, bigotry and hatred, is labeled as going too far, what to say?" (p. 32).

Lots of critique inside and outside the field of ethnomathematics provoked reaction and gave the opportunity for furthermore development mostly regarding theoretical/epistemological perspectives of the field. In their 1997 paper entitled *The end of Innocence: a critique of 'ethnomathematics'*, Ole Skovsmose and Renuk Vithal discuss broadly the issue of ethnomathematics definition highlighting the problems that prefix *ethno*, as a word incriminated in the context of South Africa causes, despite it is clarified that this prefix is note referred to races. Also, they notice that ethnomathematical practices that take place in a particular cultural group are not only framed by the natural and social environment but they are related to interactions with the power relations both among and within cultural groups, as well. In other words, ethnomathematics has been critiqued that, at its origins, its political dimension was not obvious, even though D'Ambrosio connects ethnomathematics with social justice very early on. In later texts, he definitely makes this link more explicit.

In his doctoral thesis, Barton (1996) explores the philosophical dimension of ethnomathematics. Exploring the connections and compatibilities between ethnomathematical perspective and several philosophical aspects concludes that Wittgenstein theory gives a philosophical framework for ethnomathematics; despite Wittgenstein has not focused on culturally relevant mathematics, the way he approached philosophy laid the foundations for the study of cultural diversity. This connection of ethnomathematics with Wittgenstein opened the door to perceive ethnomathematics under the postmodern/poststructural[4] perspective.

Bello (2010) makes also a poststructural *reading* of ethnomathematics with references to Wittgenstein and Foucault. By using the Wittgenstein's notions of language games, and Foucault's notions of discursive practice, power-knowledge, and truth games, the author discusses Ethnomathematics theory as a discursive practice whose rules establish not only ways of seeing and saying about mathematics practices of identifiable cultural groups but also as multicultural device of government when defining and reorganizing identities and differences.

Knijnik (1998) develops her argumentation about the inclusion of ethnomathematics in the postmodern scene in her attempt to respond mostly to the critique of

[4]The analysis of the two terms exceeds the objectives of this work, but it is necessary a basic reference. Many authors use alternative or complementary the terms. In short, we could say that "being postmodern indicates a historical, sociological point of view. Being poststructural indicates a strategy of analysis" (Yeaman et al., p. 26).

Taylor (1993) and Dowling (1993). Dowling supports that ethnomathematics is a project of Modernity, basing on his interpretation about monoglossism: he supports that the monoglossism of a person is transferred to the monoglossism of a cultural group in ethnomathematics framework. Taylor focusing, for his critique on the work of Walkerdine who considers as "pre-eminent theorist of *ethnomathematics* (Knijnik 1998, p. 2) stresses the strong connection between ethnomathematics and *modernity*, when speaking about a "profound ambiguity in the ethnomathematics discourse" (Knijnik 1998, p. 2).

Knijnik (1998), finally, takes into consideration the above aspects together with ideas that came from Walkerdine work and Skovsmose and Vithal (above mentioned paper) in order to justify why ethnomathematics could be seen as included between Modernity and Post-Modernity. She speaks about her own experience developing empirical projects with the *Landless People*, in Brazil, in an ethnomathematical perspective. Exploiting her experience she highlights the role of the power relations produced within that social movement, which led her to define what kinds of knowledge among those practiced by their members are instituted as regimes of truth, in the words of Foucault.

My personal experience in my previous involvement with Roma communities led me to the realization that the enrichment of ethnomathematics with perspective of postmodern/poststructural could provide us more tools for understanding and interpret complicated situations that are connected with Roma children schooling and particular their mathematics learning; learning of school mathematics. Here I feel the need to speak personally, since it is appeared for me a contradiction: on the one hand, I totally respect the culture, the cultural differences, and the mathematics knowledge of each group, and at the same time I am strongly interested in promoting formal mathematics education for Roma children.

It is difficult to separate cultural differences and cultural choices from the social and political situations that are related to them. This has generated a number of questions to me: To what extent, what appears as cultural feature is really cultural? Why not connected with the fact that this group is been located out of the boundaries (in both geographical and symbolic level) of the broaden community? Why is the claim that Roma parents are not interested in their children school attendance because of their culture (and how this perceptions are constructed) so strong? Why is the fact, for example, that in some cases there are not available buses for their children to move from the community to school largely ignored? Why is the fact that many non Roma families are reluctant in having their children attending common classes with Roma children, which, therefore, makes the latter consider school as a hostile space, not considered?

Just recently, Christos, a 12-year old Roma child who had moved for a few months to another place, told me that for long time he did not attend school in his new area because he was the only Roma there and other children made fun of him, calling him names like *gyftaki*, which while it literally means *gypsy boy*, it also has a negative connotation. Some researchers suggest cultural relevant pedagogy as a response to the realization that some learners need more attention due to obstacles to social change (Appelbaum 2002).

Culturally relevant pedagogy is an approach that validates students' cultural backgrounds, as ethnomathematics also does, and ethnic history, providing ways for educators to support cultural connections between the school and the community, while simultaneously challenging social injustices by identifying obvious and subtle individual, institutional, and cultural actions that perpetuate social structures (Rosa and Orey 2011). Therefore, ethnomathematics and culturally relevant pedagogy together create opportunities for mathematics curriculum relevant and meaningful for students, while the same time maintaining interest in issues of equality in and out of the classroom.

Compatible also with an ethnomathematical perspective is the diversity in mathematics education, which recognizes that "intercultural relationships are (maybe always) asymmetrical, and that intercultural education should aim to repair the damages caused by attention to what once might have been called egregious cultural differences" (Appelbaum and Stathopoulou 2016, p. 348). Instead of removing the differences, or criticizing current practices as harmful to social justice, diversity educators work with the recognition of differences and promote direct attention to how the use of cultural difference by a language of multiculturalism might be coopted for social justice. Diversity educators search for those times and places where the function of something designated as *cultural difference* serves to support a new version of racism or marginalization, as for example, in the expression, *culturally incompatible*, in order to respond together with their students (Mishra 2012).

5.4 The Project

A linguistic colleague and I had undertaken to support Roma preschool children mathematics and language learning with ultimate aim Roma social inclusion. Despite of this focus on Roma children education, being aware of the complexity of education and much more in the cases of students coming from out of main stream backgrounds, marginalized groups, we were led to broaden our fieldwork including their community of origin.

Our methodological choices as well as the fieldwork in details are presented in the next section while here the planning, the processing, and the implementation of the program (project's logistics) in both school and community are outlined. On the one hand, the teachers and we (the academics) conducted research on the spot. Following a *Collaborative Inquiry* framework, together with teachers we explored the possibilities for incorporating the knowledge and experience that Roma students bring to schools (funds of knowledge). On the other hand, through a community of practice we had the opportunity of continuous interaction: teachers had every week to present a report, addressing to the whole community, regarding the design of their teaching and its implementation in the classroom while all members (teachers and academics) of the community had access to it, providing feedback. Furthermore, regular face-to-face meetings were including in the community's actions. The communication and the interaction in the framework of this community of practice, informed our next pedagogical activities.

For our project's purposes, ethnographic research in their community of origin was of great interest in order to improve our understanding and our interface, through real contact and relationships, aiming to their education and social inclusion. Searching for funds of meaning in the students' everyday life, we drove down the streets; we observed the neighborhood, the surrounding area, and the external markers of what identifies this as a neighborhood. We visited student's homes and we engaged in conversations with their parents and other family members regarding their habits, what they do, where they go every day, how they are involved with their children education; creating for them the opportunities to express their expectations regarding education, their problems connected directly or not with education. For data collection we used three techniques of CCM: communicative life story, communicative focus group, and communicative observations. Besides, participation in festive activities (for example celebration of cycle live) helped challenge the (symbolic) boundaries of the two communities.

5.4.1 In the Community

In the community we selected a critical ethnography together with CCM (Gómez et al. 2011), since it has proven successful for analyzing educational inequalities in ways that generate real transformation towards social justice. This methodology derives from Habermas' communicative action theory in which he speaks about the universality of capacity for language and action: "All human beings can communicate and interact with others, regardless of their cultural, ethnic, or academic background" (Gomez et al. 2011, p. 237).

In a framework of a CCM, researchers co-participate with the members of the community creating situations of equal ability and responsibility to evaluate and criticize research. Both a researcher and a participant work on the same level and go beyond the dichotomy of a researcher and an informant; researchers and participants are engaged in common practical projects aimed at social transformation (Caterino 2013).

The research cannot exclude those who play ordinary roles in society aiming to understand and transform social reality; in the same way, communicative research techniques aim to generate reflection among people who want to better understand the topic they are investigating as a first step to promote change. Indeed, communicative research with cultural minorities, and people who have low incomes and no academic background, has proven that when their voices are included in research, they help advance the state of the art, and the research moves from an ethics of interest to an ethics of responsibility (Gomez et al. 2011).

The above theoretical and methodological perspective and the realities we faced in the field guided some of our next initiatives. Although originally the project had an orientation from top to bottom, we turned it to bottom-up actions. These choices were detected by our fieldwork in the students' community of origin where we participated in several activities, e.g. professional, live cycle events etc. Thus, we had the opportunity to explore on the one hand, their collective and subjective perceptions and representations about education, and on the other hand, the funds of knowledge, practices, and discursive practices related to mathematics learning, exploitable for our educational interventions.

At the same time we were organizing meetings with other institutions, involved in Roma issues, where Roma people's *voice* was asked and dialogue developed in these meetings led to our next steps. Despite our focus on child education a lot of other needs, explicitly or implicitly connected with education, emerged. For example, the factor of poverty affected directly school attendance: there were families without the capacity to give a snack to their children for the time they were at school or pocket money to buy something from school cantina. After the head of the Church in Volos undertook this expenditure school attendance was increased.[5]

For data collection, we used three techniques of CCM: communicative life story, communicative focus group, and communicative observations. Part of our data is presented through the five episodes/situations that follow, highlighting how complicated education/mathematics education is for unprivileged groups such as Roma in Greece. A main issue that emerged through our communicative observations and participation in Roma community of Aliveri[6] was the fact that Roma people experience long-term social processes of discrimination and separation from the broader (non-Roma) community. This is even represented literally, as a marginalized district at the outer border of the city of Volos (Greece), with distinctly real, railway lines, and symbolic boundaries from the rest of society. The differentiation of the two communities' settlements symbolizes the lasting, maybe less visible, production and reproduction practices of cultural distance on behalf of the non-Roma community members. The fact that the two communities appear as two parallel universes has consequences on Roma children's education.

The first of the situations presented here, concerns the occasional strong resistance from the part of the non-Roma parents against the efforts made towards a smooth operation of classes with students from both communities. Using as a pretext for their pro-discrimination arguments, the cultural difference and the insufficient knowledge of the Greek language on the part of the Roma students, the parents of the dominant community seem to shape and perpetuate. Among other things, hegemonic perceptions and positions of distancing vis-à-vis the value of the

---

[5]In some other cases we took into consideration the needs that were emerged through our communication in the next project's application (Inclusion and education of Roma children in the region of Thessaly-code 5001369 and IIS integration Judgment A.P.17556/10.14.2016) that has started in November, 2016). For example, among other initiatives we established a school for the parents of Roma students for familiarizing them with the school structure and whatever constitutes the school in order to be able to support their children.

[6]The main pragmatological material used here it is coming from this particular community of Aliveri.

other culture in a public space such as the school and hence, seriously undermine Roma children access to formal education.

The communicative procedures helped us to identify issues of discrimination and exclusion. Through dialogue with parents (from both parts) and formal structures (school, administrative services etc.), a subversive discourse was developed challenging the ones of the dominant society resulting in the transformation of the exclusionary situation. However, our attempts were not effective at all times for both communities: this school year some of non-Roma students were moved to other areas of the city because of the increasing number of Roma students. A new challenge now for both teachers and researchers is to keep non-Roma students in school, since the aim is the coexistence, and not to make the school a school for Roma, a ghetto. This recent reality makes obvious the complexity of the support of Roma students' education, it looks like a Lernaean Hydra; killing one hand two are appeared.

The second episode concerns the material derived through a communicative focus group; members of a family and their friends constituted the group. Issues like insurers topics, permits (for street vendors), housing conditions (houses without electricity), unemployment, poverty, etc. emerged, that more or less, obviously are strongly connected with education, since it cannot be considered as neutral; attitudes and at the same time life conditions are strongly involved. The voice of a Roma father brings this dimension to the fore. Antonis, a (35-year-old) father of one of the school students:

> School for us is a kind of luxury. For us, the one who has the money can attend school and the one who has no money can't attend (....). The children need food, clothes and hundreds of other things, and all of them affect the school (he means the attendance of the school) (...) we have understood that only school (education) will change us but some help by (...) who are in charge, is needed. (...) The fact that breakfast, since last year has been provided (by Metropolis and other Municipality's structures) is very important to me (...). I'm doing sacrifices (...) if children go to school; but they could have a better life".[7]

A friend of his also highlighted the problems they face regarding getting permission for open markets and lower taxes: "In order for Roma have their children at school they need better life conditions (...) because of poverty children do not go to school but are on the streets begging.

The third episode emerged by using the technique 'communicative daily life store'. Although the part of the narration presented here concerns not a real life story but a tale narrated by an old Roma, it reveals their collective representations about their education shedding thus light on their perception about Roma education in the past, in the present and in future expectations.

---

[7]The above communicative incidence took place the last school year. At this time, his family was living in a house and his children attended school. This school year, because the retailer license is on the name of his wife, they have to move together all around the country, making children's school attendance problematic. The response to this situation was to send the boy to his grandparents and to stop girl's attendance.

Once upon a time, there was a King. The Queen that used to give birth to girls. The King disappointed by this situation threatened the crones midwives who helped the Queen to give birth that if the next child was a girl again he will kill them.

(…)

The day of the next birth came and, for one more time, the Queen gave birth to a girl. Beside the palace at the same time, a Roma woman gave birth to a boy. In order to save their lives the crones midwives changed the girl born by the Queen with the young Roma boy.

(…)

Both children reached the age of 15 years and the young Queen did very well in school. She was excellent while the young Roma boy although he was 15 years old he continued to attend a first grade class; he became 15 years old but he continued to attend first grade.

One day the girl followed her mother who was going to sell baskets. During their trip they met a group of soldiers that were inviting people to discover an encrypted message that was hidden on a coin and to become the winner. Several educated people tried but nobody managed to give the right answer.

The girl asked to try and soon she decoded the message: if in January, February, March, April there is no snow we will not have a good summer (implying difficulties in crops). After that the girl became the winner, the King who was next to the group of the soldiers invited the girl to visit the palace. The King was impressed very much by her and wondered: how could a Roma girl be so smart and so educated and to be in a so high level.

Among other questions, he asked her about her birthday and realized that she was born on the same date with his son. Then he started to wonder how that has happened that his boy continued for so many years to attend the same class and the girl to be so perfect. The girl t he suspected that was so well educated was not possible to be a Roma girl. Finally, he forced the crones midwives to admit what had happened.

The Gypsy boy turned 15, while still in the first grade, was the selected epilogue by the old Roma for this story telling.

The fourth communicative episode comes from my communication with two high school Roma students, Christos and Stelios who among other things were asked to talk about possible difficulties they mainly confronted in the first years at school and how those affected their school trajectory. Part of this is here:

Charoula: *Which were the main problems that you faced the first time at school?*

Christos: *We confronted a lot of problems coming for first time at school.*

Charoula: *Like what? Do you remember?*

Christos: *The language, lady, because we have learned from the time, we were little children to speak another language ... and then (...) lady it is like the Greek people learn English: we confront the same difficulties to learn Greek. It is as a second language but at the same time it is a necessary language (...) the first time lady, I came here at school I could not talk at all!*

Charoula: *But, why? Do you not speak Greek at home?*

Christos: *No, we used to talk, but I was embarrassed to talk to the other children* (he means not Roma children); *I was scared that I would make mistakes ... after a couple of years had passed, after it (...) it was ok (...). I know now* (to speak Greek).

Charoula: *Now, do you face any kind of difficulties with language?*

Christos: *No.*

Charoula: *You mean while you talk; but when you write.*

Christos: *When I have to write, yes, but just to speak with the children I have no problem.*

Charoula: *What about you Stelios, growing up what kind of difficulties did you face;*

Stelios: *Growing up I had no problems, on the contrary; growing up lady the situation was better (…). I learned better the language; I had no problem. I started mainly to behave like the others (not Roma students), having not forgotten my language, but (…).*

Charoula: *You do not need to forget your language!*

Stelios: *But I was learning the language better and better; just in the beginning (…).*

Charoula: *Did you face difficulties adapting to the classroom?*

Stelios: *(…) at the beginning, yes!*

Charoula: *Only because of the language or because of other things, too.*

Stelios: *The behaviors lady because we were taught other modes of behaving.*

Charoula: *What things, regarding behavior, did stress you; what was hard for you to do the first time; what was difficult for you in the classroom?*

Stelios: *(….) in general lady, (I did not know) how someone should address the other people (…) what means "hello", what means "how are you", while at the same time the other children (non Roma) knew from the time they were born what 'hello', 'how are you' mean. How is this thing, how is the other, they were able to speak comfortably.*

Charoula: *You mean in the classroom; which was your behavior in the classroom? Had you attended preschool education?*

Stelios: *No, I had not attended. I was ashamed because I thought the other children were something else, something different, something elusive, because lady my father, since I was a little boy, used to say to me: you must be like the Greek people, not to be like the other Roma children that are used to getting married very young, you must be different, to have your own work to go out in the society* (he means to be a member of a broader society). *So, lady, when I went to school I thought that the non-Roma people were so different, so important, something that I could never reach.*

Fifth episode is the narration of Panagiotis, a young Roma boy who studies at the University in Volos; the only person of the community that has reached University level. Since he represents a positive facet of the community, his experience from school that resulted in a successful trajectory is important.

Charoula: *To what extent did the language cause difficulties in school performance for you?*

Panagiotis: *And now is difficult, someone who is bilingual always faces problems, regardless of speaking good Greek.*

Charoula: *But, you speak well.*

Panagiotis: *My writing is much more poor.*

Charoula: *How did this difficulty in language affected your performance in mathematics.*

Panagiotis: *In a word problem I could not understand what the problem was asking.*

Charoula: *And (…).*

Panagiotis: *I needed the teacher to read 2-3 times and explain it.*

Charoula: *What do you think we need to know in order to help effectively the younger Roma children that attend now school?*

Panagiotis: *I believe that, since the young Roma are used to speaking during the majority of the day Romani, teachers should speak basic Romani, not fluently, but just as a bridge to come closer to the child. Last year when I visited the kindergarten, the teacher spoke to me about her experience of a Roma child that was crying for a long time saying 'pani'. This word in our language means water, so the child was crying for water. After this, the teacher started to learn our language. This is very important for the children. When children listen to you speaking in their language they not only feel happy but also welcomed and accepted. In this way, you come closer to the child. I think this is an important tool; teachers to have the basic knowledge of our language.*

(…)

Charoula: *How do you experience the boundaries of the two communities?*

Panagiotis: *Speaking about the boundary of the two communities (...). For two years, I lived in a students residence. An uncle of mine, an old man, asked me: what has happened to you? Have you forgotten us? Have you become Balamos (non-Roma)? I explained to him that although I communicate a lot with non-Roma during day, I have not became Balamos and I'm not going to became, I'm Rom. (...) the boundaries and the lines, yes I have experienced them. But I think you decide where you would like to belong to. Me, as Panagiotis I have transcended the boundaries.*

Charoula: *How the people in your community perceive what you are doing?*

Panagiotis: *Some of them are thinking positively about the fact I am in the University, some others can't understand what I'm doing in the University.*

Charoula: *How was your experience the first time you attended the kindergarten? Were you the second child of your family at school?*

Panagiotis: *I was the third but the first one at the kindergarten. It was so unprecedented for me, completely different from my community. I could understand Greek because my father had Greek friends, but not everything. I did a clever thing: I followed what the other children did. At that time (1996), I was the only Roma child at school.*

(...) In the community, I did not have many friends. Because of my disability I was mocked by children in the mahala (community).

Charoula: *What kind of difficulties did you face regarding mathematics when you were a school student?*

Panagiotis: *The exercises on the blackboard that teacher used to write. He wrote down an exercise and then he moved and seat in his desk. I copied what was on the blackboard and left it unsolved. He never came to me to see what I was doing, he never asked me: did you solve it? Teachers are usually interested in the students that are good, they do not devote time to all children. Every child is different with different needs. The first two years at primary school, it was difficult for me to understand. But, in the 3rd grade I was attending a special class (a class for integration of students with learning difficulties)*[8] *and a teacher, Vasilis K. helped me very much.*

Charoula: *How did he help you?*

[8]It happens very often to be included in this classes students that have no learning difficulties; students that have just a different cultural background, illegally.

Panagiotis: *First of all, he gave me a test in order to see which my level was. (...) for example, I had a problem to learn the multiplication tables. He did not ask from me to memorize it; he gave me some materials, some blocks and explained to me how to use them for understanding and not for memorizing.*

The main common characteristic that is depicted in the above situations, more or less evident, is the notion of power relations that permeate all aspects of life for this marginalized group. The effect of power relations is so strong that it manages to dictate even their collective representations, which are internalized and adopted as their own. The structures and the people of the dominant culture have the power and *possess the truth*. As Foucault has written (1979), power and truth are intimately intertwined. Truth is perspectival: it is the mere creation of the strong. Roma people marginalization even in formal education depicts a situation of inequalities.

Through the communication with the young Roma the complexity of education and particularly their school mathematics learning is obvious. They do not only confront problems because of the language, a written language while their own is only oral, but also problems to understand the cultural codes in communication and the norms of a classroom. It is difficult for them to understand the norms and to adapt accordingly. They come to school context conceiving themselves as inferiors and expect through the school to become something (socially) *better*.

Working in an ethnomathematical and CCM framework, we faced the challenge to pursue ways of improving social situations aiming at social justice and people's dignity, a common aim of both approaches. Our response addresses both: the community and the classroom. In order to support the individual and social emancipation of the Roma community of Aliveri, we incorporated intracultural and intercultural comprehension processes and actions/initiatives that counteract the existing border-making practices, thereby creating and offering opportunities for public participation to all inhabitants of the area.

As part of this process, we organized events together with other organizations and structures and the members of the community, in order to challenge the boundaries of the two, separated, communities. The interaction of all contributors, academics, teachers, students, members of the Roma community, other local institutions or structures that act within the community, emerged as a necessity for responding to the complicated issue of (Mathematics) Education for Roma students; research data were exploited in the classroom, concerning both understanding cultural conflicts and learning mathematics.

5.4.2 Our Action in the Classroom

Preschool teachers, supported by us, in mathematics and language teaching, used to teach in several and different structures in both the Roma community and in typical schools. Aiming to develop a continuous and egalitarian dialogue and interaction, we developed a community of practice, as it has already been mentioned earlier, comprised of the academics involved in the project and the preschool educators.

Our educational interventions were informed by a comprehensive and interdisciplinary range of socio-cultural approaches that support educational practice: culturally responsive and intercultural pedagogical orientation (Gay 2010; Govaris 2013); approaches that consider the school as a place to create relationships and identities for students with cultural and linguistic diversity (Moje et al. 2004; Gutiérrez 2013), ethnomathematics (Stathopoulou 2005). Therefore, we implemented a *bottom-up* curriculum, encouraging teachers to exploit students' knowledge for mathematics and language teaching.

5.4.2.1 Examples of Our Bottom up Curriculum

According to an ethnomathematical perspective, we explored students' funds of knowledge, identified through our fieldwork on the community of Roma, and incorporated relative ideas in our bottom up curriculum. We tried, on the one hand, to challenge the dichotomy of informal-formal knowledge and on the other hand, to create a *space* in the classroom encouraging Roma students' participation and their voice strength in the classroom. The two activities that follow are based on the pragmatological material selected on the spot.

On the one hand, because traditionally they had no permanent residence and on the other hand, because of the poverty many houses are still shanty houses that do not respond to the people's needs, part of their activities take place outdoors. So we had the opportunity to observe Roma people's activities, either household or entertainment, happening outside home. Playing games with cards was an activity of this kind. Groups of adults, groups of young Roma (even younger than five years old), mixed groups of young Roma and adults were very often playing together. It should be mentioned here, that playing cards, often aiming to win money, is considered a negative activity by the broader community. By incorporating games with cards in our *bottom up* curriculum, we transformed an everyday activity, familiarizing Roma students to a mediation tool that facilitated the teaching of mathematical concepts and number sense.

Regarding the second activity, while the 'pretext' was the narration of a fairy tail about a trip of a lion, identifying it with the experience of Roma students that are used to travelling very often, was the aim to be investigated here. Since Roma traditionally are travellers for professional purposes, and the whole family is used to participating in these journeys Roma children have strong travelling experience. In this activity, we tried to integrate their experience to the development of space notions and other mathematical ideas as the distance, the direction, the comparison of magnitudes (length) like the strait line, and generally the line. These ideas as well as the experientially understanding of the role of the note, and consequently the symbolic representations as well as the intuitive approach of the notion of *scale* would re-inform their out of school experience.

The first activity, as it has already been mentioned above, concerned the card game. The learning objectives of the activity were to recite, read and write numbers up to 10, recognize numerical amounts using direct identification strategies, count objects up to 10, find the previous and the next of a number up to 10, and compare quantities and numbers to approach the operation of addition and subtraction.

Roma students, due to their familiarity were willingly involved, gained *voice* in classroom, could count up to 10, and with the help of their teammates they could recognize the symbolic representation of the numbers appearing on the cards. Some quotes from the unfolding activity. From the teacher's question: "Which is bigger? 2 or 8; and why?" the student answered: "8", "It's much more!" (Showing symbols on paper, one by one and counting mentally) and to the question: "Which is bigger? Five or seven?", the student answered: "7". Lipizune 2" (Lipizune: sounds like the appropriate Greek word (leipoun) meaning: missed two), and the teacher said: "Yes, missing 2" (accepting the developing interlanguage). During the negotiation of this activity, Roma students presented a significant difference in both, the involvement in the learning process and in the classroom interaction (with teacher and peers).

The other indicative activity, presented here, is part of a cycle dealing with the development of spatial thinking of our young students. In this cycle of activities, the questions were: (a) which is the understanding of our students of fundamental mathematical concepts associated with navigation in the space (compare sizes, perception of the scale, location and description of routes and spaces); (b) how the teaching would promote the expansion of the symbolic repertoire and transformation of space's representations? (Valai et al. 2015).

In order to plan the activities the logic of Multiliteracies was used: it refers to various channels and forms of representational embodiments (multimodality) and aims at different ways (cognitive processes) in which students can take ownership of knowledge. The methodological framework allows the incorporation of culturally acquired knowledge of students (Gutiérrez and Dixon-Román 2011). In terms of implementation, initially, students were introduced to the theme of navigation in space through reading a tale, which was about a lion wandering in an area. The map (see, Fig. 5.1) of the area in which the hero wandered constituted an organic component of the text, it was available to students and its usability as a practice of representation was exploited during interactive exchanges that accompanied the story telling.

Fig. 5.1 Students working on the map. *Source* Personal file

Students were actively involved in a meaningful for them activity: the map was an attractive multimodal text incorporated into the story telling and simultaneously it was understandable as it was based on their experiences in visiting different geographic locations. Although they were not familiar with the use of a map, their travel experience, following their parents in their professional work, contributed to the understanding of the map functionality as a spatial representation tool. Some parts of the dialogues in the classroom:

Teacher: *How do you think we could design your homes?*

Student: *So (...)* (make a circle).

(...)

Disciple: *Here, over the mountain* (a remote cycle).

Teacher: *How?*

Pupil: *Circle!*

Teacher: *Which house is close to our school, the Paraskevoulas's or Chrisovalanto's?*

Student: *Uh!*

Teacher: *Which could you reach faster?*

Pupil: *Of Paraskevoula's.*

(...)

Teacher: *And then, which way the lion will get if he likes to go for swing?*

Student: *He will go strait and then to this way here (showing on the map) and he will go on the swings.*

Teacher: *So, turn right and then straight.*

Student: *Yes, right, straight.*

Throughout the above dialogue, students realized both the potential to represent the various natural objects by using symbols, buildings here, as well as to navigate through their imagination on the routes that are marked on the map. They could understand that the places and routes depicted on the map reflected potential ways of transition from one place to another. Furthermore, other concepts associated with the determination of space such as the distance, the direction and the comparison of magnitudes (length) were used successfully in this context, while the students could experientially understand the role of the note, and consequently the symbolic representations and could approach intuitively the sense of scale.

It seemed that the association of extracurricular activities, compatible with an ethnomathematical perspective, in conjunction with the strategic support of every communicative resource their hold (e.g. mixture of languages and other types of semiotic resources like the design, music etc.) created a hybrid learning space that supports student's involvement in the learning process. Accepting and validating what students considered as their 'own' improved their self-image and supported the renegotiation of their identity as equal participants in classroom interaction. The overall practice constituted a small step to the development of a framework where cultural backgrounds are a tool for change.

5.5 As an Epilogue

Being involved in young Roma (mathematics) education if you stay in the classroom observation you could only perceive the 'small picture'. An ethnomathematics perspective, being a holistic approach, gives the tools for studying the broader picture. According to my reading ethnomathematics, among others:

- Challenges the dichotomy of formal and informal mathematics knowledge/ education;
- Supports all children's mathematics learning;
- Responds to the acculturation process that students (mostly those outside of mainstream) experience during schooling;
- Values any kind of mathematics knowledge working in its context.
- Challenges eurocentrism of knowledge and any kind of central or local authority/power (and the same time); and
- Fights indignity and injustice (it is a force for social justice).

Taking into account the original purpose of this project, which was, to support Roma (preschool) students' mathematics and language teaching, we conducted research in their community of origin. Our methodological approach based on ethnomathematical perspective gives us information about issues that explicitly or implicitly are connected with Roma students (mathematics) education, while post structural ideas regarding power help to attempt to offer some interpretations.

The episodes from the research on the spot come to the same scene with different roles. The first one constitutes an example of the problems that Roma community face because part of the broader community is negative towards the Roma children's inclusion in school. Basing on the *power of whom has the knowledge* very often people of the dominant society exploit the deficits of egalitarianism despite the rhetoric for the contrary, and they use various pretexts for exclusion. The second, the narration of a Roma father, expresses the difficulties, they confront to facilitate their children's schooling; difficulties that very often are real obstacles for them, preventing the implementing a policy of egalitarianism. The third episode, the story telling of the old Roma man, depicts the collective representations they have developed regarding their school performance; representation detected by the rest of society and adopted by them as realities. Since the others *have the power, they possess the truth.*

Similar collective representations emerged through the two next episodes with the Roma highs school students and the Roma University student. Education and the culture of the others are considered by them as something alien, unfamiliar. The students of the mainstream appear like the ideal they potentially are called to reach. Roma students come to school, an acculturation process, with these kinds of representations and so low expectations that are followed by low performances resulting in a vicious cycle of failure.

Despite the fact that this framework dictates behaviors all the time there is a place for subjectivities to select a different way of life, paying of course the cost.

Panagiotis, for example, seems to challenge the boundaries, in geographical and symbolic level, selecting to study and to live out of the community, but, trying at the same time to keep balance between the two strong identities.

Concluding, our attempt to respond to both learning and social justice issues was addressed to both community and classroom. Working with real people the situations are not linear; their complexity needs a combination of methods and methodologies to understand and to suggest changes. Ethnomathematics it is a perspective that can incorporate all of them and to respond to complicated situations keeping its main aim: to social justice and peoples dignity.

In the community, through a continuous and open dialogue with the members of the community and the co-existing with them in several situations and activities, we contributed to create situations of expressing their own 'voices' and of renegotiation of their symbolic boundaries, more strong than the geographical ones, with the broaden community contributing in overcoming inequalities. The knowledge we acquired through the fieldwork in the community contributed to the understanding of how and why inequalities are constructed through discursive practices making the inclusion of Roma and other marginalized inclusion just a rhetoric.

Also, the ethnomathematical perspective, together with the exploitation of funds of knowledge, informed our practices/our interventions in the preschool education. The design of the two activities in the classroom was based in students' culturally acquired knowledge, valuing this knowledge. As Sousa Santos (2012) notices, in *Epistemologies of the South*, the hierarchy of the knowledge, is not based on an intrinsic value of knowledge itself, but on the dominant social and economic structures. Thus, the dichotomy of formal-informal knowledge was challenging, giving to students a personal meaning. As a result, the students had access and actively participated in school mathematics knowledge, a dimension of Social Justice according D'Ambrosio (2012) and the teaching becomes more effective and the knowledge and skills more easily integrated (Ladson-Billings 1995; Stathopoulou 2005).

References

Appelbaum, P. (1995). *Popular culture, educational discourse, and mathematics*. Albany, NY: State University of New York Press.

Appelbaum, P. (2002). *Multicultural and diversity education: A reference handbook*. Santa Barbara, CA: ABC-CLIO.

Appelbaum, P. (2008). *Embracing mathematics: On becoming a teacher and changing with mathematics*. New York, NY: Routledge.

Appelbaum, P., & Stathopoulou, C. (2016). Critical issues in culture and mathematics learning. In L. English & D. Kirshner (Eds.), *Handbook of international research in mathematics education* (pp. 336–358). New York, NY: Routledge.

Barton, B. (1996). *Ethnomathematics: Exploring cultural diversity in mathematics*. Doctoral dissertation. Auckland, New Zealand: University of Auckland.

Becker, R., & Tremel, P. (2011). Bildungseffekte vorschulischer betreuung, erziehung und bildung fuer migranten im deutschen schulsystem. In R. Becker (Ed.), *Integration durch Bildung* (pp. 57–70). Wiesbaden, Deutschland: VS Verlag.

Bello, S. E. L. (2010). *Language games, discursive practices and truth production: contributions to contemporary (mathematical) education*, Special Edition (Vol. 18, pp. 545–588). Zetetike.

Bishop, A. (1988). *Mathematical enculturation: A cultural perspective on mathematics education*. Boston, MA: Kluwer Academic Publishers.

Butler, J. (1999). *Gender trouble*. New York, NY: Routledge.

Borba, M. C. (1990). Ethnomathematics and education. *For the Learning of Mathematics, 10*(1), 39–43.

Brown, T. (2010). Cultural continuity and consensus in mathematics education. Special Issue on Critical Mathematics Education. *Philosophy of Mathematics Education Journal, 25*, Online.

Caterino, B. (2013). Phronesis and the participants' perspective. *The British Journal of Sociology, 64*(4), 739–745.

Carraher, T. (1988). Street mathematics and school mathematics. In A. Borbás (Ed.), *Proceedings of the twelfth annual meeting of the international group for the psychology of mathematics education* (pp. 1–23). Veszprem: Hungary International Group for the Psychology of Mathematics Education.

D'Ambrosio, U. (1985). Ethnomathematics and its place in the history and pedagogy of mathematics. *For the Learning of Mathematics, 5*(1), 44–48.

D'Ambrosio, U. (2012). Mathematicians, mathematics educators and the state of the world. *Journal of Research in Mathematics Education, 1*(1), 5–28.

D'Ambrosio, U. (2007). Peace, social justice and ethnomathematics. *The Montana Mathematics Enthusiast, Monograph, 1,* 25–34.

Dowling, P. (1993). Mathematics, theoretical "totems": A sociological language for educational practice. *Political Dimensions of Mathematics Education, 2*, 35–52.

Foucault, M. (1977). *Discipline and Punish* (p. 191) (A. Sheridan, Trans.). New York: Vintage, 1979.

Foucault, M. (1978). *The history of sexuality: An introduction* (Vol. 1) (R. Hurley, Trans.). New York, NY: Vintage.

Foucault, M. (1979). *Discipline and punish: The birth of the prison*. New York, NY: Vintage.

Freitas, E., & Walshaw, M. (2016). *Alternative theoretical frameworks for mathematics education research: Theory meets data*. Switzerland: Springer International Publishing.

Fuchs-Rechlin & Bergmann, C. (2014). Der abbau von bildungsbenachteiligung durch kindertagesbetreuung für unter 3-Jährige—zwischen wunsch und wirklichkeit. In K. Maaz, J. Baumert, & M. Neumann (Hrsg.), *Herkunft und bildungserfolg von der frühen Kindheit bis ins erwachsenenalter*. Zeitschrift für Erziehungswissenschaft, Sonderheft 17/2014 (pp. 95–118). Wiesbaden, Deutchsland: VS Verlag.

Gana, E., Stathopoulou, C., & Govaris, C. (2016). *Expanded pedagogical spaces: enhancing Roma students involvement to school*. Paper presented at the Conference Rethinking Language, Diversity and Education, 2015. Rhodes, Greece University of the Aegean.

Gay, G. (2010). *Culturally responsive teaching: Theory, research, and practice*. New York, NY: Teachers College Press.

Geertz, C. (1973a). Religion as a cultural system. In: *The interpretation of cultures*: *Selected essays* (pp. 87–125). NewYork, NY: Basic Books.

Geertz, C. (1973b). Thick description: Toward an interpretive theory of culture. In: *The interpretation of cultures selected essays* (pp. 3–30). New York, NY: Basic Books.

Gutiérrez, R. (2009). Embracing the inherent tensions in teaching mathematics from an equity stance. *Democracy and Education, 18*(3), 9–16.

Gerdes, P. (1994). Reflections on ethnomathematics. *For the Learning of Mathematics, 14*(2), 19–21.

Gogolin, I. (2009). Förderung von kindern mit migrationshintergrund im elementarbereich. In Roßbach, H-G., Blossfeld, H-P. (Hrsg.), *Früpädagogische förderung in institutionen. Zeitschrift für Erziehungswissenschaft,* Sonderheft 11/2008 (pp. 79–90). Wiesbaden, Deutchsland: VS Verlag.

Gómez, A., Puigvert, L., & Flecha, R. (2011). Critical communicative methodology: Informing real social transformation through research. *Qualitative Inquiry, 17*(3), 235–245.

Govaris, C. (2005). Intercultural pedagogy: Recognition relationships in multicultural school. In Aluffi-Pentini, A. (Ed.), *Intercultural workshop: Reception and interaction in a multicultural learning environment* (pp. 9–24). Athens, Greece: Atrapos.

Govaris, C. (2013). Intercultural pedagogy: Thoughts and positions from the perspective of the theory of education. In: Govaris (Ed.), *Teaching and learning in intercultural school* (pp. 17–30). Athens, Greece: Gutenberg.

Gutiérrez, R. (2013). The sociopolitical turn in mathematics education. *Journal for Research in Mathematics Education, 44*(1), 37–68.

Gutiérrez, R., & Dixon-Román, E. (2011). Beyond gap gazing: How can thinking about education comprehensively help us (re)envision mathematics education? In B. Atweh, M. Graven, W. Secada, & P. Valero (Eds.), *Mapping equity and quality in mathematics education* (pp. 21–34). New York, NY: Springer.

Hasselhorn, M., & Kuger, S. (2014). Wirksamkeit schulrelevanter Förderung in Kindertagesstätten. *Zeitschrift für Erziehungswissenschaft, 17*(2), 299–314.

Knijnik, G. (1998). Ethnomathematics and postmodern thinking: Conver/divergences. In P. Gates (Ed.), *Proceedings of the First International Mathematics Education and Society Conference* (pp. 248–252). Nottingham, England: Nottingham University.

Ladson-Billings, G. (1995). Toward a theory of culturally relevant pedagogy. *American Educational Research Journal, 32*(3), 465–491.

Lave, J. (1988). *Cognition in practice: Mind, mathematics and culture in everyday life*. Cambridge, MA: Cambridge University Press.

Lerman, S. (2000). The social turn in mathematics education research. In J. Boaler (Ed.), *Multiple perspectives on mathematics teaching and learning* (pp. 19–44). Westport, CT: Ablex.

Marcus, G. E., & Fischer, M. M. (1986). *Anthropology as cultural critique: An experimental moment in the human sciences*. Chicago, IL: University of Chicago Press 176.

Mesquita, M., Pais, A., & François, K. (2014). Communitarian mathematics education: Walking into boundaries. *Em Teia: Revista de Educação Matemática e Tecnológica, 5*(1), 1–18.

Mishra, V. (2012). *What was multiculturalism?* Melbourne, Australia: Melbourne University Publishing.

Moje, E. B., Ciechanowski, K. M., Kramer, K., Ellis, L., Carrillo, R., & Collazo, T. (2004). Working toward third space in content area literacy: An examination of everyday funds of knowledge and discourse. *Reading Research Quarterly, 39*(1), 38–70.

Moschkovich, J. (2007). Examining mathematical discourse practices. *For the Learning of Mathematics, 27*(1), 24–30.

Noula, I., Cowan, S., & Govaris, C. (2015). Democratic governance for inclusion: A case study of a Greek primary school welcoming Roma pupils. *British Journal of Educational Studies, 63*, 47–66.

Pais, A., & Valero, P. (2012). Researching research: Mathematics education in the political. *Educational Studies in Mathematics, 80*(1–2), 9–24.

Papachristou, D. (2014). *Literacy practices and orality of Roma students.* Master Dissertation. Mytilene, Greece: University of Aegean.

Rosa, M., & Orey, D. C. (2011). Ethnomathematics: The cultural aspects of mathematics. *Revista Latinoamericana de Ethnomatemática, 4*(2), 32–54.

Sousa Santos, B. (2012). Public sphere and epistemologies of the South. *Africa Development, 37* (1), 43–67.

Stathopoulou, C. (2005). *Ethnomathematics: Exploring the cultural dimension of mathematics and of mathematics education*. Athens, Greece: Atrapos.

Stathopoulou, C., & Appelbaum, P. (2016). Dignity, recognition, and reconciliation: Forgiveness, ethnomathematics, and mathematics education. *International Journal for Research in Mathematics Education, 6*(1), 26–44.

Stathopoulou, C., & Kalabasis, F. (2007). Language and culture in mathematics education: Reflections on observing a Romany class in a Greek school. *Educational Studies in Mathematics, 64*(2), 231–238.

Stinson, D. W., & Bullock, E. C. (2015). Critical postmodern methodology in mathematics education research: Promoting another way of thinking and looking. *Philosophy of Mathematics Education Journal, 25th Anniversary Issue* (29), 1–18.

Taylor, N. (1993). Desire, repression and ethnomathematics. In C. Julie & D. Angelis (Eds.), *Political Dimensions of Mathematics Education* (Vol. 2, pp. 130–137).

UNESCO (2014). Inclusion from the start-Guidelines on inclusive early childhood care and education for Roma children. *UNESCO Council of Europe*. Retrieved from http://unesdoc.unesco.org/images/0022/002275/227503e.pdf.

Valai, Ph., Gana, E., Stathopoulou, C., & Govaris, C. (2015). Roma students use and design maps in the kindergarten. In D. Desli, I. Papadopoulos & M. Tzekaki (Eds.), *Proceedings of the 6th National Conference of EN.E.DI.M: Mathematics with Distinction and without Discrimination* (pp. 279–288). Thessaloniki, Greece: ENEDIM.

Vithal, R., & Skovsmose, O. (1997). The end of innocence: A critique of ethnomathematics. *Educational Studies in Mathematics, 34*(2), 131–157.

Walkerdine, V. (1988). *The mastery of reason*. London, England: Routledge.

Walshaw, M. (2007). *Working with Foucault in education*. Rotterdam, The Netherlands: Sense Publishers.

Zaslavsky, C. (1994). Africa counts and ethnomathematics. *For the Learning of Mathematics, 14* (2), 3–8.

Chapter 6
Mathematical Ideas in Chundara Culture: Unfolding a Nepalese Teaching and Learning System

Jaya Bishnu Pradhan

Abstract Chundara is one of the occupational caste groups in Nepal that possess a unique knowledge generation system. This research was carried out setting up the objectives to explore approaches to Chundara teaching and learning and to uncover mathematical knowledge used in constructing wooden artifacts. The methodological procedures included in-depth interviews and the observation of activities in the Chundara workplace. This includes daily activities at their places of work and exploration of the mathematical knowledge found in various sectors of life, which have been analyzed using written documents, photographs, and videos. This chapter presents the findings of this research project which has drawn four major components as exclusive approaches to teaching and learning: observation, imitation, estimation, and practice. Chundaras' teaching and learning activities involved participatory and cooperative approaches in which they learned with the help of more knowledgeable seniors. In addition, it was observed that the Chundaras make wonderful wooden artifacts by applying implicit high level of mathematical concepts, knowledge, and skills.

Keywords Chundara · Culture · Ethnomathematics · Ethnography · Indigenous knowledge

6.1 Brief History of *Chundara*

Nepal is a rich land, comprised around 59 indigenous communities speaking more than 125 recorded languages (NEFIN 2006) with distinctive cultural characteristics. There are many groups of people of diverse ethnicities and castes. A caste group is problematic distinction which requires a study of diverse cultures of different ethnic groups. Even the government classification system is open to question (NDC 2006). Among some caste groups are the *Dalits*, more than twenty Dalit caste groups exist

J.B. Pradhan (✉)
Tribhuvan University, Kathmandu, Nepal
e-mail: jaya@mrctuktm.edu.np

M. Rosa et al. (eds.), *Ethnomathematics and its Diverse Approaches for Mathematics Education*, ICME-13 Monographs,
DOI 10.1007/978-3-319-59220-6_6

in Nepal alone. The Dalit are discriminated people under this caste system and suffer from many kinds of restrictions including the use of public amenities, deprivation of economic opportunities, and are neglected by the state and society. The number of Dalits counted in the 2001 population census was 2,946,652, which represented 13.02% of the national population.

Bishwakarma is a Dalit community which is considered to fall under Indo-Aryan ethnic group. Some Bishwakarmas are blacksmiths and some others are goldsmiths who have been scattered in almost all hill districts of Nepal. The blacksmiths are famous for the making *Khukuri* (knives) used by the Nepal Army and used by Gurkha military. They have about 26 surnames among which one is the Chundara (NDC 2006). According to the 2001 population census, the total population of Bishwakarma in the country was 895,954 out of which 96.69% were Hindu and 2.21% were Buddhists and the remaining followed other religions (NDC 2004).

According to the religious legends, various work divisions were made in which Asbini Kumar was considered as the Doctor of the Gods, Kuber as Finance Director, and Brihaspati as the teacher of Gods. The Bishwakarmas had skills and expertise in technology, so, they were labeled as the engineers of the Gods. In Nepalese culture, it is believed that they constructed and designed various war machines and vehicles for the Gods at that time.

There is another story related to Lanka Darbar, considered a highly sophisticated articulation of the time, and was also made by Bishwakarmas on behalf of Shiva and Parbati. Currently, they are contributing in various segments of social life in many Nepalese contexts. They served as technicians playing significant roles by making different war machines during the period of Nepali unification about two hundred fifty years ago. Even today, they are affluent in their *ethnoknowledge*. Which is also known as the *ancestral knowledge* transmitted across the generation orally and practically.

Orey and Rosa (2010) mentioned that *ethnoknowledge* is acquired by people of a particular cultural group through culturally relevant teaching and learning processes. Both culturally and religiously, there is a custom of praying to Bishwakarma Baba as both the inventor and the saver and conductor of machines. This implies that the Bishwakarmas are the drivers of machinery goods. Though they have many wonderful skills and talents, they are regarded as low and the untouchable caste in the Hindu custom. That is a shocking paradox of our community life. Because the Bishwakarmas do not have their own specific language, Nepali is spoken amongst them with different disguising accents depending upon the place and region. The Terai Bishwarkarmas are not regarded as untouchable as hill Dalit of Bishwarkarmas. They also speak a local language like Maitheli and Abadhi etc. They use the Devnagari script for writing purposes.

Chundara is a caste group having indigenous occupational knowledge and skills. They are involved in making various wooden artifacts like theki, mana, paathi, dhyangro, madal, khaijadi, flower cases, pencil cases, and damaru. These are forms of both wooden kitchenware and musical instruments with decorative arts. There are a wide range of mathematical components of knowledge that the artifacts involved in making these implements. They include measurement, geometrical shapes, size, symmetry, transformational geometry etc. The creativity of *Chundara* that include attention to decorative arts encompasses and relies on mathematical concepts as well.

It is a basic tenet of ethnomathematical practices and research that all cultures use mathematics to manage their social and economic life. So it is that mathematics is an important tool for many occupations that are often hidden in indigenous cultural contexts. Chundara are no different and have artistic skills and use implicit mathematical knowledge in the production of their wooden artifacts. They make use of this implicit mathematical knowledge, which they have been practicing to fulfill their everyday needs. This knowledge, skills and unique ways of understanding the world have been transferred from generation to generation both verbally and empirically.

Regarding indigenous knowledge, Ezeife (2002) states that the knowledge embedded in an indigenous community is often unique to a given culture. This indigenous knowledge covers traditional beliefs, perceptions, practices, customs, rituals and worldviews. It also refers to alternative and informal forms of knowledge rooted in the indigenous community. Vygotsky (1978) affirms that "learning is a necessary and universal aspect of the process of developing culturally organized, specifically human, psychological functions" (p. 90). Indigenous knowledge generation follows participatory processes, and is communal and experiential form of knowledge with the reflection of local geography. It refers to systems of knowledge and skills that have been developed to fulfill everyday needs. Their teaching and learning is based on their own ways distinct from formal educational systems that enable communities to survive.

In this context, this researcher has made an attempt to look into Chundara indigenous knowledge of mathematics and explore it using different theoretical lenses. The dominance of a global knowledge system has largely led to a prevailing situation in which indigenous knowledge is largely ignored and neglected, and in danger of being lost. Thus, the research of mathematical practices by the Chundara and their ancestral knowledge has brought a new insight in mathematics education research in Nepal.

Ethnomathematics is the mathematical concept inherent and practiced in the culture of a certain group of people. D'Ambrosio (2006) says "Ethnomathematics is the mathematics practiced by the cultural groups, such as urban and rural communities, groups of workers, professional classes, children in a given age group, indigenous societies and so many other groups that are identified by the objectives and traditions common to these groups" (p. 1). The Chundara people, as a caste and worker group have their own way of understanding and using mathematical knowledge. They practice distinct mathematical concepts in their everyday lives which are unspoken and informal mathematical components of knowledge.

Further, Orey (2000) explained the term ethnomathematics as defined by D'Ambrosio (2006) as the motive by which the members of specific cultures *(ethno)*, in this case, the Chundara, developed over time, the techniques and the ideas *(tics)* to learn how to perform their everyday work regarding ideas of measuring, calculating, comparing, and classifying in the construction of woodenwares.

They have the ability to model natural and social environment in which they use to (*mathema*) explain and understand the phenomena. In this regard, it is necessary to understand how far Chundara mathematical knowledge: counting, locating, measuring, designing, playing and explaining (Bishop 1990) are practiced in their cultural activities.

As stated earlier, Chundaras are considered a primary occupational caste group in Nepali culture. They are known for making different instruments of wooden craftwork and form a part of Nepali cultural traditions. They are known for their creative and artistic skills and use implicit mathematical knowledge in their daily work place. I realized an immediate need to explore the unique perspective and mathematical ideas, knowledge and concepts of Chundara people as there are very few specific studies so far about their traditional mathematical knowledge.

This chapter explores the pedagogy of how Chundaras generate new knowledge. Furthermore, this researcher shares his experience in how he attempts to find the knowledge transformation system of the Chundaras and what kind of mathematics and geometry found in a particular kind of their daily activities. This research shows connections between a special wooden craft making by the Chundara and the mathematics studied in relation to them. Thus, the research about the ways of learning and generating new knowledge from the Chundara's context is of paramount importance. More specifically, this chapter has an aim to achieve the answers of the questions: "What are the mathematical ideas embedded in the Chundaras production of wooden artifacts?" And "how do they share, teach and learn these mathematical ideas in their own cultural setting and context?"

6.2 Literature Review and Conceptual Understanding

The term ethnomathematics was coined with different perspectives. Rosa and Orey (2010) view ethnomathematics as the mathematical ideas and concepts embedded in diverse cultural contexts. For Ascher and Ascher (1997), ethnomathematics is the study of the mathematical ideas of non-literate people. And, others believe that it is the methodological postures in the learning process of mathematics (Ferrerira 1989). D'Ambrosio (2006) views ethnomathematics as a research program about the history and philosophy of mathematics and it is also the program of the way in which cultural groups understand, articulate, and use the concepts and practices, which we describe as mathematical, whether or not the cultural group has developed a concept of mathematics.

For us in Nepal, ethnomathematics is a relatively new research field linked to the domain of mathematics education, and is described as the study of mathematical ideas and activities embedded in their cultural context. Further, Bishop (1991) believes that mathematics is a cultural product developed as a result of various activities and that *counting*, *locating*, *measuring*, *designing*, *playing*, and *explaining*, are all part of that cultural product (Powell and Frankenstein 1997). Everyday life is impregnated in the knowledge and practices of a culture. Common to all peoples everywhere, individuals compare, classifying, quantify, measure, explain, infer, generalize, and evaluate, using materials and intellectual instruments that belong to their culture (D'Ambrosio 2006). The cultural activities invented by diverse peoples always use some kind of the mathematical knowledge and concepts.

The ethnomathematics existing in different cultural practices of *Chitwan's Tharus* (Paudel 2008) support the preservation of their cultural identity, and highlights a certain discontinuity between traditional and modern measurement systems, with similar practices found among illiterate groups. *Tharus*' ethnomathematical practices support the need to preserve call for a developing links to formal education of the community connecting to the local corpora of knowledge with the global corpora of knowledge (Paudel 2008).

Similar results are also found in the literature of ethnomathematical studies done in other indigenous communities in Nepal. For example, Dahal (2007) demonstrated how measurement systems, numerical systems, geometrical applications and other mathematical concepts are preserved by the Gopali people living in Chitlang village in the Makawanpur district and were different from school mathematics. Before school enrollment children have some mathematical concepts developed in their mind in a way that is totally different from school teaching practices. Gopali children certainly were no different. Dahal (2007) concluded that the ethnomathematics emerged from cultural practices he found in Gopali culture, and is more applicable and can be used easily in practical life contexts. Primary level school mathematics education can be related to the ethnomathematics which was the major concluding remarks of the study.

According to UNESCO (2008), it is necessary to conduct research to develop culturally contextualized mathematics curriculum resource materials for Nepalese lower secondary schools in order to foster a culturally pluralistic society. This study found that numerous local practices were identified as being linked to formal mathematics at the lower secondary level. In particular, practices of farming, local business, household activities, children's games, cultural practices, artifacts and social events were found to be strongly linked to school mathematics concepts. These contexts were incorporated into the curriculum resource materials in order to enhance children's learning of school mathematics, especially girls, whose experience and family social role could be legitimated in formal education by such contextualization.

Analysis of Nepali lower secondary school mathematics curriculum revealed formal geometric, arithmetic, algebraic and set theory concepts that could be linked directly to local practices and day-to-day life activities documented in the fieldwork data. The field study indicated how parents and elders can be trained in order to help their children's learning of mathematics at home by engaging them consciously in local activities linked directly to school mathematics, such as wages, expenses and cost calculation and estimations. Given that children cannot escape household tasks in rural Nepal, a possibility observed in the fieldwork was that children can enrich their mathematical understanding by linking classroom learning to these daily household activities.

Gurung (2009) conducted a study that explored the knowledge generation, continuation, distribution and control of Pariyar people and compared it with the ways that they learn and teach their everyday activities from school pedagogy. He observed that conventional curricula and achievement tests, however, did not support students' learning based on indigenous knowledge. They prescribed school pedagogy and curriculum as alien to the rich local knowledge of the Pariyars. The school pedagogy was developed in Kathmandu for students of affluent families and was used in the remote villages of Gorkha district was the same. Their local environment and other cultural aspects were largely ignored.

This was when this writer realized that learning environments need to adapt to indigenous community knowledge and recognize and respect diverse cultural value systems. The different studies done in the context of an indigenous community identified that teaching and learning activities involved participatory and cooperative approaches in which they learn with the help of their parents as Vygotsky (1978) seeing learning as an activity in which shared mathematical meanings are constructed socially.

Millroy (1992) found that a group of carpenters used the concept of tacit knowledge of mathematical ideas. This concept manifested that the people use mathematical ideas in their activities implicitly rather than spell out the mathematical concept. Thus, tacit knowledge displays itself through their activities in the process of constructing different artifacts, which are not necessarily expressed in written or spoken form. According to Millroy (1992), "the physical act of designing and building furniture would involve tacit mathematical knowledge" (p. 13). The results of my study showed that indigenous knowledge acts as a powerful tool in a learning environment to teach students. These same experiences have been observed in the *Chundara* community, where they are using implicit mathematical knowledge in their everyday activities.

The studies that this researcher reviewed depicted many indigenous people of different places around the world who have the same basic mathematical ideas of counting, measuring, playing, sorting and deciphering methods which they use to perform their everyday activities. Another study presented by Mosimege (2000) about mathematical knowledge and its use in the daily activities of workers at South African villages. His study intended to observe the use of mathematical knowledge in the work of indigenous people. The study identified how artifacts and cultural activities of the villagers provided an opportunity to explore mathematical concepts that are used in their everyday context.

The focus of each village is on a particular cultural group that is dominant in the province in which that group resides in a large number. In each of these samples, villages employ the use of elderly people who are knowledgeable about cultural activities, and show activities properly in their historical and socio-cultural backgrounds. The elders were involved actively in making the various artifacts and they used implicit mathematical ideas in the process of the construction of various types of artifacts.

Gerdes (1999) indicated how the people south of the Sahara Desert constitute a vibrant historical and cultural mosaic, extremely rich in its diversity. The indigenous people he studied have demonstrated a variety of geometrical and mathematical concepts and ideas in the construction of dwellings, wooden materials, ivory carvers, potters, painters, weavers, mat, and basket makers. Both the men and women of this region contribute to the construction of artifacts, which was in contrast to Nepal, where only male Chundara are directly involved in the construction of wooden wares. The works regarding different artifacts possesses the varieties of geometrical shapes. Further, Gerdes (1997) mentioned that the sophisticated mathematical ideas are embedded in their artifacts of decorated handbags, coiled baskets, mats, pots, houses, fish traps, decorated pottery, grass brooms, tattooing and body painting, bead ornaments, mural decorations and cultural games like string figures.

Mosimege and Lebeta (2000) conducted an ethnographic study of mathematical concepts in the cultural activities at the Basotho cultural village reported that the inhabitants of the village used sophisticated mathematical knowledge. They found extensive use of a variety of mathematical concepts in the traditional baskets, traditional hats, and miscellaneous items involving *Motlhotlho* plaited rope used for binding, and of grass artifacts. This illustration of mathematical concepts used in various cultural activities showed that most indigenous people are knowledgeable about a variety of artifacts and activities in which mathematical concepts are used extensively and they were dealing with mathematical concepts like estimation, patterns, geometry, and symmetry in the construction of different grass artifacts. It is clear that such mathematical knowledge and concepts are used regularly in their work, even though they do not necessarily know the mathematical terms as described in mathematics literature (Mosimege 2000).

After a review the literature related to the ethnomathematics and the research done in the various dimensions on ethnomathematics, and having gone through the available resources on ethnomathematics in and around the globe, it was observed that indigenous people contexts are often different in respect to their knowledge, languages, political situation, aims of education, and access to resources, however, most indigenous peoples are relatively culturally homogenous (Barton 2008). This researcher realized an immediate need to explore the knowledge generation system of the Chundara, including insights to their ways of teaching and learning, the hidden mathematical ideas, knowledge and concepts. There has not been any specific study so far about their traditional mathematical knowledge gives special important to this work.

In this chapter, I have developed an innovative approach that fills the void in our Nepali context. The study of literature on ethnomathematics provides a new conceptualization of ethnomathematics which avoids some of the difficulties which emerge from the literature. Being a relatively new field and with many scholars conducting research in equally diverse locations worldwide, there is no consistent view of ethnomathematics in the literature. In this case, often the relationship with mathematics itself has been ignored, or the philosophical and theoretical background is missing. The literature also reveals the ethnocentricity implied by ethnomathematics as a field of study based on culture which has mathematics as a knowledge category. In particular, this work has studied the mathematical ideas embedded in the everyday activities of indigenous people of Nepal. The need of incorporation of ethnomathematical ideas of other people has been highly emphasized in school curriculum.

As mentioned earlier, my work has been to identify the knowledge generation system of the occupational caste group of the Chundara people. Also, my interest was how they learn their ancestral work and what are the specific ways of teaching and learning and its commonalities/disparities with the school pedagogy. Further, this researcher was equally interested in seeing the mathematical knowledge embedded in their work in constructing artifacts.

Regarding the teaching and learning approaches of the community members, Vygotsky (1978) viewed that the cognitive development does not happen just in the head of the children. He emphasized the learning as the social aspect and maximum learning can occur only with the guidance or collaboration with more capable peers. In this regard, Vygotsky (1978) supplied the term Zone of Proximal Development (ZPD) and "it is the distance between the actual developmental level as determined by independent problem solving and the level of potential development as determined through problem solving under adult guidance or in collaboration with more capable peers" (p. 86).

In my research it was observed that the teaching and learning approaches to the Chundara cultural setting involves concepts of ZPD. The senior members who are more knowledgeable about the construction of wooden stuffs help their children to construct the same with assistance. The Chundara children work as an apprentice worker under the guidance of the more knowledgeable adult. After some time, their children become able to construct wooden stuffs independently on their own. The knowledge sharing and supportive teaching and learning situations in the Chundara culture possess the sociocultural theory of Vygotsky.

According to this context, this researcher produced a conceptual understanding of qualitative inquiry about the ways of knowledge generation and distribution of the Chundara. Regarding the teaching and learning process in the Chundara cultural setting, the empirical evidence that this researcher gathered are related to motivation, imitation, and observation of the other's work, drill, practice, exercise, and estimation. Figure 6.1 represents the conceptual framework for knowledge generation and distribution system of the Chundara in their work places.

There are complex structures in regard to social and cultural manifestations in the indigenous community. Their perceptions, understandings, beliefs, love and affection cannot be observed directly in their social and cultural setting (Neuman 2008) and hence they are often invisible. However, their surface reality possesses how the Chundara people work in their workplace, what they produce, how they participate in the construction of woodenwares, and what role does the senior knowledgeable member play in the process of teaching and learning in their cultural settings and so forth are more often visible. In this process, this researcher visited their workplace and patiently observed the social reality and context of the Chundara on the basis of the surface reality and established how they formulate and generate knowledge from their daily lives.

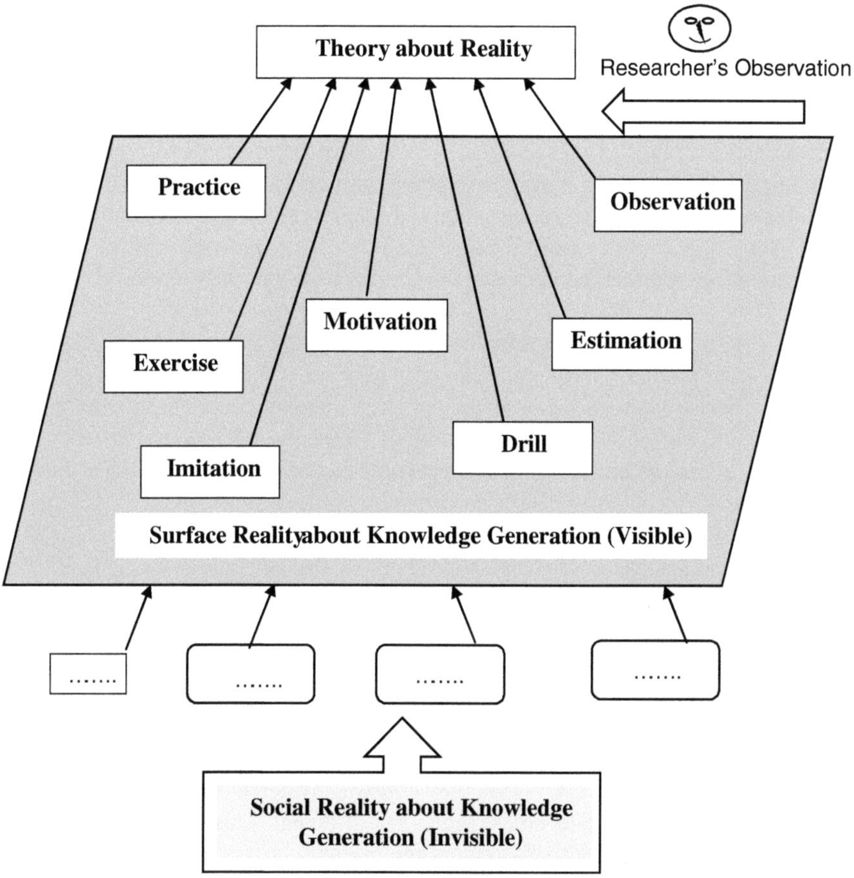

Fig. 6.1 Chundaras' social reality about knowledge generation. *Source* Neuman (2008)

Chundara indigenous knowledge regarding the construction of their wooden vessels comprises a complex set of technologies developed and sustained by their community over time. Often, most forms of indigenous knowledge are transmitted from one generation to another using both empirical and verbal methods often oral and symbolic, and is transmitted and passed on to the next generation through modelling, practice, and animation rather than through a written form.

Here, modelling means that the Chundara's cultural knowledge which was transmitted from one generation to another verbally. The modelling process focus on critical analysis of the generation and production of knowledge in the cultural settings of particular group of people, their social mechanism of teaching learning and transmission of culturally embedded knowledge to the new generation (Orey and Rosa 2010). Gerdes (2005) argues that all knowledge is first of all a local knowledge. Thus, all schooled knowledge is the explicit product of the culture, and may have different ways of teaching and learning approaches to perform their ancestral work.

6.3 Methods and Procedures

In this study, I used qualitative research methods to make sense of the complex world of the implicit mathematical knowledge of the Chundaras and their way of teaching and learning. *Qualitative researchers* have been described by Glesne and Peshkin (1992) as meaning makers who draw on their own experiences, knowledge and theoretical outlooks, to collect data and to present their understanding to the world. Thus, the current research became a process of meaning making with the assistance of my research participants, the Chundara, rather than reporting objective reality.

As my study is based on the ethnomathematical knowledge of the Chundaras, it was necessary to make frequent visits to where the Chundara were making the wooden artifacts under study. As stated earlier, Chundara people are spread around the different parts of Nepal; however, the Chundaras of Gorkha have a great recognition about the construction of wooden artifacts. Their production is famous not only in their local market, but also they are exported to the international market as well. Thus, I chose the Deurali Village Development Committee of Gorkha district where there were substantial numbers of the Chundaras carrying on this ages-long occupation.

Currently, most of Chundara have moved to other places where they can find wood in sufficient quantity. Hence, I tracked down the Chundaras and found them residing in the Aabu Khaireni Village Development Committee as this place was convenient for them to find the materials they needed and it was easier for the marketing of their products, where they make the wooden artifacts popular in their surrounding communities. Similarly, they construct different forms and varieties of the materials as compared to other Chundaras of Nepal. This was the reason why this researcher chose this site to conduct the research.

Purposive sampling is a non-random sampling used in ethnographic research (Neuman 2008). It was suitable for me because I wanted to identify the mathematical knowledge of the Chundara people. In this research a purposive sampling was adopted because it is one of the most common sampling strategies in qualitative research and group participant according to preselected criteria relevant to a particular research question (Ary et al. 2010). This study intended to explore the knowledge generation system of the Chundara people and perhaps unearth their *hidden* mathematical knowledge in relation to their work.

The sample sizes in qualitative research may or may not be fixed prior to data collection. It depends on the resources and time available as well as the study's objectives (Denscombe 2010). Purposive sample size is often determined on the basis of consistencies of the data to be gathered from the field. This is because of the same specific pattern of behavior emerged over and over again (Fetterman 2010) in the activities of the members of cultural groups. Purposive sampling is most successful when data review and analysis are done in tandem with data collection.

Since the objective of this study was to uncover the hidden mathematical knowledge of the Chundara people, the writer purposively selected five of them as research participants. Approximately two weeks were spent with them. The researcher followed them to their work places and patiently observed their activities. As well, the writer took part in every activity of their families and we had meals together. This eased the situation and they considered the researcher positively and did not hesitate to talk on any issue or question raised.

Neuman (2008) has said that ethnography is the "field study research that emphasizes providing a very detailed description of a different culture from the viewpoint of an insider in the culture to facilitate understanding of it" (p. 381). The research also intended to understand and describe a social and cultural scene (Fetterman 2010) of the Chundara, so, ethnographic methods were adopted for the collection of data. In this regard, and to be more specific, in-depth interviews were used to collect views and interpretations of certain social actors and their knowledge about the specified social context, their accounts of that social arena which are significant in informing my research (Dunne et al. 2005).

An observation check list was prepared so that it could be easier for the researcher to generate the data from the field observations. Meaning was generated by experiencing their lives, daily patterns of human thought and behavior (Fetterman 2010). While analyzing the data, the writer read and scrutinized the transcripts of interviews. The data thus obtained from the field were categorized according to themes regarding to the Chundara teaching learning system and the mathematical knowledge developed in their everyday activities. Then, the field data was triangulated with theory and literature and then a sense of meaning was generated without dismantling the essence of the participant views.

6.4 Emic and Etic Perspective in Ethnomathematics Research

The emic perspective in research sees the reality of cultural elements in terms of the insider's perspective. In this regards, Rosa and Orey (2012) viewed that the emic approach seeks to understand a particular culture based on its own reference. Further, Fetterman (2010) said that "the insider's perception of reality is instrumental to understanding and accurately describing situations and behaviors" (p. 20). The emic method is used to investigate how the members of particular cultural groups think, how they perceive, how they explain things and what it means for them.

Thus, emic perspectives and interpretations within a culture are determined by local custom, meaning, and belief and best described through the lens of the indigenous culture. The emic approach to research in mathematical ideas in Chundara culture investigated mathematical phenomena and their interrelationships through their lens, their perspective. In this vein, Orey and Rosa (2015) mentioned that the primary goal of an emic approach in ethnomathematical research is to emphasize the uniqueness of mathematical ideas, procedures, and practices developed by the members of particular cultural groups.

An etic perspective in research views cultural elements through the eyes of the outsider. Etic knowledge is a description of a behavior or belief of the members of cultural groups by a scientific observer through the culturally neutral and outsider's perspective. The etic perspectives investigate lawful relationships and casual explanations valid across different cultures (Orey and Rosa 2015). Thus the etic method applies an outsider's interpretation of events, customs, beliefs from decontextualized perspectives (Neuman 2008). Further, an etic approach is a way of examining the emic knowledge of the members of cultural groups and it is universal in the international community of scientific observers, researchers and investigators (Orey and Rosa 2015).

Fetterman (2010) highlighted that the best ethnography requires both emic and etic perspectives. The emic approach to research tries to apply overarching values to a single culture from native's lens. However, the etic approach helps in enabling researchers to see more than one aspect of a culture and to apply the observations to cultures around the globe. Thus, by combining these two approaches in study, a richer view of cultural elements of group members can be better understood. In this research both emic and etic perspectives were used to increase understanding of the knowledge generation and distribution, and the ethnomathematical ideas embedded in Chundara cultural setting.

6.5 Findings and Discussions

This section deals with the findings of the study in terms of the data gathered from the field. The analysis of the data in this study was based on research participants' own words, direct quotations and the images of the different artifacts of their own community. The writer is confident that a connection was established between the literature, theory, and empirical data collected from the field for the analysis and interpretation of the results of this study. During the analysis of field data, two vital research questions were answered: (1) How do the Chundara people generate new knowledge? and (2) Are there mathematical ideas embedded in the production of their wooden artifacts? While doing so, some major findings were listed from the study by two consequent subheadings on the basis of the themes that emerged from the field.

6.5.1 Knowledge Generation and Distribution in Chundara Culture

Ascher (2002) views how indigenous people she observed have had their own ways of looking at and relating to the world, the universe, and to each other. The

Chundara people also have their own ways of generating knowledge. Regarding the attainment of skills to make wooden wares, one of my research participants, Mangal Bishwakarma (age not known, but he was familiar with the massive earthquake of 1934 AD) says:

> In the hope of getting Ghaiya ko bhat ra pina ko tiun (rice and vegetables- better foodstuffs in metaphorical sense); I followed my maternal uncle from Baseri of Dhading to Chyangli of Gorkha District. I just looked very closely and observed my uncle working with the wooden things. This is how I learnt to work on this (…) when I made a mistake he used to hit badly on my hand, and I always tried to do it consciously without making any mistake.

This tells us that the Chundara gained knowledge through observation and imitation. In their culture there is strict system of reward and punishment in the process of learning in the work places. To understand the ways of knowledge generation, the other research participant, Namaraj Chundara (16) says:

> My father used to go to the work place early in the morning. So, I had to carry food for him in the same place. I was interested in his work and used to observe closely when he worked. I wished I could do this work some days. I was curious what it would be like if I worked that way. This became a regular routine. After some time, I started working with him. This is how I learnt to make wooden things.

People learn in the natural setting with the help of their family members or seniors as mentioned by Namraj above. All knowledge is created as individuals and groups adapt to make sense of their experiential world. Vygotsky (1978) argues that the construction of knowing is not a matter of individual, solitary construction of understanding, but a dialectical process firmly grounded in a system of social relations. Following this Vygotskian position novice workers in the Chundara workplace begin to follow with guidance of the elder (teacher) to construct different wooden artifacts as they focus on individualized approaches in acquiring mathematical knowledge. With the senior's function in the role of expert, the novice was encouraged to develop individual thinking, learning and practical approaches. This established a collaborative teaching learning setting that enhances the knowledge generation and transmission of the Chundara community.

Regarding specific ways of making wooden artifacts, the research participant, Namaraj Chundara said "there are no hard and fast rules for construction of the wooden stuff. With an experience when we see the timbers, we estimate and then make particular wooden stuffs. Estimation is vital skills in all stages of constructions". This researcher further asked him to elaborate what was the basis of his estimation, he said "to make wooden vessel pathee (eighty handfuls) a cylindrical timber with the length of one bitta (a span) five amals (digits) and of circumference as double of the height is required. The hollow part is made with the help of bako (a form of drill machine)".

The research participants agreed with the view and idea of Namraj Chundara. They told the researcher that the same process can be used in the construction of *theki* (wooden pots) and *theko* (small wooden pots). They exhibited a high level of understanding in hollowing out the timber for a required volume. One of the research participants, Mangal Bishwokarma also said:

> We need a cylindrical timber of height two bittas (span of the hand) and five amal (a finger's breadth) and base circumference of three bittas to make a theki (wooden vessels) of two pathees. We put this timber in the machine to give the external shape and the internal hollow part be made with the help of bako. For estimating the theki of two pathee, the hollow part of the theki should be loosely measurable by bitta. We put the dust of the timber on outer surface of the theki and then started knocking the inner surface by thumb to verify whether the thickness of the theki is all right. If the dust falls down; we understand the theki is made of two pathee. If the dust doesn't fall, we keep hollowing the inside of the theki.

All the participants in my study also agreed with the Mangal Bishwokarma's views. Further, Gam Bahadur Chundara said that "the wooden stuffs of equal capacity can be estimated by comparing the dust of the two instruments". From the observation of their activities regarding the selection of log of timber and construction of different wooden artifacts, they were using a non-standard (for formal educations viewpoint) system of measurements. The height, diameter, circumferences of circular cylindrical log are measured with *haat* (one *haat* = 18 inch), *bitta* (one *bitta* = 10 inch) and *amal* (digits). The capacity of wooden vessels is measured with *mana* (10 handfuls), and *pathee* (8 *mana).* They also use smaller units like *chauthai (*quarter of *mana*), *chakhanti* (quarter of *chauthai*) and *chimti* (substance held between three fingers thumb, index and middle).

Once again, the Chundara people have their own way of teaching and learning and have developed their own knowledge generating system. They have their own indigenous way of knowledge generation and distribution of acquired knowledge to new generations. The field data that was collected revealed that they have the ability to select and estimate the wooden timber needed to construct artifacts having a particular size and capacity. All of my research participants had frequently come up with fairly accurate and consistent estimations used to construct the needed wooden artifacts.

The experiences and practices needed to perform accurate and consistent estimation for the construction of wooden vessels includes the learning and teaching approaches in the process of making wooden vessels involved with the individual efforts of observation, imitation, and practices. Chundara used non-specific units for measuring the dimensions of length, volume and capacity in their everyday activities. The specific units involved body units such as *amal, kuret, bitta, haat* were made frequently accurate and consistent in their production of wooden wares.

Chundara ways of teaching and learning in their own setting act can be a powerful tool in the learning environment for their children. Reflecting and integrating both indigenous and formal/school knowledge generation systems. It was found that there is no predetermined condition to learn or to construct Chundara wooden artifacts. Those who are interested in it can learn from their elders. There is no test conducted nor are they discouraged when they fail to learn in the initial stage. The final product is the evaluation.

However, in formal school education, Jarlais (2008) mentioned that "competency is often assessed based on predetermined ideas of what a person should know, which is then measured indirectly through various forms of objective tests. Such an approach does not address whether that person is actually capable of putting that knowledge into practice" (p. 43). In this regard, when this writer talked to the children of the Chundara, many of them liked the ways of learning to make different wooden artifacts as compared to going to school and learning from teachers. Naresh Chundara a fifth grader said:

> My father always encourages me to construct *mana* and other decorative instruments; I don't get punishments for my mistakes. But at school, the teachers tell me to do homework on my own. If I fail to do so, they punish me. So, I have no interest in doing mathematics, I want to be like my father in the future.

This shows how the boys enjoy the learning process at home, but shows their resentment about school pedagogy. When this researcher lived with them, he had a great time talking to them. They were innocent and spoke the truth, whatever the question asked. From the interview with the Chundara children, it was felt that they genuinely and honestly spoke out. This researcher learned that they could very easily copy or imitate others in terms of constructing the different wooden artifacts. They learned easily by observing, imitating, and participating. The more knowledgeable senior members of the community played supportive roles and there was frequent interaction between senior members and the children during the construction of wooden artifacts.

In this researcher's observation, their physical gestures, their elation and smiling faces during the work exhibited how much they enjoyed the learning process in their cultural setting. To the contrary, the researcher asked them if they could say what type of this shape they were using, and, it showed a problem related to the measurement and volume they had learned at school. In all cases, they could not explain or connect the two. This shows how our school pedagogy is indifferent to indigenous approaches to knowledge generation (Hammond and Brandt 2007).

This writer has discussed the above major findings and other general findings in connection with theoretical review and framework, and field reflections. When reflecting on the field data and the literature, it was found that knowledge generation systems of the Chundara people have, in accordance to Ascher (2002), had their own ways of looking at and relating to the world, the universe, and to each other for eons. Knowledge is a negotiated product with the people individually and collectively to make sense of our world emerging from the interaction of human consciousness and reality. Their traditional education processes were carefully constructed around observing natural processes, adapting modes of survival and using natural materials to make their tools and implements (Bernhardt and Kawagley 2005).

All of this was made understandable through demonstrations and observations accompanied by thoughtful stories in which the lessons were imbedded. The sources of indigenous knowledge and the ways of obtaining it have very distinct approaches. The intersection of their beliefs, perceptions, and experience facilitate the construction of their knowledge. From the data collection and observation in the field study it was found that Chundara teaching and learning activities involved participatory and cooperative approaches in which they learn with the help of more knowledgeable seniors using visual learning as an activity in which shared mathematical meanings are constructed socially. Here, the social aspect of learning and the interplay of speech and action in children's learning activities were emphasized as well.

The ethnomathematics is the study of the relationship between mathematics and culture. Its aim is to contribute both to the understanding of culture and the understanding of mathematics. Eglash (1997) viewed how ethnomathematics is the study of mathematical concepts embedded in the indigenous cultures. In this regard, Bishop (1988) expressed: "mathematics (small m) is a pan-cultural phenomenon, something which exists in all cultures; and Mathematics (capital M) is a particular variant of mathematics which has been developed through the ages by various societies" (as cited in Clements and Ellotron 1996, p. 87).

Mathematics with capital M can be taken as the mathematics of the global village or the mathematics that have been practiced by academic mathematicians and professors at universities. Thus, Mathematics is an internationalized discipline (Clements and Ellerton 1996) and has influenced the entire planet in some aspect. It is the so called formal mathematical system and schools and universities everywhere study and practice it, develop it, as part of the mainstream academic Mathematics. In this context, Mathematics is a system that enables people to count, calculate, measure, and use the various forms of knowledge related to it to fulfill their daily lives.

The members of every cultural group have developed and use some form of its own knowledge and skills to perform work and solve problems even though they may be unable to utilize the global Mathematical knowledge. Thus, ethnomathematics can be seen as the pedagogical tool that seeks to study how students have come to understand, comprehend, articulate, process and ultimately use mathematical ideas, concepts and practices to solve the problems related to their everyday activities (Rosa and Orey 2010). Chundara use mathematical ideas in the production of different wooden artifacts. The study of mathematical ideas, concepts, procedures, and practices to perform everyday activities of Chundara implicitly can be considered as ethnomathematics.

6.5.2 Chundara Ethnomathematics

Here, an episode is presented in relation to the mathematical practices in the process of hollowing out a Chundara *dhyangro*. The Dhyangro, is a wooden vessel forming a musical instrument especially being used by Gumba monks and Lamas. Chundara artisans have traditional ways of selecting, estimating and identifying the center of circular timber for the preparation of the hollow part of Dhyangro (Fig. 6.2).

This researcher patiently observed the construction process, specifically the hollowing out of the dhyangro. The selection of diameter depends upon the size of

Fig. 6.2 **a** Hollow part of dhyangro. *Source* Personal file. **b** Finished product: a dhyangro. *Source* Personal file

timber available, as well, the timber with larger radii are selected as it is the characteristic feature of a proper dhyangro. In their workplace, two different ways of identification of the center of the circular face were observed. There arise two different cases: (i) face of log is irregular and (ii) face of log is approximately circular. In order to make the irregular face of timber exactly circular, they estimate and locate one point of the face as a center. Then, they take a piece of thread whose one end is on the center and next at the nearest point in the circumference. The thread is rotated around the center point and marked. This is similar idea of making a circle by our students using a compass. With the same basis, Chundara crop the timber into a circular cylinder.

In the next case when the timber is approximately circular face, first they make it as circular as possible. It was interesting to see how they identified the center of the circular face. For this purpose, Chundara take a piece of thread and fix one end at a point on the circumference. And, the next end of the thread is caught by another person and is moved along the circumference by making varying chords. Finally, the Chundaras assume the longest chord to have center of that circular face in it.

This longest thread is a diameter whose middle point is marked and hence the center of the circle is finalized. Likewise, they also mark a line through the longest thread and draw another line with the same thread within the circumference. Thus, the point of intersection of these lines (i.e. diameters) is identified and is marked as the center of circle.

On the basis of above episode, it was recognized that the Chundara are practicing numerous mathematical ideas to perform their everyday activities. They have developed practical mathematical ideas and concepts which are practiced in the construction of different wooden artifacts. They use their emic view of the knowledge in the construction process. The mathematical knowledge is useful and practical in their everyday lives and has passed across generations verbally. The emic approach to research emphasized the investigation of the elements of culture through the lens of the local perspective. In this approach, the researcher detaches herself/himself from theories and literature, and focuses on the field data obtained from the participants, and pays attention to behavior patterns or themes of the members of the cultural groups.

So, it is impossible for the researcher to go through the emic approach because they all have preconceived perspectives, ideas and the theories. In this regard, Fetterman (2010) viewed that "most ethnographers start collecting data from an emic perspective and then try to make sense of what they have collected in terms of both the indigenous view and their own scientific analysis" (p. 22). Further, Lett (1996) mentioned that an etic approach seeks to compare and examine cultural practices of the members of certain cultural groups by using standardized methods (as cited in Orey and Rosa 2015, p. 376). With this regards, this researcher would like to compare the ethnomathematical ideas of Chundara in their own views with the global mathematical ideas as I wanted to see their everyday activity from the lens of an etic perspective.

As a researcher in ethnomathematics, it is important that we give the proper meaning of their knowledge without dismantling their own views, ideas and their cultural identity. Table 6.1 indicates that the Chundaras have a full and robust mathematical knowledge and that they use it implicitly in the construction of various wooden artifacts. Their rules of measuring and estimating a log for a particular product and the ideas they use to make the vessels of exactly the same volume involve sophisticated mathematical concepts.

This comparison shows that the Chundara have mastered a number of geometric activities in their work. Some of the mathematical ideas and concepts embedded in their work place as observed include: frustum of cone, cylinder, and concepts of height and base perimeter of cylinder, capacity and volume, concept of transformation, and axis of rotation. This implicit mathematical knowledge is tacitly embedded in their activities. Mosimege and Lebeta (2000) reported that the indigenous people use different mathematical concepts like estimation, tessellations, and symmetry in the construction of the traditional artifacts and cultural activities. The Chundara use lots of mathematical concepts and knowledge. This knowledge is sufficient for performance that is consistent with rules, even though the person might not be aware of the rules.

Table 6.1 Chundaras' ethnomathematics versus global mathematics

Chundaras' ethnomathematics	Global mathematics
These are cut sections form a tree trunk. They estimate the size of vessel later produced by embracing the trunk with both hands 	This is a frustum of a cone 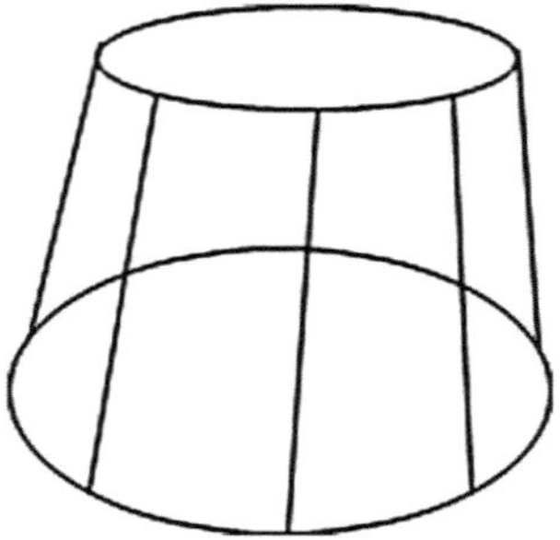
First they select the medium section of the trunk of a tree 	First we transform the frustum into a circular cylinder 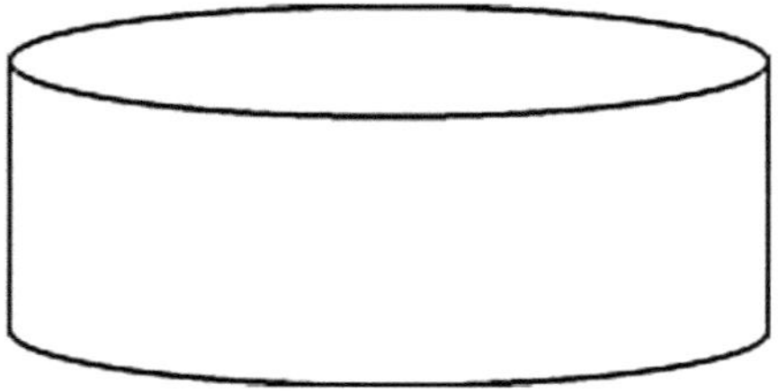

(continued)

Table 6.1 (continued)

Chundaras' ethnomathematics	Global mathematics
It is necessary to make a hole at the middle of the circular face of log. The middle point of the longest chord can be identified as the center of circular face. Square hole is made to adjust it in the machine 	The midpoint of the diameters is center of a circle. The diameter of a circle is the longest chord. The longest line segment through the center is diameter. The center is a point through the intersection of two diameters 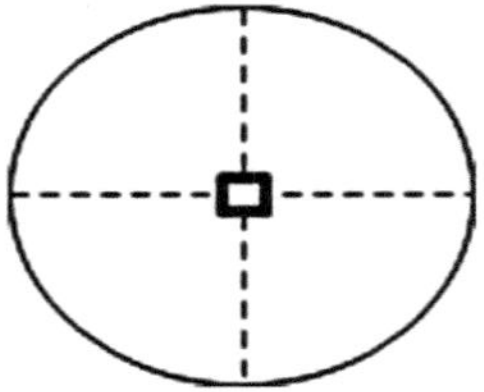

(continued)

Table 6.1 (continued)

Chundaras' ethnomathematics	Global mathematics
Making outer surface of *dhyangro* in cylindrical shape. Outer surface is finished with decorative arts that reflects the religious and cultural significances 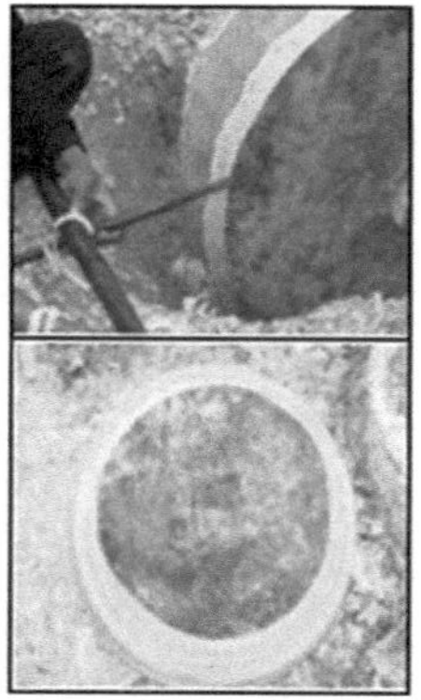	Closed circular cylinder with decorative arts in outer surface that are rich in the concepts of congruencies, similarities and symmetries. It is the culturally contextualized teaching materials for the total surface area of circular cylinder is required 
After hollowing out the closed cylindrical can. It is ready to supply into market 	Open circular cylinder. Presenting example to find the lateral surface area of circular cylinder 

Source Personal file

In the field, it was found that they used different mathematical knowledge in their activities, but they could not explain what they were. The field data also supported the findings of Millroy (1992) that the Chundara use tacit mathematical knowledge in the process of constructing different wooden artifacts as Millroy (1992) observed that the carpenters' physical act of designing and building furniture involved tacit mathematical knowledge. The tacit mathematical knowledge was manifested itself through their activities in the process of constructing different wooden artifacts in the everyday activities of Chundara.

Chundara used different mathematical knowledge in the process of constructing different wooden artifacts in their everyday activities without ability to explain written or spoken form. They estimate the size of log of tree by embracing the trunk with both hands and measure the height of trunk with *haat*. In this situation, they most definitely practice the mathematical ideas of circumference, center and radius of circle; the height and diameter of circular cylinder; and the volume and capacity of the vessels they produced.

Therefore, it was found that they have implicit mathematical knowledge with the ability to perform a task, but without the ability to explain their performance. The indigenous language pattern of the speakers reflects mathematical concepts and ideas (Owens 2010) who observed how the most indigenous languages of Papua New Guinea (PNG) possess mathematical concepts like volume, mass, area, and length. The children from indigenous communities are deeply immersed in their own cultural world. Their home language reflects the mathematical thinking of the people in own ways (Barton 2008; Owens 2010) and the researcher also observed that Chundara have tacit mathematical knowledge of circle, identification of center, diameter in the various stages of construction of *dhyangro* and in their conversation.

The Chundara use a high level of mathematical knowledge while they construct their wooden artifacts. It was found that the Chundara possess culturally embedded knowledge, which is historically excluded from schooling. From this perspective, learning is socially mediated activity and all learning takes place within social context. They have high degree of estimation power and their experiences of age long job enhance to do this better. They used a sophisticated level of mathematical concepts and knowledge in their productions though they are far from formal/academic mathematical knowledge. As an ethnomathematical researcher, one of the ethics used by this researcher is to give the proper meaning of their knowledge by respecting and valuing their own views, ideas and cultural identity. Ethnomathematics has contributed both to the understanding of culture and to the understanding of mathematics, but mainly to appreciating the connections between the two.

6.6 Implications for Teaching and Learning

Often home and school are the two different worlds where children live and learn simultaneously (Winter et al. 2009). The teaching and learning of mathematics can be enriched if the two different worlds interact with each other in a harmonious manner. However, when many traditional cultures, such as that of the Chundara, are relatively isolated from formal schooling, they still practice *haat, bitta, amal* and other bodily measurement unit for length measurement and *pathee, mana, chauthai,* and *chakhanti* for the measurement of liquid and solid artifacts. The incorporation of traditional measurement systems and mathematical ideas in Chundara culture in school mathematics curriculum has been shown to enrich the knowledge of Chundara children in learning of mathematics. It also allows them to develop positive attitudes towards mathematics and schooling in general.

The mathematical ideas gleaned from Chundara culture, the acknowledgement of their ways of knowledge generation and transmission, students' experiences should be blended with formal mathematics in the classrooms. But their mathematical ideas have generally been excluded in discussions of formal and academic mathematics. The implicit mathematical ideas situated in everyday activities of the Chundara can be linked with the school mathematics. In this study this writer observed that there are various teaching and learning approaches in Chundara culture in comparison to that of formal school pedagogy. It is true that their mathematical knowledge and learning approaches are not taken into consideration in the formal school mathematics curricula. For example, a majority of students failed to locate the center of the circle drawn through the help of a coin or a ring.

On the other hand, I observed that the Chundara are practicing and using wonderful ideas of identification of the center in the process of making dhyangro. However, their learning and teaching approaches are different from the approach taught in formal education. In everyday cultural activities; the Chundara have been practicing a diversity of mathematical ideas such as finding the frustum of a cone, total surface area and lateral surface area of cylinder, center of circle, radius and diameter of circle, and its different segments. Recognizing and valuing Chundara cultural heritage, can significantly influence the development of mathematical ideas as well as positive attitudes towards school mathematics.

In this context, D'Ambrosio (1994) argued that the cognitive power of children, including their learning capabilities and attitudes towards mathematics are enhanced by linking it to their own culture and value system. It can also be noted that Chundara ways of knowledge generation, applications, and transmission are participatory, communal, experiential, and reflective of local practices as opposed to formal classroom practices focusing on the mass teaching and learning of knowledge.

6.7 Conclusions

Chundara is a Nepalese occupational caste group who construct various wooden artifacts. These artifacts are the products of Chundara's ancient ancestral skills and knowledge. Previously, Chundara mathematical knowledge has not been formally documented or transmitted; rather it has been shared orally from generation to generation with the Chundara people themselves. Their production begins with wooden timber in the form of a truncated cone. They generally transform the frustum of a cone into a cylinder in accordance with the need of the materials to be constructed in the given capacity.

They have excellent experiences and skills related to the expertise that allows them to accurately estimate the capacity of the artifacts they create. The findings of the study on Chundara culture and their mathematical ideas were dependent on fieldwork data gleaned from the oral traditions. It was also found that they have a high degree of estimation skill in selecting timber to make particular shape and size for required volume of the wooden artifacts. This mathematical knowledge is the capital or the thesis of the knowledge developed by their ancestors, and is still valid and workable in their life. In the field, it was observed that geometrical activities as part of the implicit mathematical knowledge of the Chundara are tacitly assumed and practiced.

The Chundara people developed their own methods of teaching and learning. On the basis of findings of this study, it can be concluded that traditional Chundara teaching and learning approaches involve observation, practice, estimation, and imitation. Besides these, the Chundara wooden artifacts involve a high level of mathematics knowledge and skills. The Chundara use a relatively high level of mathematical concepts and knowledge while constructing wooden materials. They have their own unique way of generation and distribution of knowledge to the next generation.

All Chundara knowledge is derived from and passed down from their ancestors outside of the formal schooling context but it is factual, scientific and stands on a strong base for their work and living. The implicit mathematical knowledge of Chundara culture, their ways of teaching and learning approaches, and student experiences help us to enhance and understand school mathematics as well. It would be relatively simple to assist Chundara teachers to make these connections.

Appendix: Glossary of Terms

(a) **Length Measurement Units**

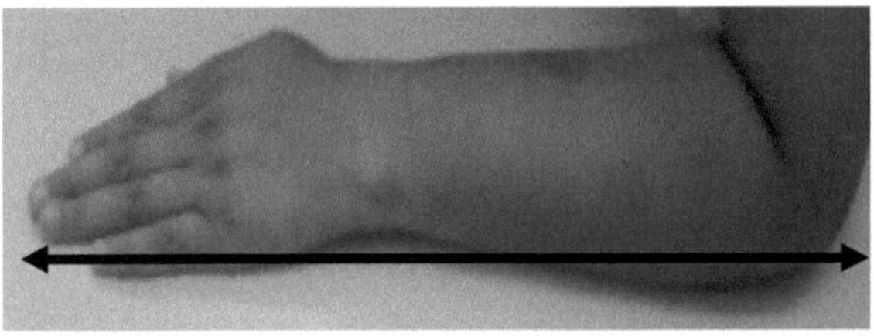	To measure the length of any object, Chundara use their hands as unit, the distance between the elbow and the tip of middle finger is called ***Haat***

(continued)

(continued)

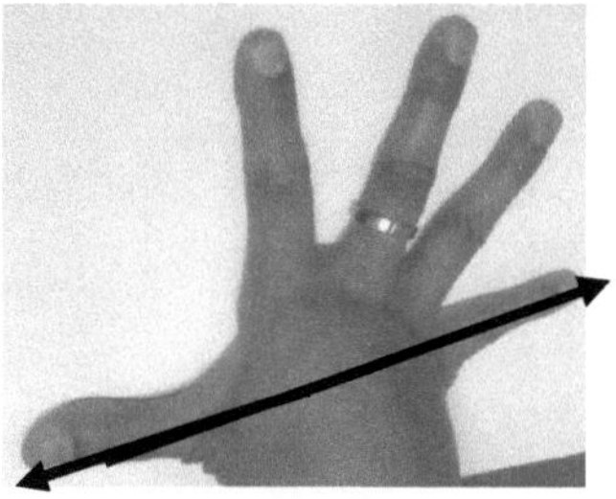	For the shorter length, Chundaras measure by the distance between the tip of the thumb and the tip of the little finger, which is called ***Bitta***
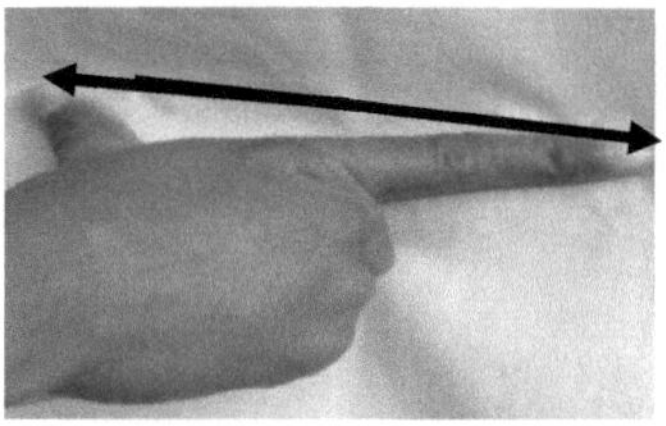	***Kuret*** is another measuring unit of Chundaras, which is the distance between tip of the thumbs and the tip of the index finger with stretched palm
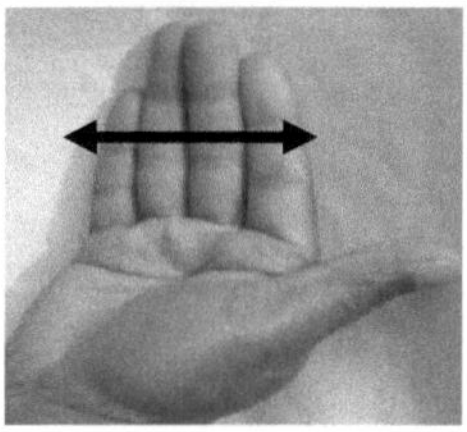	To measure even shorter distance Chundaras use finger width which is called ***Amal***

Source Personal file

(b) **Volume Measurement Units**

Mana: is the volume measuring vessel that contains ten handfuls.
Pathee: is the volume measuring vessel that contains eight *mana*.
Chauthai: is the smaller units to measure the volume and it contains quarter of *mana*.
Chakhanti: is the smaller units to measure the volume and it contains a quarter of *Chauthai*.
Chimti: the substance held between three fingers thumb, index and middle.

(c) **Exclusive Product of Chundara**

Dhyangro: It is drum like musical instrument 	Khaijedi: It is musical instrument smaller than Dhyangro in shape and size 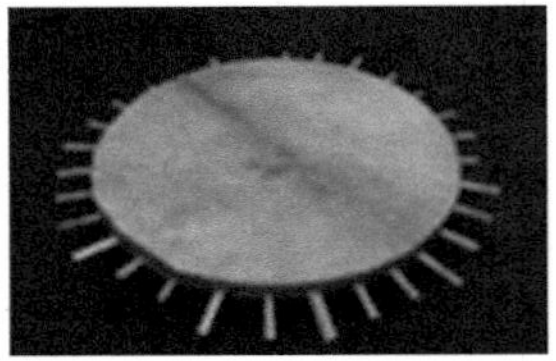
Madal: It is popular musical instrument in Nepalese culture and it is cylindrical in shape 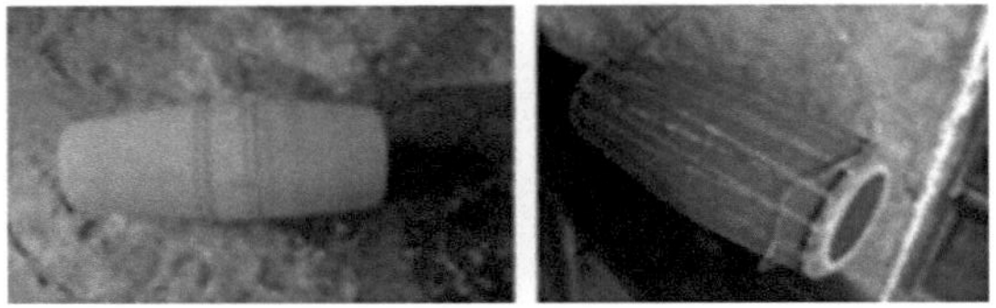	Damaru: It is musical instrument hyperboloid in shape 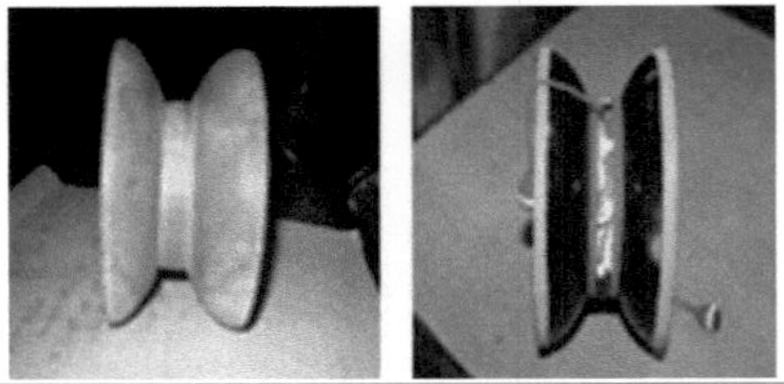
Theki: It is a wooden vessel used to make cord and related products  Source: http://nepalitreasure.blogspot.com.br/2011/12/theki-and-madani.html	Mana and Pathee: Volume measuring instruments 

Source Personal file

References

Ary, D., Jacobs, L. C., & Sorensen, C. K. (2010). *Introduction to research in education*. Belmont, CA: Wadsworth.

Ascher, M. (2002). *Mathematics elsewhere: An exploration of ideas across cultures*. Princeton, NJ: Princeton University Press.

Ascher, M., & Ascher, R. (1997). Ethnomathematics. In A. B. Powell & M. Frankenstein (Eds.), *Ethnomathematics: Challenging eurocentrism in mathematics education* (pp. 25–50). New York, NY: State University of New York.

Barnhardt, R., & Kawagley, A. O. (2005). Indigenous knowledge systems and Alaska native ways of knowing. *Anthropology and Education Quarterly, 36*(1), 8–23.

Barton, B. (2008). *The language of mathematics: Telling mathematical tales*. New York, NY: Springer.

Bishop, A. J. (1988). *Mathematics enculturation: A cultural perspective on mathematics education*. Dordrecht, The Netherlands: Kluwer.

Bishop, A. J. (1990). Western mathematics: The secret weapon of cultural imperialism. *Race and Class, 32*(2), 51–65.

Bishop, A. J. (1991). Mathematics education in its cultural context. In M. Harris (Ed.), *Schools, Mathematics and Work* (pp. 29–41). London, England: Falmer Press.

Clements, M. A., & Ellerton, N. F. (1996). *Mathematics education research: Past, present and future*. Bangkok, Thailand: UNESCO Principal Regional Office for Asia and the Pacific.

D' Ambrosio, U. (2006). *Ethnomathematics: Link between traditions and modernity*. São Paulo, SP, Brazil: UNICAMP.

D'Ambrosio, U. (1994). Ethnomathematics, the nature of mathematics and mathematics education. In P. Ernest (Ed.), *Mathematics, education and philosophy: An international perspective* (pp. 230–242). Bristol, PA: The Falmer Press.

Dahal, R. C. (2007). *Indigenous mathematical concepts of Gopali community from their cultural perspective*. Unpublished M. Phil. Thesis, Kathmandu: Tribhuvan University, Faculty of Education.

Denscombe, M. (2010). *The good research guide for small-scale social research projects*. New York, NY: Open University Press.

Dunne, M., Pryor, J., & Yates, P. (2005). *Becoming a researcher: A companion to the research process*. London, England: Open University Press.

Eglash, R. (1997). When math worlds collide: Intention and invention in ethnomathematics. *Science, Technology and Human Values, 22*(1), 79–97.

Ezeife, A. N. (2002). Mathematics and cultural nexus: The interactions of culture and mathematics in an aboriginal classroom. *International Education Journal, 3*(3), 176–187.

Ferreira, E. S. (1989). The genetic principle and the ethnomathematics. In C. Keitel, A. Bishop, P. Damerow, & P. Gerdes (Eds.), *Mathematics, education and society* (pp. 110–111). Paris, France: Division of Science and Environmental Education.

Fetterman, D. M. (2010). *Ethnography: Step by step*. Thousand Oaks, CA: Sage.

Gerdes, P. (1997). On culture, geometrical thinking and mathematics education. In A. B. Powel & M. Frankenstein (Eds.), *Ethnomathematics: Challenging eurocentrism in mathematics education* (pp. 223–247). New York, NY: State University of New York.

Gerdes, P. (1999). *Geometry from Africa: Mathematical and educational explorations*. Washington, DC: The Mathematical Association of America.

Gerdes, P. (2005). Ethnomathematics, geometry and educational experiences in Africa. *Africa Development, 30*(3), 48–65.

Glesne, C., & Peshkin, A. (1992). *Becoming qualitative researchers: An introduction*. New York, NY: Longman.

Gurung, G. B. (2009). *Unveiling the ways to knowledge generation, continuation, distribution and control of the Pariyars: Commonalities and points of departure from school pedagogy*.

Unpublished M. Phil. Dissertation, Faculty of Education. Kathmandu, Nepal: Tribhuvan University.

Hammond, L., & Brandt, C. (2007). Defining an anthropological approach to science education. *Studies in Science Education, 40*, 1–47. Retrieved from: http://www.csus.edu/indiv/h/hammondl/anthscifinal4-19-04.htm. Accessed in January 10, 2017.

Jarlais, C. W. D. (2008). *Western structures meet native traditions: The interfaces of educational cultures*. Charlotte, NC: Information Age Publishing.

Lett, J. (1996). Emic-etic distinctions. In D. Levinson & M. Ember (Eds.), *Encyclopedia of cultural anthropology* (pp. 382–383). New York, NY: Henry Holt and Company.

Millroy, W. L. (1992). An ethnographic study of the mathematics of a group of carpenters. *Journal for Research in Mathematics Education Monograph 5*. Reston, VA: National Council of Teachers of Mathematics.

Mosimege, M. D. (2000). *Exploration of the games of malepa and morabaraba in South African secondary school mathematics education*. Unpublished Doctoral Thesis. Cape Town, South Africa: University of the Western Cape.

Mosimege, M., & Lebeta, V. (2000). An ethnographic study of mathematical concepts at the Basotho cultural village. In S. Mahlomaholo, M. Nkoane, & K. Smit (Eds.), *Proceedings of the 8th Annual Conference of the Southern African Association for Research in Mathematics and Science Education*. Port Elizabeth, South Africa: University of Port Elizabeth.

NDC. (2004). *National Dalit's Commission: An introduction*. Kathmandu, Nepal: National Dalit Commission.

NDC. (2006). *Dalit in Nepal*. Kathmandu, Nepal: National Dalit Commission.

NEFIN. (2006). *Involvement of indigenous nationalities in the state mechanism*. Kathmandu, Nepal: NEFIN.

Neuman, W. L. (2008). *Social research methods: Qualitative and quantitative approaches*. New Delhi, India: Dorling Kindersley.

Orey, D. C. (2000). The ethnomathematics of the Sioux tipi and cone. In H. Selin (Ed.), *Mathematics across culture: The history of non-western mathematics* (pp. 239–252). Dordrecht, The Netherlands: Kulwer Academic Publishers.

Orey, D. C., & Rosa, M. (2010). Ethnomodelling: A pedagogical action for uncovering ethnomathematical practices. *Journal of Mathematical Modelling and Application, 1*(3), 58–67.

Orey, D. C., & Rosa, M. (2015). Three approaches in the research field of ethnomodelling: Emic (local), etic (global) and dialogical (glocal). *Revista Latinoamericana de Ethnomathematica, 8* (2), 364–380.

Owens, K. (2010). Papua New Guinea indigenous knowledge about mathematical concepts. ICEM4. *Journal of Mathematics and Culture, 6*(1), 20–50.

Paudel, K. P. (2008). *Tharu culture*: *An ethnomathematical perspective*. Unpublished Master Degree Thesis. Lalitpur: Nepal Kathmandu University.

Powell, A. B., & Frankenstein, M. (1997). *Ethnomathematics: Challenging eurocentrism in mathematics education*. New York, NY: State University of New York.

Rosa, M., & Orey, D. C. (2010). Ethnomodelling as a pedagogical tool for the ethnomathematics program. *Revista Latinoamericana de Ethnomatheamtica, 3*(2), 14–23.

Rosa, M., & Orey, D. C. (2012). The field of research in ethnomodelling: Emic, etic, and dialectical approaches. *Educação e Pesquisa, 38*(4), 865–879.

UNESCO. (2008). *Developing culturally contextualized mathematics resource materials: Capturing local practices of Tamang and Gopali communities*. Kathmandu: United Nations Educational, Scientific and Cultural Organization Office.

Vygotsky, L. S. (1978). *Mind in society: The development of higher psychological processes*. Cambridge, MA: Harvard University Press.

Winter, J., Andrews, J., Greenhough, P., Hughes, M., Salway, L., & Yee, W. C. (2009). *Improving primary mathematics: Linking home and school*. New York, NY: Routledge.

Chapter 7
Meaningful Mathematics Through the Use of Cultural Artifacts

Toyanath Sharma and Daniel Clark Orey

Abstract This chapter offers a perspective on how to make mathematics meaningful for students by using a cultural artifact named dhol, which is a musical instrument constructed and played by the members of the Rai cultural group in Nepal. The authors discuss how the use of culturally contextualized mathematics found in these drums may help students to connect school mathematics with their own home cultures by elaborating ethnomodels. The theoretical basis of ethnomathematics and culturally relevant pedagogy could be an appropriate response for education in Nepal. In this context, the authors invite the readers to connect mathematics with their own community and cultural practices in order to make mathematics meaningful for students.

Keywords Cultural artifacts · Culturally relevant pedagogy · Dhols · Ethnomathematics · Rai culture

7.1 Introduction

Traditionally, education in Nepal was not accessible to everyone; it was originally limited to the ruling families. Over time, the wealthy and other people of means have an increasingly better chance for an education. However, the educational condition of the minority groups continues in decline, where many of the children cannot take part in their primary education because they must work and earn money to maintain their family or help out at home.

T. Sharma (✉)
Center for Activity Based Instruction,
Kathmandu University School of Education, Patan, Nepal
e-mail: toyanath@kusoed.edu.np

D.C. Orey
Departamento de Educação Matemática/Instituto de Ciências Exatas E Biológicas,
Universidade Federal de Ouro Preto, Ouro Preto, Brazil
e-mail: oreydcead@gmail.com

M. Rosa et al. (eds.), *Ethnomathematics and its Diverse Approaches for Mathematics Education*, ICME-13 Monographs,
DOI 10.1007/978-3-319-59220-6_7

However, currently, the Nepali government is committed to the progress, development and growth of education for all. Nepal has introduced school education from preschool education (kindergarten level) to school education (primary and secondary). At the same time, educational opportunities are disseminated unfairly according to gender and geographical location. Nepal has three different regions (Himalaya, Mountain and Tarai). Each region has its own way of seeing the world, religion, and celebrating life, and distinct ways of living.

Nepal is a country rich in sociocultural aspects, but there are still several educational issues and challenges to overcome, such as: (a) low literacy rate, (b) quality of education standard lower than expected, (c) the state of sociocultural and educational infrastructures have not been substantially developed, (d) relevancy of education since the human resource produced by the existing education system could neither find proper place in the employment market nor is able to generate self-employment, (e) literacy rate of women, marginalized and oppressed group, people with disability, and geographically disadvantaged is not satisfactory, and (f) adequate educational opportunities are not available to children, women, marginalized and oppressed group, people with disability, and geographically disadvantaged people (Nepal 2007).

According to the previous assertions, in relation to school education, the major problem relates to both access and equity, which makes learning experiences very different from region to region in Nepal. However, although access to primary education has increased considerably, increasing this access and improving equity in education still remain a formidable challenge (Mathema 2007, p. 47). For example, there are many reasons for the unsatisfactory access and equity at school level. For example, while primary school is free, parents have to pay many direct and indirect costs for their children's education.

Among other barriers that exclude children from educational opportunities in Nepal is their distance from school, the negative attitude of teachers toward poor children's ability to succeed, and language, and cultural factors. In the context of Nepal, there over 123 languages with Nepali as the most widely spoken. Over 44.6% of Nepalis have Nepali as their mother tongue. Learners who do not speak, or have Nepali as a second language, have many difficulties in schools (Mathema 2007; Civil 1995, 1998). As well, participation in school education is unequal across social, cultural, gender, regional, linguistic, and income groups (Mathema 2007). Educational challenges are still in existence due to geographical, social, and cultural aspects (Nepal 2007).

Despite all these challenges, Nepali education at present is in the process of transformation including recent policies on student assessment. With the development of new teaching and learning theories, new technologies, the teaching and learning process of mathematics in Nepal is transforming from a teacher centered to use of cultural artefacts as a teaching material for mathematics classes, which is related to the innovative pedagogy named *student centered teaching* that helps teachers to create and develop attention of the students and respect for their cultural background and develop understanding of their mathematical knowledge (Luitel 2009).

In this regard, the existing domination of teacher centered teaching methods in school contradicts the essence of student assessment and ultimately, letter grading. The idea of letter grading can only evaluate student performance, if the teaching and learning process is student centered. Addressing the issue of ownership and identity of the students is the key issue for educators in Nepal.

Since it instantly overcomes many boundaries and language barriers, globalization has acquired a new meaning for Nepali educators. The new mathematics activities must respond to new meanings of globalization and will, necessarily, be transdisciplinarian and transcultural. Therefore, we agree that the theoretical basis of ethnomathematics (D'Ambrosio 1985, 2000) could be an appropriate response for education in Nepal.

7.2 Issues of Culture

The students' culture has been identified as one of the factors that influence mathematics learning, particularly in respect to indigenous learners as quoted in defines culture as the body of learned beliefs, traditions, and guides for behavior that are shared among members of any human society (Barrett 1984; Hollins 1996). Similarly, Erickson (1986) states: "Culture, as a social scientific term, refers to learned and shared standards for ways of thinking, feeling, and acting" (p. 117). However, it was Hall (1976) who concisely described the function of culture, thus:

> Culture is man's medium; there is not one aspect of human life that is not touched and altered by culture. This means personality, including how people express themselves (shows of emotion), the way they think, how they move, how problems are solved, how their cities are planned and laid out, how transportation systems function and are organized, as well as how economic and government systems are put together and function (pp. 16–17).

Moreover, Huntington (1993) stated that individuals of different cultural groups have different worldviews that are a product of centuries, which will not disappear rapidly because they are far more fundamental than differences among political ideologies. In addition, culture is expanded to include also the cultures of differing professional groups and age classes (D'Ambrosio 1985) as well as social classes and gender. On the other hand, Bullivant (1993) defined culture as a social group's design for surviving in and adapting to its environment. Despite these definitions of culture, according to Banks and Banks (2002) culture is considered as the ideations, symbols, behaviors, values, knowledge and beliefs that are shared by a community (Civil 1995, 1998).

Looking closely at the above discourse, we agree with Rosa (2010) who states that culture may have a pervading influence on how a group of people live and learn. The culture of local populations, for instance, would influence how students learn and retain what they are taught in schools. It is the culture itself that shapes learning styles in mathematics, determining to a large extent, to what use they put the mathematics knowledge they acquire in schools. Because culture provides the essence of who we are and how we exist in the world, and because it is derived from understandings acquired by people through experience and observation about how to live together as a community, how to interact with the physical environment, and knowledge or beliefs about their relationships or positions within the universe, culture can be a an essential tool in learning mathematics (Hollins 1996; D'Ambrosio 2006). Hence, culture plays a vital role in the development a worldview of teaching mathematics.

Within this context, D'Ambrosio (1985) affirms that culture considerably affect the way in which people understand mathematical ideas, procedures, and practices. Hence, ethnomathematics has demonstrated how mathematics is made of many diverse and distinct cultural traditions because each cultural group has developed unique ways of incorporating mathematical knowledge and has often come to represent given cultural systems, especially in ways that members of cultural groups quantify and use numbers, incorporate geometric forms and relationships, and measure and classify objects.

For example, Rosa (2010) affirmed that if a mathematical problem is not contextualized students may not be able to use their prior knowledge to develop learning activities. This is especially true if this knowledge is based on their histories, experiences, and interactions with the environment where they live. Many educators forget that context plays an important role in the teaching and learning process. Thus, context is the most powerful and influential factor in this case because it focuses on and relates to prior knowledge to the mathematics of the classroom. From this point students find mathematics in their everyday life, they feel mathematics as flexible, and they increase their conceptual understanding in order to solve mathematical problems.

If teachers are aware of the cultural context of their students, they will encourage understanding of mathematical concepts (Sharma and Neupane 2016). As suggested by Shirley (1995), ethnomathematics is a key to finding connections between students everyday lives and school, that is, within mathematics, the members of distinct cultural groups blend two or more mathematical areas to other subject areas such as art, geography, economics, etc., in order to meet their needs as they look at the *culture of the others*; and to the local culture in order to incorporate local mathematics. Therefore, they develop their own mathematical language and symbols to understand phenomena, and to empower themselves.

Hence, Hofstede (1986) argues that culture influences mathematics through its manifestations such as cultural traits, geometric shapes, values, artifacts, and symbols. His premise was that the values preferred by a group of people separate them from other groups and, thus, cultures can be compared with each other using values as a standard. Under this circumstance, local mathematical manifestations focus on the meaning of objects such as cultural artifacts in the lives of the members of different cultures have also applied values to explain how people organize information in their own environment, which are the ideas according to which phenomena, such as mathematical, are organized, constructed, evaluated, and diffused.

For example, cultural artifacts such as musical instruments themselves can be considered as the solid reminders of the commitment of Nepali people to a musical and spiritual life, connections between its cultures, signs of great creativity, and physical emblems of the sounds they create. Music is about expression, communication, release, celebration, worship, ritual and enjoyment, and musical instruments support all of these *cultural traits.*[1]

Therefore, Dunipace (2010) stated that Nepali musical instruments are understood according to the purposes of the members of the cultural groups who use them. Some of these instruments are used in Nepali classical music, for folk music, by the Gandharba and Damai musician castes, for religious or spiritual ceremonies and processions, and some for specific purposes in the life of Nepal's many ethnic groups. Most of the instruments used in Nepal originated in India and to a lesser extent Tibet, but all have developed in uniquely Nepali ways.

7.3 Cultural Artifacts

Cultural artifacts are objects created by the members of cultural groups, which give cultural clues and information about the culture of its creators and users (D'Ambrosio 1993). They have some significance in the daily life of distinct cultural groups because "the language of the shapes, the designs, the myths, and the colors, confirm the community's sense of reality and give it control over its own time and its own space" (Voltz 1982, p. 45). In this regard, these artifacts are made to adorn walls, ceilings, baskets, utensils, clothes, jewelry, and even the human body itself as well as to serve as religious purposes (Onstad et al. 2003).

However, it is important to emphasize here that the essence of cultures is not only the development and use of their cultural artifacts, tools, or other tangible cultural elements, but the distinct ways that their members develop in order to interpret, use, and perceive their main characteristics of their own culture (Banks and Banks 2002). Cultural artifacts such as language, myths, and literature influence the representational system of diverse civilizations may be used in different cultures in distinct ways as well as for different purposes.

Hence, members of distinct cultural groups develop models that represent systems taken from their own reality in order to help them to understand and comprehend the world by using small units of information named *ethnomodels*, which link cultural heritage with the development of their mathematical ideas, procedures, and practices. According to D'Ambrosio (2016), humanity:

---

[1]Cultural traits are related to the appreciation of features developed by the members of a specific culture such as religion, language, government, customs and traditions, social organization, and arts as well as the establishment of relations between the members of that group. In this context, it is important that educators, investigators, and researchers understand the cultural roots of other cultures in order to value the ideas, procedures, and mathematical practices used by students from distinct cultural contexts (Rosa and Orey 2016).

> (…) develops the perception of past, present and future, and their linkage, and the explanations of facts and phenomena encountered in their natural and imaginary environment. These are incorporated to the memory, individual and collective, and organized as arts and techniques, which evolve as representations of the real (models), as elaborations about these representations which result in organised systems of explanations of the origins and the creation of myths and mysteries (mentifacts). Some of the representations materialize as objects, concrete representations and sophisticated instruments (artifacts) (p. 41).

Through the elaborations of models, humanity try to explain phenomena that occur in their daily lives. These explanations are organized as arts, techniques, theories, as strategies to understand and deal with daily facts and events. These strategies, have been historically organized, in different groups, in different spatial and temporal contexts, which are the support of cultures as systems of knowledge frequently represented by the elaborations of models (D'Ambrosio 2016). In general, a model is a representation of an idea, a concept, an object, or a phenomenon (Gilbert et al. 2000).

Therefore, ethnomodels is defined as cultural artifacts that are the pedagogical tools used to facilitate the understanding and comprehension of systems taken from reality of the members of distinct cultural groups (Rosa and Orey 2009). In this regard, according to Rosa and Orey (2013a, b), ethnomodels are considered as external representations of local phenomena that are precise and consistent with the scientific and mathematical knowledge socially constructed and shared by members of specific cultural groups. The main objective for the elaboration of ethnomodels is to *translate* the mathematical ideas, concepts, and practices developed by the members of distinct cultural groups into school mathematics.

This approach helps the organization of pedagogical actions that occur in classrooms by using local aspects of mathematical thinking through the development of cultural artifacts as well as "provides an important opportunity for educators to link current events and the importance of these artifacts in the context of ethnomathematics, history, and culture" (Rosa and Orey 2008, p. 33). Consequently, D'Ambrosio (2008) states that representations given by these elements enable the expansion of students' reality by incorporating the artifacts regarding mathematical simulations and models.

7.4 Connecting Cultural Artifacts with Mathematics

Early knowledge acquired by youngsters is inter-twined with the cultural milieu and environment in which the children are born and grow up. Hence, by the time children get to school, they already have a considerable amount of experience and prior knowledge ingrained in them through their home and peer group interactions. If the school culture reinforces the students' home cultures, by respecting prior knowledge and learning experiences, then learning is facilitated for these students; if not, learning is drastically inhibited.

The situation described here holds true, to a greater or lesser degree, for all students, but is more cogent for local students, most of who live in different regions of Nepal. For example, important objectives of Nepali education is to "develop positive attitudes towards the norms and values of democracy and diverse culture of the nation" and to "familiarize with the national history, culture, geography, economics, ethnic and cultural diversity and environment for nation's development by promoting national unity, cordiality and peace" (Nepal 2007, p. 42).

When students attend schools and encounter abstract and academic subjects such as mathematics, they find themselves disadvantaged because the mathematics curriculum currently in use in most schools is alien to them and so are the teaching and learning styles adopted in classrooms. Shirley (1995) addressed this issue by stating that the problem for many teachers is that most of the mathematical content in our academic curriculum has been derived from the developments in European mathematics and many educators have difficulty finding examples that do not seem to be Eurocentric.

Since our schools have increasingly heterogeneous populations from many different cultures around the world, an apparently Western-based (without connection to Nepali culture) curriculum can be counterproductive to our interest in recruiting members of underrepresented groups into mathematics. In accordance to Sharma (2012), if our children learn to see mathematics only from a foreign perspective, they may believe that our culture is less powerful and might not work with mathematics or, even worse, that they cannot work with the mathematics that they see in schools, as it is for *others*.

If children come from a South Asian cultural heritage, this belief may be dangerously extended to "I cannot work with mathematics". Thus, a culturally relevant pedagogy can go a long way towards rectifying errors that currently exist, and making mathematics attractive to all students, especially students from the under-represented indigenous groups. This kind of pedagogy was developed out of concern for serious academic achievement gaps experienced by low-income students, students of color, and students from culturally diverse environments (Gay 2000).

Culturally relevant pedagogy uses cultural knowledge, prior experiences, frames of reference, and unique learning styles of ethnically diverse students to make learning more relevant and effective with the objective to strengthen their connectedness with schools and as a consequence reduce behavior problems and enhance learning (Klotz 1994). In this regard, educators can benefit from being culturally relevant by contextualizing instruction and schooling practices while maintaining academic rigor and helping students to achieve their academic potential (Ladson-Billings 1995). The use of a culturally relevant approach to teaching has resulted in success in some recent cases (Rosa and Orey 2015a).

The most important assertion in the previous paragraph is to remind us that it is necessary to discourage teaching mathematics to students from indigenous cultures through instructional methods that emphasize the abstract, as opposed to the concrete, the imaginary rather than the real (Davison 1992) because the "abstract, decontextualized teaching of mathematics has affected many American Indian students' school success" (Davison 1992, p. 243). In order to redress this situation, it is important to recommend that:

> Wherever possible, mathematics concepts should be presented in a culturally relevant manner, using situations that the students find interesting and familiar. Above all, the presentation of mathematical ideas needs to be consistent with how students learn. The use of tactile or visual approaches assists students to form meaningful images (Davison 1992, p. 243).

For instance, Simard (1994) reported the impressive outcome of an integrative, culture-based approach employed at a Winnipeg high school in Canada, the children of the Earth High School, an example of an urban school that teaches the standard curriculum from a cultural perspective. The school has been very successful in graduating students from the grade 12 program into post-secondary institutions. The use of community people and their wide array of talents helped them put back meaning into education. Several researchers and pioneers in this field (Cajete 1994; D'Ambrosio 1980; Davison 1992) have advocated the use of holistic and multisensory approaches to mathematics and science teaching.

For indigenous students in particular, teaching mathematics through the use of methods and instructional designs that glorify the familiar explain-example-exercise pattern of expository teaching (Matang 2001) has led to high dropout rates from mathematics and associated science courses. For such students, *rules without reasons* frequently used in mathematics teaching as exemplified by *change side, change sign* (Backhouse et al. 1992) has often led to confusion, not comprehension.

7.5 Mathematical Practices with Culturally Relevant Pedagogy

Nepal is a tiny country made up of numerous multicultural and multiethnic communities that also contribute to a large diversity in the quality and understanding of the teaching and learning process. Because children learn better in their own contexts, the diversity of religions, cultural practices, geographies, castes, ethnicities, languages, and histories contribute to the learning process. It has a national curriculum framework, which demands specific skills, behaviors and knowledge for all children.

Thus, a *National Curriculum Framework for School Education in Nepal* (Nepal 2007) was developed to play the role of the core document of Nepali school education. In accordance to this document:

> It is expected that it will provide a long term vision of school education. It has presented the policy and guidelines on contemporary curricular and other important aspects, issues and challenges, vision of school level education, basic principles of curriculum development, objectives and structure of school education, student assessment and evaluation policy, and open education (Nepal 2007, p. iv).

Even though Nepal has a national curriculum framework, a single and unique standardized teaching process has not been able to address the context of distinct cultural groups nor to reach expected learning achievements for all learners. One of the reasons for this problem is that, often, the "teaching in all levels is examination oriented" (Naidoo 2004, p. 55).

At the same time, Nepali children have their own ways of learning. Moreover, living in extremely diverse contexts such as Nepal is highly dependent on cultural practices. In order to address this issue, in relation to the teaching methods, the national curriculum framework of Nepal states that the process of teaching and learning should be:

> (...) practical and effective in order to transform the learning achievements set by the curriculum. Schools and classroom environment as well as the activities conducted in classes are considered as the key elements for the successful implementation of the present day formal school curriculum. The relation between school and community, teacher development and management, education materials, and the evaluation system bring about great effect on instructional approach. Similarly, instructional approaches are considerably significant from the angle of teaching and learning because a teacher has to play the role of a communicator, colearner, facilitator, motivator and an agent to make learners inquisitive in learning (Nepal 2007, p. 21).

This assertion may suggest that culturally relevant pedagogy play an important role in the teaching and learning of mathematics because it can help teachers to make mathematics learning contextual and create activities where students can easily relate to their own culture and everyday life. A culturally relevant pedagogy approach provides ways for students to maintain their cultural identity while succeeding academically. It is designed to fit school culture with students' culture to help them understand themselves and their peers, develop and structure social interactions, and conceptualize mathematical knowledge (Ogbu and Simons 1998).

Therefore, ethnomathematics encourages the use of and study of cultural aspects of mathematics. It presents mathematical concepts of the school curriculum in a way in which concepts are related to the students' cultural backgrounds, thereby enhancing their abilities to make meaningful connections and deepening their understanding of mathematics (D'Ambrosio 1985). In accordance to this context, Rosa (2010) argues that there is a need to examine the embeddedness of mathematics in its cultural context by drawing from a large body of literature that takes on the students' cultural roots of knowledge production in the context of the mathematics curriculum.

The application of ethnomathematical approaches and culturally relevant pedagogy into the mathematics curriculum are intended to make school mathematics more relevant and meaningful to students and to promote the overall quality of their education. In order for teachers to implement a sense of cultural connection, they need knowledge of and respect for the various traditions and languages in their communities (Rosa and Orey 2011). Consequently, schools need to be "encouraged to conduct practical classes by developing appropriate teaching materials and made them capable in the development and management of teaching materials by using local means and resources" (Nepal 2007, p. 56).

In this regard, our research asks, "How can we make mathematics teaching and learning processes more relevant with our cultural practices, realities, and our children's prior knowledge as well as we share a number of questions related to school mathematics in Nepal, which include:

1. How we might assist students to become multidimensional thinkers, rather than linear thinkers;
2. How can we empower and encourage students to seek mathematics in their everyday life;
3. How cultural practices makes mathematics meaningful; and
4. If empowerment can happen, how does it make school mathematics meaningful to children?

In order to find the answers to these questions, we agree with Irvine and Armento (2001) that a culturally relevant pedagogy allows teachers to provide and use meaningful learning materials; create environments, which include cultures, customs, and traditions that are different from their own; and include lessons that assist students in making meaningful connections between their lives and school-related experiences. In this direction, Rosa (2010) states that in a culturally relevant mathematics pedagogy, teachers construct bridges between the home culture and school learning of the students, where it promotes the background, experience and knowledge of learners.

In this case, local cultural artifacts provide a natural means for students to access the framework for their own conceptual understanding of mathematics. Thus, it is important to emphasize that "Human development is a cultural process. As a biological species, humans are defined in terms of our cultural participation" (Rogoff 2003, p. 3). Therefore, culture is one of the most important resources for teaching and learning of mathematics (Rosa and Orey 2011).

This context allows us to state that teaching through cultural aspects increases student conceptual understanding. In this regard, Ladson-Billings (1995) describes culturally relevant pedagogy as a pedagogical approach that empowers students intellectually, socially, emotionally, and politically by using cultural referents to impart knowledge, skills and attitudes. If a form of culturally relevant pedagogy is used in the classrooms, it can help to develop a student's whole personality (intellectual, social, emotional, and political).

In relation to the pedagogical work developed in schools, the "views of pedagogy within the literature on ethnomathematics are compatible with work on culturally relevant pedagogy" (Hart 2003, p. 42), which examines the cultural congruence between students' communities and schools. However, in order to implement the main ideas of cultural congruence in schools, Zeichner (1996) emphasizes that teachers must have knowledge of and respect for the various cultural traditions and languages of students in the mathematics classrooms.

Thus, Rosa (2010) recommends that teachers also need to develop general sociocultural knowledge about their students' development as well as the ways that socioeconomic circumstances, language, and culture shape their mathematical performance. Finally, teachers should develop a clear sense of their own ethnic, social, and cultural identities in order to understand and appreciate those of their students.

7.5.1 Comments on Ethnomathematics and Culturally Relevant Pedagogy

In the context of this study, culturally relevant pedagogy focuses on the role of mathematics in the sociocultural context of the students, which involves ideas and concepts associated with ethnomathematical principles related to geometric concepts. According to Rosa (2010), this process of teaching mathematics through cultural relevance and ethnomathematical perspective by using cultural project-based learners help students to know more about their own reality, culture, and society by providing them with mathematics content and approaches that enabled them to master academic mathematics.

Thus, when practical or culturally-based problems are examined in proper sociocultural contexts, students realize that mathematical practices developed by the members of different cultural groups reflect themes that are profoundly linked to their daily lives (Rosa and Orey 2015a). It becomes connected to objects and places they literally walk by everyday. Thus, students may be successful in mathematics when their understanding of it "is linked to meaningful cultural referents, and when the instruction assumes that all students are capable of mastering the subject matter" (Ladson-Billings 1995, p. 141). Consequently,

> Students investigate conceptions, traditions, and mathematical practices developed by members of distinct cultural groups in order to incorporate them into the mathematics curriculum. Teachers learn to engage students in critical analysis of the dominant culture as well as the analysis of their own culture (Rosa and Orey 2015a, p. 902).

This approach aims to draw from the students' cultural experiences in using them as vehicles to make mathematics learning meaningful as well as to provide students with the insights of mathematical knowledge as embedded in their sociocultural environments (Rosa and Orey 2013a, b). This is one important element of the use of ethnomathematics and its connection with cultural relevant pedagogy for mathematics teaching and learning process since it most certainly includes indigenous populations, but also cultures of labor and artisan groups, communities in urban environments and in the periphery, farm communities, and all types of professional groups. According to D'Ambrosio (2016), all these groups have specific strategies of a mathematical nature, that is, they have developed their own strategies for observing, comparing, classifying, ordering, measuring, quantifying, and inferring.

7.6 Dhols as Cultural Artifacts in Nepal

Dhols are double sided drums, which are traditional musical instruments in Nepal, mostly, in communities in the Himalayan belt. People from diverse communities in Nepal also construct the dhols as they need for the festivals or other functions such as artistic events. These musical instruments can be easily found in stores in Nepal. Figure 7.1 shows one of the Nepali dhols.

The dhols play an important role in the Nepali culture because it has emerged as a musical instrument in which the members of distinct cultural groups use them as a symbol of their ethnic identity. Culturally, the dhol is very important to these members because they use it for their celebrations and other events.

Many different and diverse occasions are associated with these musical instruments that are played across the Indian subcontinent, which includes festivals, jatras, religious ceremonies, folk dramas and other circumstances of celebration such as weddings or children birth. It is also widely used on diverse events ranging from wrestling bouts to folk dances as well festive occasions. Different Nepali musical instruments such as dhols are used for diverse purposes. For example, the dhols are mostly used for marriage and other social ceremonies (Basu and Siddiqui 2011).

These musical instruments are made in diverse varieties and materials, which give them colour and rhythm to any music they are associated with during the performance of its players (Barthakar 2003). In accordance to Schreffler (2010), throughout history, diferent rhythms were played on the Nepali dhols. However, with the decline or disappearance of some cultural practices, recent generations of dhol players have become unfamiliar with many of these rhythms.

There are considerable variations in techniques to play the dhols, which are usually played with free hands or by using two wooden sticks and various combinations of both. The most well-known style is the *bhangra* that uses sticks on both sides of the dhols. There is also variation as to whether the low pitched side should be played on the left side or on the right side. However, throughout most of South Asia, it is more common to play the lower pitched side with the left hand (Courtney 2016).

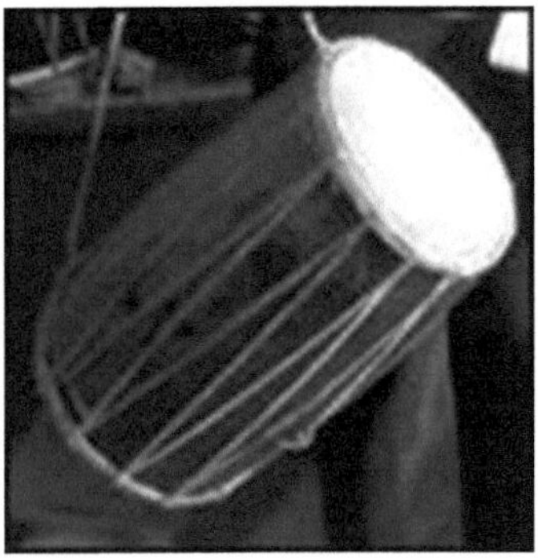

Fig. 7.1 Dhol drum. *Source* Personal file

For example, people from the Newar community use two different sticks to beat the dhol, one for each head of the drum, but the members of the Rai cultural group only use one side for beating. In fact, for them, it does not matter which side, left our right, they beat the dhol to produce the sound. It is importante to state here that sound is not produced because of the size of the drums, rather, it is the quality of the wood and the leather that influence the sound produced by the dhols.

Nepali people who construct these drums, generally, use leather that come from the ox[2] and the wood they call *utis* (alder), but currently they are using wood from the *Siris* and other locally available trees. Mostly, they are able to get these materials from the mountain area of Nepal. They may also purchase them from the local suppliers.

Historically, Schreffler (2002) argues that the dhols are drums that date back to the 15th century. It was probably introduced to the Indian subcontinent via the Persian drum type dohol (duhul) (Thakur 1996; Nabha 1998). Evidence of this fact is found in the *Ain-i-Akbari*, which is the *Constitution of Akbar*, a 16th-century, detailed document recording the administration of Akbar's empire, written by his vizier, Abu'l-Fazl ibn Mubarak (Majumdar 2007) who describes the use of dohol in the orchestra of the Mughal emperor Akbar (Schreffler 2002).

However, the Indo-Aryan word dhol only appears in print around 1800 in the *Treatise of Sangitasara* (Tarlekar 1972). However, Schreffler (2002) states that images of dhol players appear to be present in the bas relief carvings on Indian temple walls from the earliest times. It is possible that both the instrument as well as the name have some deep Indo-European connection.

Musical instruments in Nepal have a very strong relationship with Nepali religion, culture, and society. The musical traditions of Nepal are as diverse as the various ethnic groups of the country. In the course of the past 2000 years, the musical culture in the Himalayas has absorbed mostly Indian influences in shaping a unique musical tradition (Shankar 2011).

In the Nepali society, special groups play musical instruments in specific events such as the Bratabandha, which is a *Vedic Brahmanic* ritual, and welcoming ceremonies and in any other festivals in order to perform their rituals and traditions (Shankar 2011). Hence, musical instruments as cultural artifacts are very important in Nepali culture and society, mainly, for the members of the Rai cultural group.

Mostly, the Rai people live in the Eastern mountain region of Nepal. They are one of the most ancient indigenous ethnolinguistic groups, with their own unique cultural practices. They possess a rich and unique cultural heritage and are known for the construction of the dhol musical instruments, which is one of the cultural artifacts that represent the Rai community.

---

[2]Since more than 80% Nepali people are Hindu by religion and they do not eat cow and ox, it is getting difficult to provide these kinds of leather. Currently, they are using buffalo leather that has less quality than the ox leather. However, people from the Rai community can eat ox and this kind of leather is provided to them.

Fig. 7.2 Dhol players playing the instrument. Retrieved from: https://en.wikipedia.org/wiki/File:Sakela_dhol.JPG. Accessed on December 29th, 2016

In the Rai community dhols are played in their festivals such as the *Sakela Sili*, which is one of a few rituals in Nepal that are performed collectively, and that is found among all the Kirati people. It gives a noble sense of feelings such as awareness of we, togetherness, motivations and mystical harmony with a deified nature. The main characteristic of this festival is the *Sakela* dance performed by large group of Rai people (Schlemmer 2004).

The beating of the *Dhol* and jhayamta (drums and cymbals) accompanies the rhythms and the styles guided by the *Silimangpa* and the *Silimangma* that remind them of their devotion towards their *God* and *Goodness*, ancestors, home, village, and the *Mother Nature*. The small size of the dhols they play reflects their tendency to dance and run around with the drum during performances. They use the *Dhole* in *Chandi Sili* on the important full days of *Ubhauli* and *Udhauli* (Rai 2012). Figure 7.2 shows dhol players beating the instrument.

Even though the Rai people perform different ways of dancing, Rai (2012) argues that the dance moreover symbolizes the folkloric practice with the aim of requesting agricultural prosperity from the ancestors. It also helps to preserve traditional musical instruments such as different types of cymbals, drums and different types of Sili, which have been practiced throughout history.

7.7 Mathematical and Geometric Manifestations in the *Dhol Drums*

If we connect cultural artifacts such as musical instruments (dhols) in the classroom teaching process, children may be able to enjoy learning and also come to explore their own mathematical and geometric concepts as well as its application in their day-to-day life. For example, the results of diverse studies have revealed sophisticated mathematical ideas, procedures, and practices developed by the members of distinct cultural groups, which include geometric principles in craft work, architectural concepts, and practices in the activities and cultural artifacts of many local cultures (Eglash 1999; Orey 2000; Rosa and Orey 2009).

These mathematical ideas, procedures, and practices are related to numeric relations found in measuring, calculation, games, geometry, divination, navigation, astronomy, modeling, and a wide variety of other mathematical procedures and cultural artifacts (Eglash et al. 2006). For example, the Rai people developed their own concepts about counting numbers and unique ways of calculations that they use in their real life contexts such as the measurement used in the construction of the *dhol* drums.

Here, the authors have taken advantage of this cultural artifact to explore the dhols and their relation to mathematics and geometry. Therefore, it is necessary to state that in "relation to the pedagogical work in schools, mathematical curricular activities must be relevant to the students' cultural backgrounds" (Rosa and Orey 2015a, p. 900).

7.7.1 Making the Dhols

Typically, Nepali dhols have diferent names in relation to the distinct cultural groups that use them. In general, the size of the drums depends on the leather and wood accessed from local sources in the process to making these musical instruments. Lengths and sizes of the heads of these drums vary from region to region, which may be the reason that there are no common standards to make these drums (Barthakar 2003).

Informants from dhol-makers cultural group in the Rai community in Nepal describe the practices they acquired from their ancestors to construct the dhols. For example, in an interview, a representative from the Folk Music and Instrument Museum of Nepal informed the researchers that they have a special *myth* related to the construction of these drums.

In this story, a shaman sent a tika to find a tree in a forest to make a dhol. According to this tradition, before starting the process of making the dhol, the drum makers pray at home with a shaman. As part of this custom, they are blessed by the shaman with a *tika.*[3] On the next day, they go in search for the tika in the forest. It is believed that when they see a tika on a tree they have found the right tree that provides the best wood for them to construct the musical instrument they want to make. Again, they pray to and bless the tree before cutting it down to make the drums.

---

[3]Tika is a small spot of bright orange colored rice that people place on their forehead for blessings.

Similarly, they have a special myth or story related to the quality of the leather. People of that community say that only an instrument with a special process can work for blessing or fulfillment of their desire. In this regard, different instruments may have special reasons, and special blessings and power associated with them.

The museum informant stated that, currently, it is very difficult to determine the correct process of making dhols or any such musical instruments, but they also believe that dhols are special musical instruments that help them to develop their spiritual and religious traditions. For this reason, they are not able to conform that they are using the same measures or standard sizes that their ancestors used to make these instruments.

According to the informant, the skills required for the proper construction of these musical instruments are highly refined and it is the main differentiating factor in their quality. Therefore, in relation to the quantifiable aspects of the drums, these makers do not apply standardized mathematical procedures and techniques to make them, yet they have developed from generation to generation a sense of measurement as they estimate the amount of strings, wood, and leather they need to make the dhols. In this context, D'Ambrosio (2006) affirms that mathematical ideas, procedures, and practices are developed in different cultures in accordance to common problems that are encountered within a cultural context.

According to this informant, the skills required for the proper construction of these musical instruments are highly refined and it is the main differentiating factor in their quality. Therefore, in relation to the quantifiable aspects of the drums, these makers do not apply standardized mathematical procedures and techniques to make them, yet they have developed from generation to generation a sense of measurement as they estimate the amount of strings, wood, and leather they need to make the dhols. In this context, D'Ambrosio (2006) affirms that mathematical ideas, procedures, and practices are developed in different cultures in accordance to common problems that are encountered within a cultural context.

In order to make a dhol, wood is found that allows for a hollow to be carved inside of a wooden cylinder that will be shaped for the body of the drum. More often than not, a solid piece of wood is used that can be carved from a single block or a tree trunk. The drums they construct have, in general, a average length of about 22 in. and a diameter of 18 or 19 in. for the two drum heads. In general, both ends of the drum will have the same size or measure. The outer part of the body of the instrument measures about 0.75 in. of thickness. Thus, the total diameter of the dhols decreases by 0.75 in. +0.75 in. or about 1.5 inches.

The members of this cultural group do not have a scientific reason or formula that is equal to what we use in academic mathematics (circumference or diameter) to fix the measurement of the dhol, but they are used to working with from a sense of estimation which produces a quality sound. Earlier, people from the *Rai* community did not use current measurement units, rather they estimate when constructing the dhol. People from the dhol user community told the researchers that, dhol makers used their fingers for estimation. For example, the distance from an elbow to nail of middle finger measures about 1.5 ft. Likewise half of a hand from elbow to middle finger's nail measures a tighter or maximized length of palm.

In the traditional village areas we still find dhol makers who use the process of estimation rather than standard units. Other communities in Nepal, have slightly different kinds of practices and measurement scales. In future work with the Center for Activity Based Instruction (ABI) we make use of these differences and the diversity of measurement. A deeper discussion about this issue will have to wait for future research since it is part of future ABI curriculum project work. According to Kandel (2007) most communities use about 1.5 times diameter for length of a classic or traditional dhol. This shows that there are different perceptions and practices of dhol making. However, in the present context, many dhol makers we worked with use standard units for measurement like inches or centimeters.

It is important to emphasize here that, for musical purposes, the volume of the drum is irrelevant for Rai people, yet it is important to name different dhols such as dhaa, dhime, madhime, and tainai because they are constructed in diverse sizes, which implies that they have distinct volumes. However, it is possible to determine the volume of a dhol by using an etic approach. For example, if the drum has a given length, the volume of the dhol can be determined by developing the process: from 18”–1.5” (0.75” + 0.75”) = 16.5 in. diameter and length 22 inches.

For academic purposes, it is possible to find it by using the formula volume cylinder, which is: $V = \pi r^2 h$. This etic ethnomodel can be used as mathematics classes teaching material. If a dhol has a diameter of 18 in., then they use 0.75 in. thickness for drum’s wood for round. For the smaller ones, they only use 0.5 in.. Figure 7.3 show an etic representation of the outer and inner measurement of the dhol.

Etic constructs are considered accounts, descriptions, and analyses of mathematical ideas, concepts, procedures, and practices expressed in terms of the conceptual schemes and categories that are regarded as meaningful and appropriate by the community of scientific observers, researchers, and investigators (Rosa and Orey 2013a, b).

Similarly, the determination of the lateral area of the dhol does not make sense to the drum makers, but they know that if drums have bigger sizes, then they have bigger areas, yet they have to estimate the amount of leather to cover the dhol and

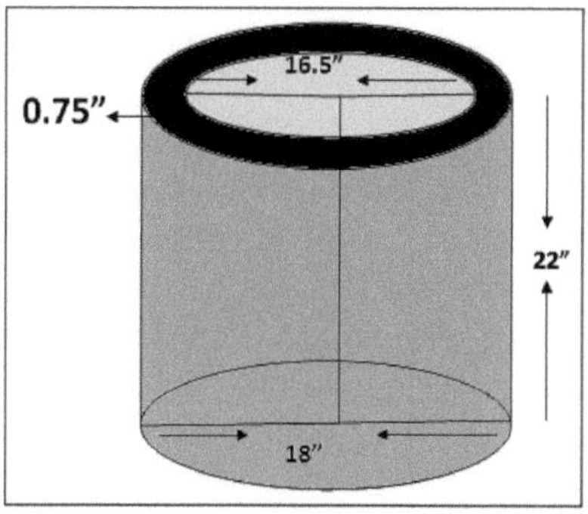

Fig. 7.3 An etic representation of the outer and inner measurement of the dhol. *Source* Personal file

its two heads. For this, they do not have standard skills and myth for estimating of leather to cover the dhol and its two heads they just see the hole of two heads and estimate it in square form.

Their process is still very traditional; they just put wood on leather then they cut it form bigger sized leather. Thus, they just estimate the total coverage area and the two heads and, and then, cut the amount of leather they need. There is no standard size for the amount of leather they use to make the dhol, but generally they cut it from bigger pieces of leather. In this context, according to their tradition a wooden frame is covered with an animal skin. Thus, they need to estimate the amount of leather they need to make the dhols. Similarly, the two open ends (heads) of the dhol are covered with goatskin, with the latter being tightened with the strings that stretch along the body of the instrument (Basu and Siddiqui 2011).

There is a variety of tightening arrangements for the dhol. There are rawhide lacings, rope, and, screw turnbuckle systems. Sometimes they are laced with rope, or rawhide, in which case, a series of metal rings are often used to pull and tighten the instrument. Sometimes metal turnbuckles are also employed (Courtney 2016). They can be retained and tightened in a direct manner by means of continuous loops of single twisted cord passing through both heads at several points on their circumferences. The cord passes through the head and around a wooden collar which is wrapped in the skin. Every two strands of the cord pass through a movable brass ring, which serves to regulate the tension of the heads (Dupree 1973).

The skins can be stretched or loosened with a tightening mechanism made up of either interwoven ropes, or nuts and bolts. Tightening or loosening the skins subtly alters the pitch of the drum sound. The stretched skin on one of the ends is thicker and produces a deep, low frequency (higher bass) sound and the other thinner one produces a higher frequency sound (Dupree 1973). However, dhols with synthetic, or plastic, treble skins are also very common.

Usually, the thickness of the skin is from 1/8–1/10th of an inch. The skins are laced together with one piece of cotton rope threaded through the edges of both skins. This is the more authentic way of setting up the skin in the drum. The skin on both the heads is stretched round leather hoops fastened to the skin and kept taut by means of interlaced leather thongs or thick rope. A leather band passed round the skin and over the braces serves to tighten the 2 heads to the pitch (Barthakar 2003; Dupree 1973). They use same leather but still cannot say its thickness because they also do not have exact measures of thickness. It depends on leather they find in the community.

In relation to the material for making the dhols, their focus is on the use of leather and wood. For example, for better sound they dry and wet the leather 4 (four) or 5 (five) times during the completion of the process of sound fixing. They use cold water to wet and heat the leather to make it flexible. After the regular process of making it tight and flexible they get the sound they have expected. After getting expected sound they stop the process for dry and wet. For fixing its sound dhol makers or user do experiment on dry and wet process.

Dhol user said that, differences in flexibility and tightness of the leather produces different types of sound. When leather becomes tight (cold), it gives bad sound then they dry its surface for few minutes to set their used sound. Moreover, in order to produce rough sound and piercing sound they use string to tight leather they use on both heads of the dhol. In the percussion instruments, the vibrating:

> (…) element is usually a wooden, sometimes shaped in thickness to tune its second mode to a harmonic of the fundamental. Apart from necessary mechanical hardness and durability, the main acoustic features desired are high density, so that considerably vibrational energy can be stored, and low internal losses, so that the sound rings for a relatively time. Percussion instruments made with wooden material have a mellow tone and shorter ring time because of the greater internal losses of wood, particularly at high frequencies (Fletcher 1999, p. 8).

The membranes of drums are "traditionally made from animal skins, for want of any alternative, and these suffered from uneven thickness, only modest strength, and sensivity to both temperature and humidity. The damping of a drumhead is almost enrirely caused by viscous losses to the air and by sound radiation, so that, the requirements on the membrane are almost purely mechanical" (Fletcher 1999, p. 8).

On the other hand, even spaced holes are cut around the circumference of the heads of the drum. A rope is them passed through the holes of the two skins, bracing them together, which fixes both heads with lather. In order to fix the lather on both heads, dhol makers use spaces holes on both sides of heads. For that they have common practices to make holes on the circumference of the heads to use string to tight them. If a drum has a diameter of 18 in., then there are 36 holes for the strings they use on the lateral surface area. To find the length of the strings they use a common formula to determine it.

For example, in this emic ethnomodel, a drum with 18 in. of diameter has 36 holes on the circumference of each head, then, there are $36 \times 2 = 72$ holes on both heads. They estimate for 36 holes on circumference. Interestingly, they know the relationship between measurement of circumference and diameter. They estimate circumference by multiplying diameter by 3 times. After getting estimation measurement of circumference, they divide it by 36 and get the space gap between holes in the circumference. So, they need 72 ft of string for one drum, which is also made with ox leather. Thus, they tight strings are tightening in order to produce better sound

In this context, emic approaches are concerned with differences that make mathematical practices unique from the *insider's* view point. Emic ethnomodels are grounded in what matters in the mathematical ideas, procedures, and practices that matter to the insiders' views of the world being modeled (Rosa and Orey 2013a, b). They represent how people who live in such worlds think these systems work in their own reality. Therefore, emic constructs are the accounts, descriptions, and analyses expressed in terms of the conceptual schemes and categories that are regarded as meaningful and appropriate by the members of the cultural group under study. This means that an emic construct is in accordance with the perceptions and understandings deemed appropriate by the insider's culture (Rosa and Orey 2013a, b).

Fig. 7.4 Geometrical patterns of the dhol. *Source* Personal file

Since, the surface of the dhols is decorated with engraved or painted patterns (Barthakar 2003). Since Rai children play the dhol in numerous festivals, the elaboration of ethnomodels may help them to understand mathematical and geometric concepts in classrooms while connecting school mathematics with their own home cultures. Figure 7.4 shows the geometric patterns of the dhols.

Thus, Rosa and Orey (2015b) argue that by using ethnomodels, students try to understand the world by means of explanations that are organized as procedures, techniques, methods, and theories, as it aims to explain and deal with daily facts and phenomena. These strategies are historically organized in every culture as knowledge systems. The elaboration of ethnomodels of cultural artifacts tends to privilege the organization and presentation of mathematical ideas, notions, and procedures developed by the members of distinct cultural groups by the elaboration of local ethnomodels. Therefore, the:

> (…) elaboration of models that represent these systems are representations that help the members of these groups to understand and comprehend the world by using small units of information, named *ethnomodels*, which link their cultural heritage with the development of the mathematical practice. This approach helps the organization of the pedagogical action that occurs in classrooms through the use of the local aspects of these mathematical practices (Rosa and Orey 2015b, p. 140).

The *etic*[4] ethnomodel below attempts to investigate and understand phenomena and their structural interrelationships through the eyes of the members of the Rai community as well as by considering aspects of mathematical procedures practiced in the mathematics school curriculum. In this context, the "etic ethnomodels represent how the modeler thinks the world works through systems taken from reality" (Rosa and Orey 2013a, b, p. 70). Hence, students can relate school mathematics with home mathematics.

[4]The etic approach corresponds to an outsider's interpretation of the mathematical ideas, procedures, and practices developed by the member of distinct cultural groups (Rosa and Orey 2013b).

In general, dhols have a cylindrical form and its sides are in circular shape. There are different line patterns and angles on these drums. They can be used as cultural artifacts for teaching geometric concepts such as cylinder, circle, lines, and angles as well as mathematical concepts such as measurements. In the mathematics classrooms, it is possible to show that a dhol resembles a cylinder and as such we are able to derive the formula to determine its total area and volume.

(a) Determining the Total Area of a Cylinder

Area of circular face = πr^2
Area of circular face = πr^2
Areas of curved surface = $2\pi rh$
Total area of a cylinder = Area of circular face + Area of circular face + Areas of curved surface
Total area of a cylinder = $2\pi rh + \pi r^2 + \pi r^2$
Total area of a cylinder = $2\pi rh + 2\pi r^2$
Total area of a cylinder = $2\pi r\ (r + h)$
Figure 7.5 shows the etic ethnomodel of the total area of the cylinder that represents the dhol.

(b) Determining the volume of a Cylinder

V = Base area times height
V = $\pi r^2 h$
Figure 7.6 shows the etic ethnomodel of the volume of the cylinder that represents the dhol.

However, it is necessary to point out that this method presents an approximated calculation for the area and the volume of the dhol as employed by the students. In this context, Rosa and Orey (2013a, b) states that ethnomodels are etic in the sense that they are built on an outsider's view about the mathematical world being modeled.

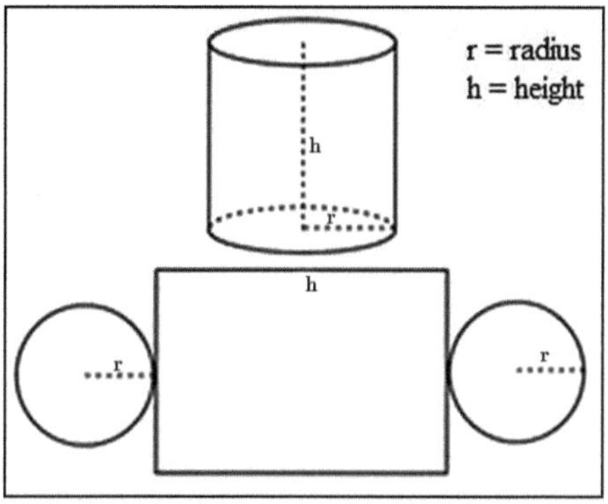

Fig. 7.5 The etic ethnomodel of the total area of the dhol. *Source* Personal file

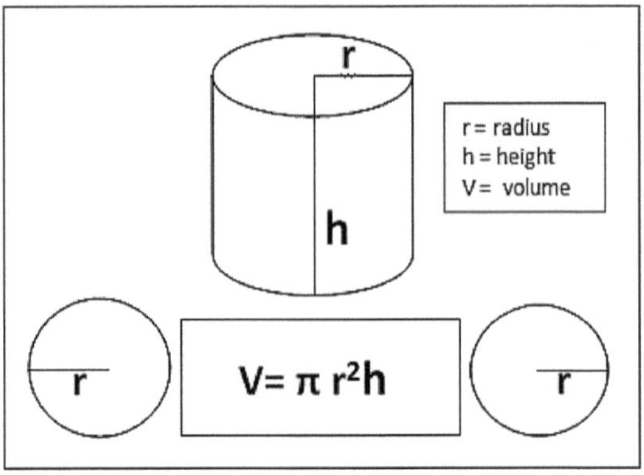

Fig. 7.6 Ethnomodel for the volume of the dhol. *Source* Personal file

7.7.2 Some Considerations About the Ethnomodel of the Dhols

The primary goal of this ethnomodel is to develop a descriptive idiographic orientation that describes the effort to understand the meaning of contingent, unique, and often-subjective mathematical phenomena since it emphases the uniqueness of mathematical practices developed by the members of distinct cultural groups (Rosa and Orey 2016).

Consequently, Rosa and Orey (2007) stated that this approach to the mathematics curriculum helps students to develop their cultural intelligence as it relates to the discovery of a hidden mathematical knowledge applied in this cultural artifact, which motivates them to rediscover their own cultural background. With this activity, students are able to incorporate abstract geometric concepts implicit in the dhol in order to help them to explain, understand, and comprehend organizational principles of their own culture.

It is important to state here that this "pedagogical approach is achieved through dialogue when community members, teachers, and students discuss mathematical themes that help them to reflect about problems that are directly relevant to their community" (Rosa and Orey 2015a, p. 902). During the development of the dhol activity, Rai students may be able to relate their everyday life with formal school mathematics as they begin to seek mathematics in their real world situation. It is important to emphasize that, this "perspective matches teaching styles to the culture and home backgrounds of their students, which is one of the most important principles of culturally relevant pedagogy" (Rosa and Orey 2015a, p. 908).

It seems to us that the cultural component in the mathematics curriculum is critical because it helps to emphasize the "unity of culture, viewing culture as a coherent whole, a bundle of [mathematical] practices and values" (Pollak and Watkins 1993 p. 490) that often appears incompatible with the rationality and the elaboration of traditional mathematical activities in schools. Thus, ethnomathematics attempts "to establish a relationship between the local conceptual framework (emic) and the mathematics embedded in relation to local designs" (Rosa and Orey 2013a, b, p. 66). This pedagogical work examines the "cultural congruencies between the community and school. This means that cultural congruence indicates the teachers' respect for social, cultural, and linguistic backgrounds of their students (Rosa and Orey 2015a, p. 899).

However, Orey and Rosa (2014) argue that in the context of mathematical forms of knowledge, what is meant by the *cultural component* varies widely and ranges from viewing mathematical practices as learned and transmitted to and from members of diverse groups to mathematical practices viewed as abstract symbolic systems with a deep internal history and logic that provides a symbolic system to its mathematical structure.

This context allowed the authors to describe this mathematical practice (emic) by using the members of the Rai cultural group own mathematical knowledge (etic) in a dialogical way between these two approaches. This means that they strive to compare, interpret, and explain the type of mathematical knowledge they observe and that the members of this particular cultural group are experiencing.

The results of the study conducted by Orey and Rosa (2014) shows that the dialogical observation of this mathematical practice tries to understand it from the perspective of the internal dynamics of the member of the Rai community (emic) while providing cross-cultural comparisons in order to comprehend it from the point of view of individuals from different cultural backgrounds (etic).

Thus, this approach is necessary to understand and explain this particular mathematical practice as a whole from a dialogical point of view, which is the dynamic of the encounters between two different cultures. In this regard, D'Ambrosio (2000) states that ethnomathematical practices can be considered as corpora of knowledge that derives from quantitative and qualitative practices, such as counting, weighing, measuring, sorting, inferring, classifying, and modeling.

7.8 Final Considerations

These are just a few of the many mathematical activities that have been developed in Nepal, which are helping to promote our own diverse cultures and respect for our diversity. Adding meaningful contexts for local activities makes for easy and sensible connections with classroom mathematics and has been shown to bring a change in the attitudes of our teachers and students in the teaching and learning of mathematics (Kathmandu University 2008). The core idea of the work presented here was to assist mathematics teachers in making it easier to help their student come to understand content in a way that they can connect to.

We are developing a series of activities as part of an ethnomathematics lab for teaching and learning with different kinds of artifacts. As the collection grows, we are engaging teachers in various regions of the country to begin to develop their own artifact-centered ethnomathematics. There are many more benefits from which student, school and communities can work together to form and use of cultural artifacts for meaningful learning. For example: an ABI school can create a ethnomathematics lab activities by using cultural artifacts found in the student's community and which directly match standardized curricular objectives.

It is important to understand that the use of the student's own unique social and cultural contexts increases possibilities that enable them to play a vital role in constructing their mathematical knowledge process. This is increasingly more important for us during this complex and difficult transition to democracy, after two tremendously disastrous earthquakes. The above presented artifacts are just a few of the examples we are developing by Rai community teachers. There is no doubt that our work is linking school mathematics to diverse cultural practices. We are learning to do mathematical modeling in ever-sophisticated levels; both our teachers and students are learning to solve algorithmic problems with the help of modeling and connecting the real life context to that of particular cultural concepts. Moreover, unique and diverse cultural practices can come to play a vital role for understanding a phenomenon.

When we teach mathematics without linking context; it ultimately does not help to create and develop a meaningful uses of mathematical ideas; and it may devalue the cultural context of our diverse communities (Rosa and Orey 2011). We have found that this work also helps to reduce mathematical anxiety because when children start seeing classroom mathematics outside of in their own cultural practices they start enjoying and teaching becomes meaningful. In schools with little financial resources, culturally relevant pedagogies are inexpensive yet powerful resources.

In the context of classroom culture, mathematics and culture are not separate issues. This is especially true in highly diverse contexts such as found in Nepal, because every student comes to the classroom with their own cultural roots and prior knowledge in relation to particular phenomena. ABI seeks to connect this with curricular objectives and trains teachers to help students to be successful by teaching mathematics that can be meaningful with cultural artifacts.

References

Banks, J. A., & Banks, C. A. M. (2002). *Handbook of research on multicultural education*. San Francisco, CA: Jossey-Bass.

Barthakar, D. R. (2003). *The music and musical instruments of North Eastern India*. New Delhi, India: Mittal Publications.

Basu, S., & Siddiqui, N. (2011). *Musical instruments: The forgotten heritage of the Himalayas*. Pragya, India: Research & Advocacy Team.

Bullivant, B. (1993). Culture: its nature and meaning for educators. In J. A. Banks & C. A. McGee-Banks (Eds.), *Multicultural education issues and perspectives* (pp. 29–47). Boston, MA: Allyn & Bacon.

Cajete, G. A. (1994). *Look to the mountain: An ecology of indigenous education*. Skyland, N.C.: Kivaki Press.

Civil, M. (1995). Connecting home and school: funds of knowledge for mathematics teaching. In L. Meira & D. Carraher (Eds.), *Proceedings of the XIX International Conference for the Psychology of Mathematics Education* (pp. 154–161). Recife, PE: UFPE.

Civil, M. (1998). *Bridging in-school mathematics and out-of-school mathematics: reflection*. San Diego, CA: AERA Annual Meetings.

Courtney, D. (2016). *Music of India*: dhol. Houston, TX: Chandra & David's Homepage. Retrieved from http://handrakantha.com/articles/indian_music/dhol.html.

D'Ambrosio, U. (1980). Uniting reality and action: A holistic approach to mathematics education. In L. A. Steen & D. J. Albers (Eds.), *Teaching teachers, teaching students*. Boston: Birkhauser.

D'Ambrosio, U. (1985). Ethnomathematics and its place in the history and pedagogy of mathematics. *For the Learning of Mathematics, 5*(1), 44–48.

D'Ambrosio, U. (1993). Etnomatemática: um programa [Ethnomathematics: a program]. *A Educação Matemática em Revista, 1*(1), 5–11.

D'Ambrosio, U. (2000). A historiographical proposal for non-western mathematics. In H. Selin (Ed.), *Mathematics across cultures: The history of non-western mathematics* (pp. 79–92). Dordrecht, The Netherlands: Kluwer.

D'Ambrosio, U. (2006). The program ethnomathematics: a theoretical basis of the dynamics of intra–cultural encounters. *The Journal of Mathematics and Culture, 1*(1), 1–7.

D'Ambrosio, U. (2008). Educação numa era de transição [Education in a transition era]. *Revista Matemática & Ciência, 1*(1), 8–18.

D'Ambrosio, U. (2016). The ethnomathematics program and a culture of peace. *Journal of Mathematics Education, 10*(2), 1–11.

Davison, D. M. (1992). Mathematics. In J. Reyhner (Ed.), *Teaching American Indian Students* (pp. 241–250). Norman: University of Oklahoma Press.

Dunipace, S. J. (2010). Play on: traditional musical instruments of Nepal. Lalitpur, Nepal: ECSNepal. Retrieved from http://ecs.com.np/features/play-on-traditional-musical-instruments-of-nepal#.

Dupree, L. (1973). *Afghanistan*. Princeton, NJ: Princeton University Press.

Eglash, R. (1999). *African fractals: Modern computing and indigenous design*. New Brunswick, NJ: Rutgers University Press.

Eglash, R., Bennett, A., O'Donnell, C., Jennings, S., & Cintorino, M. (2006). Culturally situated designed tools: ethnocomputing from field site to classroom. *American Anthropologist, 108*(2), 347–362.

Erickson, F. (1986). Culture difference and science education. *The Urban Review, 18*(2), 117–124.

Fletcher, N. (1999). Materials for musical instruments. *Acoustics Australia, 27*(1), 1–9.

Gay, G. (2000). *Culturally responsive teaching: Theory, research, and practice*. New York, NY: Teachers College Press.

Gilbert, J. K., Boulter, C. J., & Elmer, R. (2000). Positioning models in science education and in design and technology education. In J. K. Gilbert & C. J. Boulter (Eds.), *Developing models in science education* (pp. 3–18). Dordrecht, The Netherlands: Kluwer.

Hall, E. T. (1976). *Beyond culture*. New York, NY: Doubleday.

Hart, L. E. (2003). Some directions for research on equity and justice in mathematics education. In L. Burton (Ed.), *Which way social justice in mathematics education? International perspectives on mathematics education* (pp. 25–50). Westport, CT: Praeger Publishers.

Hofstede, G. (1986). Cultural differences in teaching and learning. *International Journal of Intercultural Relations, 10*, 301–320.

Hollins, E. R. (1996). *Culture in school learning: Revealing the deep meaning*. Mahwah, New Jersey, NJ: Lawrence Erlbaum Associates.

Huntington, S. P. (1993). The clash of civilizations? *Foreign Affairs*, 22–49.

Irvine, J. J., & Armento, B. J. (2001). *Culturally responsive teaching: Lesson planning for elementary and middle grades*. New York, NY: McGraw-Hill.

Kandel, R. P. (2007). *Musical instruments of Nepal: Nepali folk musical instrument museum*. Kathmandu, Nepal: Suvash Printing Press.

Klotz, M. B. (1994). Culturally competent schools: Guidelines for secondary school principals. *Principal Leadership*. Retrieved from: http://www.nasponline.org/resources/principals.

Ladson-Billings, G. (1995). Toward a theory of culturally relevant pedagogy. *American Educational Research Journal, 32*(3), 465–491.

Luitel, B. C. (2009). *Culture, worldview and transformative philosophy of mathematics education in Nepal: A cultural-philosophical inquiry*. An unpublished thesis. Science and Mathematics Education Centre. Perth, Australia: Curtin University of Technology.

Matang, R. A. (2001, July). The role of ethnomathematics in mathematics education: Implications for mathematics curriculum in Papua New Guinea. *Paper presented at the 11th World Congress of Comparative Education Societies (WCCES)*. Chungbuk, South Korea: Korea National University of Education.

Mathema, K. B. (2007). Crisis in Education and Future Challenges for Nepal. *European Bulletin of Himalayan Research, 31*, 46–66.

Nabha, B. K. S. (1998). *Gurushabad Ratnakar Mahan Kosh*. Delhi, India: National Book Shop.

Naidoo, J. P. (2004). School evaluation in Nepal. In A. D. Grauwe & J. P. Naidoo (Eds.), *School evaluation for quality improvement: An ANTRIEP report* (pp. 53–57). Paris, France: UNESCO.

Nepal. (2007). *National curriculum framework for school education in Nepal*. Bhaktapur, Nepal: Curriculum Development Centre.

Ogbu, J. U., & Simons, H. D. (1998). Voluntary and involuntary minorities: a cultural-ecological theory of school performance with some implications for education. *Anthropology and Education Quarterly, 29*(2), 155–188.

Onstad, T., Kasanda, C. D., & Kapenda, H. M. (2003). Ethnomathematics: a link between an abstract school subject, local culture & everyday experience. In P. Chilisa, L. Mafele, & J. Preece (Eds.), *Educational research for sustainable development* (pp. 36–56). Gaborone, Botswana: Lentswe la Lesedi.

Orey, D. C. (2000). The ethnomathematics of the Sioux tipi and cone. In H. Selin (Ed.), *Mathematics across culture: The history of non-western mathematics* (pp. 239–252). Dordrecht, Netherlands: Kulwer Academic Publishers.

Orey, D. C., & Rosa, M. (2014). How we came to use a combination of emic, etic and dialogical approaches in the research field of ethnomodeling. *Proceedings of the IX Festival Internacional de Matemática* (pp. 1–15). Quepos, Costa Rica: CIENTEC.

Pollak, R., & Watkins, S. (1993). Cultural and economic approaches to fertility: proper marriage or mésalliance? *Population and Development Review, 19*, 467–496.

Rai, D. B. (2012). *The impact of modernization on the traditional Sakawa Sili festival in the Rai Kirat community of Nepal*: a case study of the Rai community. Master of Philosophy in Indigenous Studies. Faculty of Humanities, Social Sciences and Education. Tromso, Norway: Tromso University.

Rogoff, B. (2003). *The cultural nature of human development*. New York, NY: Oxford University Press.

Rosa, M. (2010). *A mixed-methods study to understand the perceptions of school leaders about English language learners (ELL)*: the case of mathematics. Unpublished doctorate dissertation. California, CA: CSUS.

Rosa, M., & Orey, D. C. (2007). Pop: a study of the ethnomathematics of globalization using the sacred Mayan aat pattern. In: B. Atweb, A. C. Barton, M. C. Borba, N. Gough, C. Keitel, C. Vistro-Yu, R. Vithal (Eds.), *Internacionalisation and globalisation in mathematics and science education* (pp. 227–236). Dordrecht, The Netherlands: Springer.

Rosa, M., & Orey, D. C. (2008). Ethnomathematics and cultural representations: teaching in highly diverse contexts. *Acta Scientiae, 10*(1), 27–46.

Rosa, M., & Orey, D. C. (2009). Symmetrical freedom quilts: the ethnomathematics of ways of communication, liberation, and art. *Revista Latinoamericana de Etnomatemática, 2*(2), 52–75.

Rosa, M., & Orey, D. C. (2011). Ethnomathematics: the cultural aspects of mathematics. *Revista Latinoamericana de Etnomatematica, 4*(2), 32–54.

Rosa, M., & Orey, D. C. (2013a). Ethnomodelling as a research lens on ethnomathematics and modeling. In G. A. Stillman & J. Brown. (Eds.), *Teaching mathematical modeling:* connecting to research and practices. International perspectives on the teaching and learning of mathematical modeling (pp. 117–127). Dordrecht, The Netherlands: Springer Science +Business Media Dordrecht.

Rosa, M., & Orey, D. C. (2013b). Ethnomodeling as a research theoretical framework on ethnomathematics and mathematical modeling. *Journal of Urban Mathematics Education, 6*(2), 62–80.

Rosa, M., & Orey, D. C. (2015a). Ethnomathematics: connecting cultural aspects of mathematics through culturally relevant pedagogy. *Mathematics Education & Society, 8*(3), 887–897.

Rosa, M., & Orey, D. C. (2015b). Evidence of creative insubordination in the research of pedagogical action of ethnomathematics program. In: B. S. D'Ambrosio & C. E. Lopes (Eds.), *Creative insubordination in Brazilian mathematics education research* (pp. 131–146). Raleigh, NC: Lulu Press.

Rosa, M., & Orey, D. C. (2016). Ethnomodelling as a creative insubordination approach in mathematics education. *Journal of Mathematics and Culture, 10*(3), 111–134.

Schreffler, G. S. (2002). *Dhol*: king of the Punjabi instruments—an introduction. Punjab, India: Gibb's Dhol Pages. Retrieved from: https://www.google.com.br/search?q=Dhol+King+of+the+Punjabi+instruments+An+introduction&ie=utf-8&oe=utf-8&client=firefoxb&gfe_rd=cr&ei=bhRKWPP5E86m8wfI84TgBw

Schreffler, G. S. (2010). *Signs of separation*: dhol in Punjabi culture. Dissertation. Music Department. Santa Barbara, CA: University of Santa Barbara.

Shankar (2011). *Nepali musical instruments*: cultural heritage of Nepal. Retrieved from: http://culturalheritagenepal.blogspot.com.br/2011/04/nepali-musical-instruments.html

Sharma, T. (2012). *Present context of Nepalese mathematics education and critical mathematics education*. Pokhara, Nepal: Second National Mathematics Education Conference.

Sharma, T., & Neupane, R. (2016). Crafting cultural intelligence in school mathematics curricula: a paradigm shift in Nepali school education. *International Journal for Research in Mathematics Education*, *6*(1), 285–308. Retrieved from: http://www.sbem.com.br/ojs/index.php/ripem/article/view/177

Shirley, L. (1995). Using ethnomathematics to find multicultural mathematical connections. In P. A. House (Ed.), *Connecting mathematics across the curriculum* (pp. 34–43). Reston, VA: National Council of Teachers of Mathematics.

Simard, L. (1994). Curriculum adaptation: Just do it. In K. P. Binda (Ed.), *Critical issues in First Nations education* (pp. 78–86). Brandon, Manitoba, Canada: BUNTEP, Faculty of Education, Brandon University.

Tarlekar, G. H. (1972). *Musical instruments in Indian sculpture*. Pune, India: Pune Vidyarthi Griha Prakashan.

Thakur, S. (1996). *Himacal ke Lokvadhya*. Chandigarh, India: Abhishek Publications.

University, Kathmandu. (2008). *Developing culturally contextualised mathematics resource materials: capturing local practices of Tamang and Gopali communities*. Kathmandu, Nepal: UNESCO.

Voltz, M. (1982). *The voltaic masks. The Drama Review, 26*(4), 38–45.

Zeichner, K. (1996). Educating teachers to close the achievement gap: Issues of pedagogy, knowledge, and teacher preparation. In B. Williams (Ed.), *Closing the achievement gap: A vision to guide change in beliefs and practice* (pp. 55–77). Alexandria, VA: Association for Supervision and Curriculum Development.

Part IV
Purpose of Ethnomathematics in Mathematics/Mathematics Education: Cross-Cultural Situations

The ethnomathematics program has an agenda that offers a broader view of mathematics. It embraces the ideas, processes, methods, and practices that are related to different cultural environments. This aspect leads to an increased evidence of a diversity of cognitive processes, forms of learning, attitudes, and paradigms that may direct a learning process occurring in our classrooms in order to facilitate cultural encounters in cross-cultural communities by reflecting on the social and political dimensions of ethnomathematics. One important aspect of this agenda is to offer an important perspective for a dynamic and modern society, which recognizes that all cultures and all people develop unique methods and explanations that allow them to understand, act, and transform their own reality.

Chapter 8
Ethnomathematics and Culturally Relevant Mathematics Education in the Philippines

Wilfredo V. Alangui

Abstract Various Indigenous peoples' education initiatives in the Philippines, including recent efforts of the Department of Education, bring to light the continuing need to develop mathematics lessons that are culturally relevant for Indigenous Filipino students. In general, teachers involved in Indigenous peoples' education have difficulty in developing culturally relevant mathematics lessons, with the relative exception of PAMANA KA, a school for Indigenous Mangyan in San Jose, Mindoro. My current involvement with this school as well as with the Gohang National High School in Banaue, Ifugao, shows how a modified framework for an ethnomathematical curriculum model admits a wider possibility for mathematical thinking and investigation based on cultural practice. Ethnomathematics may be a useful theoretical framework for the pursuit of culturally relevant mathematics education in the Philippines.

Keywords Ethnomathematics · Indigenous peoples' education · Culturally relevant mathematics lessons · Ethnomathematical curriculum model · Mathematics and culture

8.1 Introduction

A quiet revolution has been taking place in the area of education for Indigenous[1] Peoples (IPs) of the Philippines. As of last count, there are more than twenty initiatives and programs on IP education all over the country, formal and non-formal that are being run by IP organizations (IPOs), non-government organizations (NGOs), community-based efforts and other private institutions. These IP

[1]We capitalize Indigenous when referring to people and communities as a sign of respect.

W.V. Alangui (✉)
University of the Philippines Baguio, Baguio, Philippines
e-mail: wvalangui@up.edu.ph

M. Rosa et al. (eds.), *Ethnomathematics and its Diverse Approaches for Mathematics Education*, ICME-13 Monographs,
DOI 10.1007/978-3-319-59220-6_8

education programs, with varying success and in different stages of development, are efforts to provide a culturally rooted and relevant education to Indigenous youth and adults.

A welcome development on the side of the Philippine government has been the recent efforts of the Department of Education (DepEd) to finally recognize the failure of state-administered education to provide a kind of education that is non-alienating and sensitive to the needs and aspirations of Indigenous peoples. On 8 August 2011, the DepEd issued Department Order 62 (D.O. 62), adopting the National Indigenous Peoples Education Policy Framework. In this important document, basic education is seen as an enabling right that will allow IPs to "claim their other rights, exercise self-determination, and expand the choices retrieved to them" (D.O. 62, p. 6).

Through D.O. 62, the department acknowledges that IPs remain to be among the most vulnerable and marginalized members of the citizenry, identifying access to culture-relevant basic education as the most critical of all the current disadvantages Indigenous peoples face in the country. The increase of IP education programs may in fact be seen as response to years of discrimination, marginalization and neglect by the Philippine government especially in the area of education for Indigenous peoples (Abadiano 2011).

For the DEpEd, the national framework shall pave the way for an IP education that is responsive to their context, respects their identities, and promotes the value of their traditional knowledge, skills, and other aspects of their cultural heritage. The long-term goal is to remove the barriers for the meaningful participation of IPs in the different levels and spheres of society and empower them to exercise their rights and duties as Filipino citizens (D.O. 62).

Considering that IP education covers all disciplines of knowledge, two questions arise in relation to mathematics education: How is a culturally rooted and relevant mathematics education implemented in these various IP education initiatives? What challenges, if any, do mathematics teachers face in developing lessons that are relevant to the context and reality of Indigenous students? A third question may be added, one that is relevant to the field of ethnomathematics: What role does ethnomathematics play in this renewed effort to push for IP education in the Philippines?

The next section briefly discusses the situation of Indigenous peoples in the Philippines and their experiences in mainstream educational system.

8.2 The Indigenous Peoples of the Philippines

The Philippines is home to around 110 Indigenous peoples communities numbering between fifteen to twenty million (ECIP, n.d.). Located in various parts of the country, they are concentrated in Northern Luzon (mostly in Cordillera Administrative Region at 33%) and in Mindanao (about 61%); the remaining 6% are in the Visayas.

The Igorot of the Cordillera region belong to different ethnic groups, such as Bontoc, Ibaloi, Ifugao, Isneg, Kalinga, Kankanaey, and Tingguian. The Gadang, Ilongot, and Ivatan are found in the Cagayan Valley, Isabela, Nueva Vizcaya, and Quirino. The Negrito groups, which include the Aeta and Dumagat, are found in North, Central, and Southern Luzon. The Mindoro Island has seven distinct Mangyan groups. Palawan islands have the Batak, Palawana, and Tagbanwa. The Indigenous peoples in Mindanao are collectively called Lumad and belong to the following groups: (1) the Monobo, (2) the Bagobo, B'laan, T'boli, and Teduray groups, (3) the Mandaya and Mansaka groups, (4) the Subanen, and (5) the Mamanwa (Abadiano 2011).

UNDP reports that "In the Philippines, IPs have been subject to historical discrimination and marginalization from political processes and economic benefit. They often face exclusion, loss of ancestral lands, displacement, pressures to and destruction of traditional ways of life and practices, and loss of identity and culture" (UNDP Fact Sheet 2010, p. 1).

Republic Act 8371, also known as the Indigenous Peoples Rights Act (IPRA), was enacted in 1997 to correct this historical injustice and to recognize and protect the rights of IPs. It has become the cornerstone of national policy on IPs including in education. However, the effective implementation of IPRA remains a challenge for all stakeholders.

There continues to be a lack of accurate statistics and figures giving the number of Indigenous peoples communities reached by the public school system of the country. Many of these communities are in the mountainous areas of the country where basic social services like public schools and health centers are not provided by the government.

The absence of schools and other educational opportunities may have been a major problem in many Indigenous communities, but for many of those who have had access to mainstream education the experience has been a difficult one.

8.2.1 Experiences in Mainstream Education

The experiences of Indigenous peoples in mainstream Philippine education are summarized in a report of the Episcopal Commission of Indigenous Peoples (ECIP) of the Catholic Bishops' Conference of the Philippines entitled *Indigenous Peoples Education: From Alienation to Rootedness*. In this report, the views, experiences, and insights of Indigenous communities and various Indigenous Peoples' Apostolates (IPAs) nationwide were consolidated through a process of regional and national consultations from 2004 to 2007.

The report states that for an Indigenous person who had experienced mainstream education, the school had been a *key venue* of discrimination. This came in various forms: IP students were treated differently by their classmates and teachers, mirroring stereotypes held by the bigger society; their inability to comply with certain school requirements like wearing uniforms and shoes (because of financial difficulties) became an identity marker that fuelled further discrimination; and there was tendency to label them as *slow learners*.

Mainstream education, in the Philippines and in other parts of the world, has also helped popularize the view that Indigenous peoples are backward and primitive, with knowledge and skills that are inferior to modern/scientific knowledge and skills (ECIP, n.d.; McConaghy 2000; Smith 1999). What is discussed inside the classroom (educational content), the manner by which these are delivered (pedagogy), and the whole educational management are dictated by an assimilationist model with its western-based epistemic and ontological assumptions about teaching and learning, curriculum, management, policies and assessment (Longboat 2012).

The overall impact of mainstream assimilationist education, according to the ECIP report, is the alienation of Indigenous youth from their own communities, heritage, culture and history.

8.2.2 Underachievement in Mainstream Education

Not surprisingly, significant gaps in achievement between Indigenous students and non-Indigenous students have been noted all over the world (see for example Longboat 2012; McEwan and Trowbridge 2007). In Canada, Haldane et al. (2011) report:

> In those places where reliable information on education achievement is retrieved, the data points to a significant disparity in education outcomes between First Nation and non-First Nation students and overall low education attainment by First Nation individuals (p. 1).

As noted earlier, there is no retrieved disaggregated data on Indigenous peoples' education in the Philippines, although government reports show relatively low participation and survival rates in education in regions with considerable IP population (Abadiano 2011). The situation of IP education in the country may be partly gleaned from retrieved data in the Cordillera region in the northern part of the country, where majority of the population are Indigenous peoples.

A 2011 report of the National Economic and Development Authority (NEDA) of the Cordillera Administrative Region (CAR) shows a lack of school buildings and classrooms: 21 municipalities/districts in the region have no elementary schools and six municipalities/districts have no secondary public schools. Because of lack of qualified teachers, 1129 classes resort to the multi-grade system, which combines as much as three grade levels (Grades 1–3 and 4–6) in a class.

Another disturbing trend in the basic education sector from 2005 to 2009 could be gleaned from a 2011 report of the Department of Education-CAR (DepEd-CAR). Table 8.1 shows low rates in all the indicators in both elementary and secondary level in the region. The average cohort survival rate shows that only 67.54% of students who started in Grade 1 reached Grade 6, while the average completion rate shows only 67.02% successfully finished elementary.

Table 8.1 Performance indicators of government elementary and secondary schools—CAR: 2005–2009 (in %)

Indicator	2005–2006	2006–2007	2007–2008	2008–2009	4-year average
Elementary level					
Participation rate	74.57	72.68	71.20	71.21	72.42
Cohort survival rate	86.40	60.42	62.94	60.38	67.54
Completion rate	85.98	59.45	62.53	60.14	67.02
Secondary level					
Participation rate	40.34	39.49	38.48	37.04	38.84
Cohort survival rate	54.95	61.79	58.11	64.23	59.77
Completion rate	50.78	57.84	54.07	59.78	55.62

Source Department of Education-CAR

At the secondary level, participation rate is dismal with only 38.84% of children attending public high schools. This means that 61.16% of youth who are supposed to be in public high schools are either not enrolled, in private high schools, or are out of school. The percentage of those in private high schools may be small given the higher cost of private education even in the Cordillera region. Table 8.1 shows the performance indicators of government elementary and secondary schools.

Among the First Nation people in Canada, the 2006 Census showed that only 39% of those aged 20–24 living on reserve have completed high school or obtained an equivalent diploma while the Canadian average for high school completion for non-Aboriginal people of the same age is more than 87% (Haldane et al. 2011). This is even lower than the 50.78% completion rate in high school that was registered in the Cordillera region for academic year 2005–2006.

Aside from the low participation, survival and completion rates, Indigenous students also underperform in disciplines involving science and mathematics (DepEd-CAR 2011; Haldane et al. 2011). That performance of IP students in mathematics courses is significantly lower than those of non-IP student leads educators to feel a 'sense of urgency' to find ways in which Indigenous students can achieve better in mathematics education (Donald et al. 2011).

8.3 A Culturally Responsive Education

The idea of culturally responsive education has been evolving in Canada for the past 40 years and this has become an important innovation contributing to the success of Indigenous learners in basic education (Longboat 2012). Similar efforts are found in Alaska, New Zealand and Australia and elsewhere.

In the Philippines, the various initiatives on IP education may be seen as efforts to address the alienating impact of mainstream education (perhaps the oldest of these is the Kalahan Academy, which was put up in 1972 in the province of Nueva

Vizcaya in the northeastern part of the island of Luzon). With varying success and in different stages of development, some of these initiatives are efforts to provide a Culturally Responsive Education (CRE).

Rejecting the culture deficit model which attributes academic weakness of IP youth to learning deficiencies, perceived lack of ability of Indigenous communities to actively support the academic achievement of their children, or the (prejudiced) view that IPs simply do not value education (Salkind and Rasmussen 2008), a CRE is "a way of bringing students, school and community relationships into a learning community of shared values and educational goals for more equitable outcomes for all students" (Longboat 2012, p. 70).

8.3.1 Formulating a Philippine IP Curriculum Framework

Following the adoption of D.O. 62, the Philippine Response to Indigenous and Muslim Education (PRIME), a program funded by the Australian government, commissioned a project for the formulation of the IP Curriculum Framework (IPCF).

A study led by University of the Philippines educators Teret De Villa and Wilfredo Alangui was initiated in November 2012 as requisite for the formulation of the IPCF, with the following aims: (1) review existing models/initiatives on IP Curriculum development; (2) provide inputs and insights into the curriculum framework formulation process; (3) propose guidelines in the formulation of the IP curriculum framework based on basic principles culled from classroom observations, interviews, and focus group discussions; and (4) provide a documentation of the different IP curriculum typologies.

Using a consultative participatory process, the research was done over a period of 3 months (November 2012 to January 2013). Sixteen (16) schools participated in the study. These schools were visited to gather information on their experiences (process and impact) in implementing IP education, to validate existing programs based on documents/reports and earlier studies conducted, and to consult with the stakeholders in the areas (De Villa et al. 2013).

For each school, the research team conducted Focus Group Discussions with teachers and school administrators, students, parents, and community elders and leaders. We followed community processes for obtaining consent, facilitated either by IPEd focal persons of the DepEd, representatives of partner IP organizations, or NGOs working in the area. Some Key Informants Interviews were held where appropriate.

Ensuring geographical representation, some schools came from the mainland (Luzon), and some from island groups in Visayas and Mindanao (De Villa et al. 2013). Of the 16, only 3 were secondary schools, 9 were elementary schools, 2 were non-formal, and 2 were Schools of Living Tradition (that focused mainly on the teaching of arts and culture and did not have a mathematics component in their curricula).

Two regional validation workshops were held, one for Luzon, and another for Visayas and Mindanao. Then, a national validation workshop was held in May 2013 to share key findings and get feedback from representatives of schools and communities that were visited.

8.3.2 Features of an IP Curriculum Framework

Several features or principles were found to be critical to the implementation of an IP Curriculum. Across sites and experiences, aspirations for a curriculum for Indigenous students were anchored on the following (De Villa et al. 2013):

1. Is rooted in the ancestral domain and its indigenous cultural institutions;
2. Embodies the sacredness of transmitting Indigenous Knowledge, Systems and Practices (IKSP);
3. Revitalizes, regenerates, strengthens, and enriches IKSP, the indigenous learning systems and indigenous language;
4. Affirms and strengthens IP identity;
5. Focuses on cultural competencies and includes other forms of knowledge, concepts, and skills attuned with the needed life-long learning values and life skills for the development and protection of ancestral domains, their culture and the advancement of IP rights and welfare;
6. Allows the whole community to discern new concepts that will contribute to the community's cultural integrity while building new relations with the broader society;
7. Uses instructional materials and resources that are culture-based and culture-sensitive;
8. Utilizes the whole ancestral domain as a learning space.

The idea of a Culturally Relevant Mathematics Education (CRME) that I use is grounded on the above features of an IP curriculum framework as it encourages building connections and meaning between indigenous knowledge and other knowledge systems. It is a curriculum framework that not only allows for the development of values, knowledge and skills that reinforce positive social and cultural identities of Indigenous students but also allows them to develop awareness about their social and political context.

The above features resonate with a CRE concept that is continuously evolving among the First Nations in Canada. Longboat (2012) elaborates that CRE efforts in Canada are guided by the belief that "academic performance is inextricably tied to an Education experience that is firmly grounded in the context of culture and language and founded on history, spiritual beliefs, songs, ceremonies, the land or place of origin, art, music, oratory, contemporary community customs and Nation building for First Nations citizenship" (p. 71).

The CRE elements forwarded by Longboat (2012) are:

1. Recognition and use of indigenous languages;
2. Pedagogy that stresses traditional knowledge, values, relationships;
3. Pedagogy in which teaching strategies are congruent with traditional culture of the community and contemporary ways of knowing;
4. Curriculum that is based on traditional indigenous knowledge and spirituality; Strong participation from the community;
5. Knowledge and use of the traditional values and behaviors/protocols of the community.

8.3.3 Mathematics Curriculum in Philippine IP Education Programs

The study also documented five curriculum typologies (De Villa et al. 2013) that showed varying approaches in their handling of Indigenous Knowledge Systems and Practices (IKSP). These typologies were relevant only to the 9 elementary schools, 3 secondary schools and the 2 non-formal schools included in the study.

Table 8.2 presents each of these typologies. I have included in the last column a description of how mathematics lessons are handled in each type of curriculum. The table also reflects both the DepEd and IP competencies that each curriculum needs to cover. DepEd competencies refer to the school competencies that need to be met by students per grade level as required by the Department of Education. IP competencies refer to what students need to learn about their culture (e.g. important values, cultural skills, knowledge about a practice).

The typology somehow dictated the extent and nature of culturally responsive mathematics lessons that were included in the curriculum. For the first two typologies, teachers felt the discussion of indigenous ways of counting or counting in indigenous language as well as localizing mathematics problems and examples were enough to meet the expectations of a culturally responsive mathematics education. Examples of localizing mathematics lessons included using local names for people and things as they appeared in word problems and providing a local *flavor* to examples.

The last three typologies demanded more from the mathematics teachers, and efforts ranged from localizing mathematics problems to providing context to the mathematics lessons being discussed. Not all lessons were put in context, and some lessons were not always successful in contextualizing them. Teachers in general admitted difficulty in designing mathematics lessons that are culturally sensitive or cross-cultural. In most cases, lessons revert back to the localization of mathematics problems and examples.

Table 8.2 Curriculum typology

Typology	Handling of IKSP	Handling of mathematics lessons
Insertion of cultural elements in specific subjects	IKSP are only cited or minimally discussed in relation to DepEd topics; cultural elements focus mainly on indigenous music and dance	Mathematics lessons generally revolved around counting in indigenous language and localizing mathematics problems and examples; Emphasis on DepEd-prescribed competencies in mathematics
Addition of a separate subject to cover IKSP and culture	IKSP treated as a separate subject; focus is on selected IP competencies (mainly indigenous music and dance, beliefs and practices)	Localized mathematics problems and examples; Emphasis on DepEd-prescribed competencies in mathematics
Integration in the curriculum	IKSP are integrated more comprehensively in all subject areas; Emphasis on DepEd competencies; IP competencies are also given importance	Counting in indigenous language; localized mathematics problems and examples; Emphasis on DepEd-prescribed competencies in mathematics
Indigenized Curriculum	Curriculum is strongly linked to community life; Equal focus on DepEd and IPs competencies	Mathematics lessons are given a cultural context; effort is given for cross-cultural mathematics lessons; DepEd-prescribed mathematics competencies are given importance
IPs Curriculum	Ancestral domain (AD) as foundation/basis of the curriculum; community life informs the curriculum; IP competencies are emphasized more than DepEd competencies	Mathematics lessons are given a cultural context; effort is given for cross-cultural mathematics lessons; while focus is on IP competencies, DepEd-prescribed mathematics competencies are also given importance

8.4 The PAMANA KA and Gohang Experiences

One school, PAMANA KA, stood out from among the rest in providing culturally relevant education to their students. PAMANA KA is a school for Indigenous Mangyan youth in the island of Mindoro. It stands for *Paaralang Mangyan na Angkop sa Kulturang Aalagaan* (literally, *a Mangyan school fit to the culture we value*). Established in 1999 in San Jose, Occidental Mindoro by the Franciscan Missionaries of Mary (FMM), it supports Mangyan students from all the seven (7) Mangyan groups in Occidental Mindoro, namely the Hanunuo, Gubatnon, Rataganon, Buhid, Taobuid, Alangan, and Iraya.

Located amidst a Mangyan settlement in a far-flung area of the municipality of San Jose, PAMANA KA makes education accessible to children who never had the opportunity to go school because of a terrain that renders them distant from all the other schools in town. There are two schools of PAMANA KA, one provides elementary education and the other is secondary. The campus for elementary education is located in the highly mountainous village of Balingasu. It is non-formal in the sense that it has yet to be accredited by the Department of Education.

It covers Kindergarten education until Grade 4. The secondary school is located in the village of Danlog, not as mountainous as in Balingasu, but to get there one has to cross a major river that is dangerous to navigate during the rainy season. The Danlog campus offers first year to fourth year levels, and has recently been accredited by the Department of Education. The last two elementary grades of Grade 5 and 6 are offered in this campus.

As an educational institution, the vision of PAMANA KA is to provide “A special education that is liberating, uplifts the Mangyan condition and brings about the formation of a Mangyan who has the freedom and capability to shape his/her own destiny; who is rooted in and developed by his own culture and indigenous wisdom; who will serve the interest and the welfare of his fellow Mangyan” (PAMANA KA Report 2012, p. 3). This vision has guided the management of the school for the past 17 years.

The efforts of PAMANA KA as a school that implements CRE have produced an annual average of 50 grade school and 50 secondary school students from the different Mangyan groups in 6 municipalities of Occidental Mindoro, according to a 2012 school administration report. It also describes the positive impact of PAMANA KA on their Mangyan students (PAMANA KA Report 2012):

> The students’ capacity to understand and speak English can be noted after 2 to 3 years from a starting point of very poor comprehension, inadequate vocabulary and inability to express themselves. Remarkable growth in self-confidence become evident in students who started very shy to answer or to put their chin up but eventually become able to speak and face up to people. Growth in the appreciation of their identity and culture, in the respect and value for their elders is manifest in practice (p. 23).

> Students who finished the course were able to go to College after PEPT (Phil. Educational Placement Test) or A & E (Accreditation & Equivalency tests). One outstanding student became eligible for College admission after 3 years of schooling in PAMANA KA. The performance of Mangyan students who enter college in the public/private schools in San Jose has changed the perception about Mangyans. They are commended not only for academic achievement but for their notable attitudes and behaviour (p. 23).

The PAMANA KA report states that some alumni have proceeded to college and have pursued degree programs such as Bachelor of Science in Secondary Education, Accountancy, and Agriculture. Three of the college graduates are now teachers of PAMANA KA, “continuing the tradition of culture-based education that they have received” (PAMANA KA Report 2012, p. 24) while another serves as the school’s accountant.

The report notes that the rest of their alumni are gainfully employed in other schools or offices, and have now the capacity to help their families in their needs. It also states that those who were not able to finish college or who dropped out have been capacitated enough to become leaders in their communities, ensuring that their fellow Mangyans are not taken advantage of in transactions with lowlanders.

While an independent and in-depth study on the impact of PAMANA KA as an institution that implements CRE is yet to be made, the school has already been recognized by the Department of Education and other agencies for its work in IP education. The latest award it received, in September 2012, was the Most Outstanding Literacy Program Award handed out annually by the Department of Education–Literacy Coordinating Council.

The National Literacy Awards was created in 1994 to encourage the development and replication of creative and indigenous literacy programs through awards and appropriate recognition (Jaro-Amor 2012). In accepting the award, Sr. Aristea Bautista, School Directress who belongs to the Franciscan Missionaries of Mary (FMM) said that PAMANA KA "has moved the Indigenous Peoples from the margins to the center. It has contributed to the rewriting of the Mangyan story, from being discriminated and set aside, now recognized, acclaimed, and given honor" (Jaro-Amor 2012, p. 3).

PAMANA KA implements an indigenized curriculum (typology 4) in their secondary school, where both the DepEd and IP competencies are articulated as important goals for the students to learn. The elementary school curriculum, on the other hand, exemplifies an IP curriculum (typology 5), one that is anchored on the life of the community. For example, lessons in the different subject areas (Biology, Social Studies, Music and the Arts, etc.) are developed around a yearly activity where elders and parents bring their children to the forest for several days as part of educating the young about indigenous resource management (e.g. what trees and plants to protect because of their value) as well as the important skills of hunting and fishing.

While DepEd competencies are consciously discussed, they take a backseat in favor of IP competencies. As such, even their school calendar is in accordance with the agricultural calendar of the community (other schools start the academic year in June; PAMANA KA starts in August to allow their students to participate in planting season from June to July). Learning the Mangyan way of life becomes the main objective of implementing the curriculum, but in the process, students are also trained to succeed in traditional schooling, and there is evidence that these students perform well as they progress in elementary and then high school (Jaro-Amor 2012). The typologies indigenized curriculum and IP curriculum were adopted from PAMANA KA.

PAMANA KA mathematics teachers go through a lot of effort in developing culturally relevant lessons for their students. Researching practices and beliefs, interview and validation with elders, immersion in community life are part of their process of developing all their lessons (not only in mathematics). Every August before the academic year starts, the school holds workshops for teachers with invited volunteer resource persons to help in the review and updating of lessons and modules.

For both the indigenized and IP curricula, the approach is the same—as much as possible, mathematics lessons are given a cultural context that informs the discussion. Unlike in other subject areas where IP competencies take precedence over DepEd competencies in Typology 5, in the mathematics curriculum of PAMANA KA, students need to learn both IP and DepEd-prescribed competencies. This reflects the administration's belief on the importance of mathematics as a subject that their Mangyan students need to learn.

8.4.1 Two Mathematics Lessons from PAMANA KA

Two mathematics lessons, one in a Grade 5 class and the other in a freshman high school class, are briefly presented to show how PAMANA KA teachers go about delivering culturally relevant mathematics lessons. I documented these lessons in March 2012 at the high school campus in Dalog.

The Grade 5 math lesson was on Harvesting Rice and Honey, and Fractions. The teacher is a Mangyan named Alma, herself a graduate of PAMANA KA. Teacher Alma started the lesson with the question: *What do we do in the month of October?*, resulting in a discussion on Mangyan practices of harvesting rice and honey; follow up questions allowed students to discuss their different beliefs and practices about rice and honey harvesting (students belong to different Mangyan groups) and the material culture used in the practice; further questions led to a discussion on the value of ensuring fair share among those who helped harvest, and the negative effect of being greedy and wanting more than what is necessary.

The discussion on equitable sharing led smoothly to the discussion of fractions and the operations with fractions. The following is an abridged transcript of the discussion in class:

Teacher Alma	"*Tuwing buwan ng Oktubre, anu ano ang ginagawa nating mga Mangyan ng sama sama?*". (What activities do we engage in together in the month of October?).
Students	"*Pag-aani ng palay at sitaw sa kaingin*". (Harvest rice and string beans in the garden). "*Pag-aatas*". (A method of catching fish in the river).
Teacher Alma picks up on topic of rice harvesting and asks	"*Anu-ano ang mga paniniwala at sistema natin hinggil sa pag-aani ng palay?*". (What are the different Mangyan beliefs and practices on rice harvesting?).
Students	"*Ipaalam sa pamayanan*". (Inform the community). "*Huwag magsipol o kumain ng palay, baka magkasakit o baka kainin ng daga ang palay*".

(Don't whistle or eat rice grains, you might get sick or rats might attack the rice plant).
"*Hatiin ang kaingin; ipapaani ang kalahati, ang kalahati ay maiiwan sa may-ari*".
(Divide the garden into two; let the community harvest the first half, the other half is for the garden owner).

Teacher Alma draws a garden of rice plants on the board, and divides the figure into two.

Teacher Alma	"*Bago nahati, ang kaingin ay buo – ang tawag dito ay whole; ang bahagi ng buo ay fraction*". (Before dividing into two, the garden was whole; the part of a whole is called fraction).
Teacher Alma writes '1/2' on the board, and explains	*Ang '1' ay isang bahagi ng buo; ang tawag dito ay numerator. Ang numerator ang dami ng nakuha mula sa buo. Ang '2' ay dalawang bahagi; ang tawag dito ay denominator. Ang denominator "ang dami ng pagkakahati-hati"*. (The number 1 means one part; it is called the numerator. The numerator indicates how many parts are being taken away from the whole. The number 2 means two parts of a whole; it is called the denominator. The denominator is a number indicating how the whole should be divided into parts).

To enrich the discussion further, teacher Alma shifts the discussion to harvesting honey.

Teacher Alma	"*Ano ang tawag sa lalagyan ng polot o balakwas?*". (What do we call the container for honey?).
Students	"*Fokfok*". One student draws a *fokfok* on the board.
Teacher Alma asks	"*Ano ang sistema natin sa hatian ng polot*?". (What is our system in sharing the harvest of honey?).
Students	"*Depende sa dami ng kumuha; pare-pareho*". (It depends on the number of persons who harvested; it should be equally shared).
Teacher Alma	"*Kung 5 ang kumuha?*". (If 5 people were involved in harvesting?).
Students	"*Lima ang maghahati-hati*". (5 people will have to share equally).
Teacher Alma	"*Paano isusulat ang bahagi ng bawat isa?*". (How should we write the share of each person?).

One student writes 1/5 on the board.

Teacher Alma "*Ano kaya ibig sabihin ng '2/5'?*".
(What could 2/5 possibly mean?).

The students are silent for a while, until teacher Alma probes further:

Teacher Alma "*Maaari kaya kung ibibigay ng isa sa isa pang kasama ang kanyang bahagi?*".
(Could it be a situation where one person gives his share to another companions).

Students "*Maaari!*".
(Yes, that is a possible explanation!).

From here, teacher Alma proceeded to discuss different types of fractions, including the operations of addition and subtraction, making sure to relate the discussion to honey harvesting whenever the opportunity arises.

Teacher Alma asks "*Bakit kailangang pare-pareho ang hati? Ano ang mararamdaman kung hindi?*".
(Why is it necessary to have equal sharing? What does one feel if the harvest is not shared equally?).

Students "*Inggit; galit; sama ng loob; makasarili*".
(Envy; anger; resentment; one is greedy).

Teacher Alma "*Ano ang sabi ng gurangon? Dapat pare-pareho; kung ano sa iyo, ganon din sa iba*".
(What do our elders say? We should always be fair with each other; no one should have more than what the others have.).

She continues "*Kung hindi, nasisira ang samahan; hindi na buo; magkakahiwalay. Ang Mangyan ay dapat buo*".
(Otherwise, our relationships will be destroyed; we will no longer be whole; there will be disunity. Mangyan should always be one.).

In this mathematics lesson, the discussion did not end with fractions but with a reminder about the importance of being just and fair in order to ensure unity among the Mangyan, as exhorted by their elders.

The mathematics lesson for the first year high school students focused on Algebraic Expressions. The teacher was Sister Ogie, a Franciscan nun, who started the lesson by asking the students of their knowledge about harvesting cassava.

Sister Ogie *Paano natin alam na may bunga ang tanim na kamoteng kahoy?*
(When do we know that the cassava plant is ready to be harvested?).

Students *Kung may mga bitak mula sa tanim.*
(When there are cracks on the ground emanating from the plant)

Sister Ogie *Gaano karami ang bunga?*
(How many cassava tubers may be harvested?).

Students *Hindi natin alam. Maaring 3 o 4 na lalim ay may bunga.*
(We would not know. There may be 3 to 4 layers of cassava tubers).

Sister Ogie *Kung ganon, sa pag-ani ng cassava, ang alam lang natin ay ang bitak; pero hindi natin alam kung ilan ang bunga. Tama ba?*
(In other words, in harvesting cassava, we only know the number of cracks on the ground; we would not know how many cassava roots may be harvested. Is this correct?).

Sister Ogie draws a cassava plant with cracks on the ground and with three layers of cassava tubers. She then explains the concepts of known and unknown variables in the following manner:

> *Sa algebraic expressions, may mga known at unknown values. Gaya ng sa kamoteng kahoy: yung bitak ang alam nating bilangin, yun ang known. Pero kung ilan ang bunga, yun hindi natin alam, kaya unknown yun.* (In algebraic expressions, there are known and unknown values. In harvesting cassava, we could count the cracks on the ground; the cracks are our known values. But we could not count the cassava roots below the ground, so those are our unknown values).

The teacher then introduces the idea of variables by identifying the 3 layers as *a*, *b*, and *c*.

> *Ang simbolong 5 ay nagsasaad na may limang bitak sa lupa. Sa bawat bitak, may 3 baytang na isusulat natin sa letrang a, b at c. Ang tawag dito ay variables. Yung 5 ay known value, nabilang natin ang bitak, alam natin may limang bitak sa lupa. Yung variables, yun ang mga bunga na hindi natin alam kung ilan. Maaring iba iba ang bilang ng bunga sa bawat baytang, kaya variable ang tawag sa kanila.*
>
> (The symbol 5 represents the five cracks on the ground. In each crack, there are 3 layers which we denote by the letters a, b and c. These are called variables. The number 5 is a known value because we were able to count the cracks, so we know there are 5 cracks. The variables represent the cassava in each layer, which we could not count. We do not know how many are there, and these could change in every layer. This is the reason why the letters are called variables).

From there, the class discussed more examples of algebraic expressions. In this lesson, an economic practice provided the relevant context to teach difficult mathematics concepts like algebraic expressions and variable.

These are examples of culture-based lessons in a difficult subject like mathematics. Western-trained mathematician are used to teaching these lessons in the traditional way (usually devoid of context). In PAMANA KA, the attempt is to discuss their mathematics lessons within a context (cultural/economic activity) that are familiar to the students. The goal is that in each lesson, two competencies are developed: an IP competency (reinforcing values and knowledge about cultural practice) and school (or DepED) competency (understanding of fractions and algebraic expressions).

8.4.2 The Gohang National High School Experience

By virtue of a Memorandum of Agreement (MOA) between the DepED and the UNESCO National Commission that was signed in February 2013, the Gohang National High School in Banaue, Ifugao was designated as a Special Secondary School for the Conservation of the Ifugao Rice Terraces. UNESCO provided grant support for the transformation of GNHS into this special school, covering (1) community engagement; (2) curriculum development; (3) training of school management and teachers; and (4) advocacy and communications.

According to Project Terminal Report (2014), the GNHS teachers, the school head and other implementers of the project underwent capacity building activities on culture-based education and immersion and onsite learning and exposure in selected IPEd initiatives in other areas. The community and other stakeholders were also involved in developing the school's educational philosophy and curriculum framework that is appropriate to its designation as a special school.

The curriculum framework was designed so that the four quarters of the school year coincide with the seasons in the agricultural cycle (i.e., *Ahitulu, Iwang, Lawang, and Tiyargo*). This curriculum framework, as described in the Project Terminal Report (2014), "underwent a serious process of reflection and discussion among the school staff, elders and other community representatives" (p. iii). Identifying aspects of their culture that shall become the framework for the school curriculum is not an easy endeavor. As in other parts of the Cordillera, these are serious matters for discussion especially by the elders.

Guided by the school's aim, which is "to develop an ancestral domain-based, culture-based and contextualized indigenous peoples' education towards cultural and global competence" (Project Terminal Report 2014, p. 44), it was clear early on that the context of the curriculum would have to be Ifugao's rice culture (and the practice of rice terracing agriculture). With this in mind, it made sense to all those involved in the discussions to anchor the curriculum framework on the community's agricultural cycle.

8.4.3 Personal Involvement in GNHS

In one of the capacity building workshops with GNHS teachers in November 2013, I talked about ethnomathematics and how this may be used to develop mathematics lessons in the context of rice terracing. I presented an ethnomathematical curriculum model developed by Shehenaz Adam (2004) with some modifications (discussed in Sect. 8.5), and discussed the potential offered by considering rice terracing agriculture as a system, identifying possible entry points for the development of lessons under different subject areas.

Almost a year later in October 2014, I again facilitated a workshop with all the 14 GNHS teachers (with 3 mathematics teachers) to process the experience of implementing their curriculum framework. The workshop revealed some of the difficulties they faced in applying the curriculum framework that was earlier developed with the community and other stakeholders:

1. The lessons the teachers developed continued to be dictated by the DepEd competencies: the curriculum framework was made to adjust to DepEd competencies, and elements of the framework were used depending on what DepEd competency is taught.
2. Lack of materials and resources: there are retrieved materials but are not particularly relevant to the experiences of the Gohang community; and
3. The teachers were adept with the DepEd competencies, but they have not agreed on what IPEd competencies to cover in their lessons.

In general, their problem was how to use the curriculum framework, still conflicted on what their special school was aiming to do. In processing these expressed difficulties, the teachers eventually arrived at some realizations:

1. The primary aim is a revitalization of Ifugao culture, expressed collectively thus: "Our special school for the conservation of the rice terraces is symbolic of the overall desire to perpetuate, revitalize, and enhance Ifugao culture not only for our children and our people, but also for the nation and the world" (Gohang teachers, 19 October 2014)
2. Expand awareness and understanding about other IPs in the Philippines, not to focus only on Ifugao peoples and their culture. The curriculum shall foster understanding about similarities and differences, for students to learn the importance of diversity, foster respect for other cultures, and strengthen their identity as IPs, as Filipinos, and as citizens of the world; and
3. The rice culture of the Ifugao is a fertile context that can arouse mathematical thinking and investigation, possibly through mathematical modeling. Although it may not be possible to do this for every cultural practice or activity related to rice terracing, there is a wide range of possibilities that can be explored to develop mathematics lessons. The students are already familiar, to a certain extent, with their cultural practice. Linking this prior knowledge to mathematics is giving them an opportunity to view their world in a different perspective, while learning new concepts and skills in mathematics. At the same time, important values, cultural practices and community perspectives are re-learned and reinforced.

Guided by a framework that I used (discussed in the next section), the teachers developed initial guide lessons using activities that coincided with the four seasons in their agricultural cycle. The workshop seemed to have lifted a veil of doubt among the teachers, and the framework for an ethnomathematical curriculum appeared to have provided some clarity and confidence on how to approach the development of lessons that are in line with the school's curriculum.

The teachers of Gohang are being encouraged by the DepED to go easy, to take baby steps, because realizing the IP Education Framework for their school, and fully implementing the curriculum framework will take some time. But they are anxious and they also know that they still have a lot of work to do to realize their own revolution.

8.5 The Need for Theoretical Grounding

There is an important gap in grounding efforts to provide culturally relevant mathematics in the context of IP education initiatives in the Philippines. The pursuit of IP education is bolstered by a conviction that Indigenous peoples deserve an education relevant and sensitive to their needs and realities.

The IPEd efforts by the schools involved in the study can be supported by principles of Culturally Relevant Education, although this grounding has not been articulated, except by PAMANA KA (and Kalahan Academy in northern Luzon, one of the schools also visited by the study). This may explain the relative success of these two schools compared to the others pursuing IPEd.

Not articulated in any of these IPEd schools is a theory that guides their efforts for the development of cross-cultural/culturally relevant mathematics lessons, and this lack of a clear theory may explain their less success in this area, exhibiting dilemma on what to focus on, what topics to integrate or *indigenize*, and how to go about their mathematics lessons. This has been a major challenge for the mathematics teachers involved in the various IPEd schools and programs in the country, in particular, in GNHS and to some extent, in PAMANA KA.

8.5.1 An Ethnomathematical Curriculum Model

Ethnomathematics can be described as the way people from a particular culture have common systems for dealing with quantitative, relational, and spatial aspects of their lives (Barton 1996). It follows what Knijnik (1997) describes as an ethnomathematical approach, one that is characterized by the *investigation* of the traditions, practices, and mathematical concepts of a subordinated social group and the pedagogical work, which was developed in order for the group to be able to interpret and decode its knowledge.

The aim, according to Knijnik, is to *acquire* the knowledge produced by academic mathematicians, and to *establish* comparisons between its knowledge and academic knowledge, thus being able to *analyze* the power relations involved in the use of both these kinds of knowledge. Ethnomathematics can provide the theory that is needed to support the pursuit of culturally relevant mathematics education in the various IPEd initiatives in the country.

The mathematics curriculum used in PAMANA KA is closest to the kind of learning that the ethnomathematical practice wants to uphold and propagate. In this model of learning mathematics, indigenous knowledge is an integral part of the education process.

Adam (2004) developed an ethnomathematical curriculum model in her research in Maldives, highlighting the integration of the mathematical concepts and practices originating in the learner's culture with those of conventional, formal academic mathematics. Her assumption is that classrooms and other learning environments cannot be isolated from the communities in which they are embedded, and students come to school bringing with them values, norms and concepts they have acquired as part of growing up.

In her ethnomathematical curriculum model, learners start from the experiences they have from the environment, build upon these mathematical ideas, and ultimately realizing and understanding the need for accuracy and the use of, say formula, in mathematics and real-life situations. A schematic diagram of the model is shown in Fig. 8.1.

The objective of her model is to allow students to become aware of how people mathematize in their culture, and to use this awareness to learn about formal mathematics. The students' world provides opportunities for students to think mathematically, and eventually connect these mathematical ideas to conventional mathematics.

It is also possible that the connection of the mathematical activities in the culture to conventional mathematics is straightforward; the activities are directly related to existing concepts in conventional mathematics. However, the model also allows the possibility of interrogating the activities for their inherent mathematical content, which is then followed by analyzing its relationship to established mathematics.

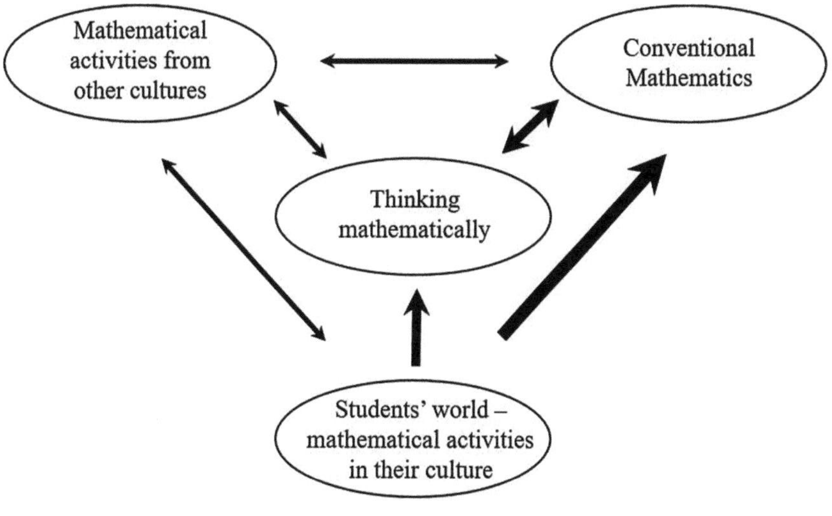

Fig. 8.1 Framework for an ethnomathematical curriculum model. *Source* Adam (2004, p. 82)

The assumption is that a curriculum of this type will motivate students: (1) to recognize mathematics as part of their everyday life; (2) enhance students' ability to make meaningful mathematical connections; and (3) deepen their understanding of all forms of mathematics (Adam 2004). Thicker arrows in the diagram signify mathematical connections (to conventional mathematics) and action (doing and thinking mathematically) that emanate from the students' context.

In the case of PAMANA KA, the activities that are used to provide context to the mathematics lessons do not immediately lend themselves to be mathematical. Harvesting honey, rice plants or cassava harvesting are activities that are at once economic and cultural: economic because it is related to the livelihood of the people, cultural because of the beliefs and values surrounding each practice. In Gohang, rice terracing agriculture and associated practices like water management, while highly developed, are not mathematical. However, these practices become the platform to discuss mathematics, inviting the students to think mathematically, and to link this mathematical thinking to conventional mathematics.

8.5.2 A Modified Framework

I provide a slight modification of the Adam (2004) framework for an ethnomathematical curriculum model to widen the scope of possibilities for mathematical thinking and linking with conventional mathematics. In this framework, we expand the source of mathematical thinking by expanding the students' world to include all cultural activities. Figure 8.2 shows a modified framework for an ethnomathematical curriculum model.

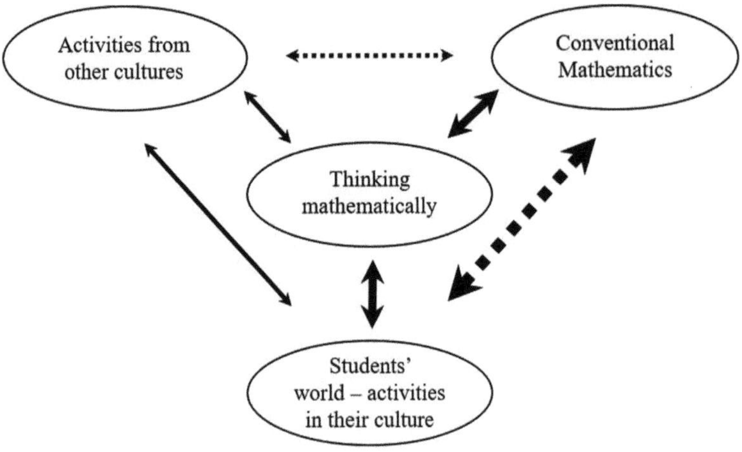

Fig. 8.2 A modified framework for an ethnomathematical curriculum model

This framework allows for a wider set of cultural practices or activities as possible take off points in the discussion of mathematical ideas. Cultural practices may not always be mathematical, hence the connection from (1) students' world to conventional mathematics, as well as the connection between (2) activities from other cultures and conventional mathematics, are no longer straight forward (represented here by the dashed arrows), but the contention is that some cultural practices, especially those that are highly systematized and have survived through time, may lend itself for mathematical analysis (Barton 1996). Similar to Fig. 8.1, the thicker arrows in this diagram (including the dashed arrow) signify mathematical connections and action emanating from the students' context.

In Alangui (2010), the practices of stonewalling and water management in Agawa, Besao, Mountain Province in the Cordillera region, northern Philippines, were investigated for possible alternative mathematical conceptions. The new model permits such kinds of investigations. Also in this new model, mathematical thinking is reflected back onto the students' world of cultural activities, establishing comparison of both kinds of knowledge (and how possibly, one can inform the other, and vice versa). In general, the aim is to determine how the ideas that are embedded in a cultural practice might relate to conventional mathematics.

8.5.3 Using the Framework to Organize Culturally Relevant Mathematics Lessons

It is possible to develop culturally relevant mathematics lessons using the modified framework (Fig. 8.2). Three examples of Guide Lessons are presented below. The first example comes from PAMANA KA, using Mangyan's knowledge about cassava harvesting and fishing to develop mathematical lessons. The last two are from GNHS, developed during the workshop in October 2014. In these lessons, both IP and DepEd competencies and standards are responded to.

Guide Lesson 1

IP Competency: Community Knowledge on Harvesting Cassava and Fishing

DepEd Competency: Variation

Harvesting Cassava (*Pag-aani ng Kamoteng Kahoy*) Fishing (*Paghuli ng Isda*)		
Students' world activities in their culture	Description of the activity; practices, beliefs, knowledge and skills needed in the conduct of the activity	Ways (*pamamaraan*) of harvesting tubers and fishing; materials and tools (*gamit*) used; associated beliefs (*paniniwala*) and traditional knowledge and practices (*talang kaalaman at kaugalian*)

(continued)

(continued)

Harvesting Cassava (*Pag-aani ng Kamoteng Kahoy*) Fishing (*Paghuli ng Isda*)		
Thinking mathematically	Qualitative, relational and spatial concepts in the activity	Knowledge harvesting cassava: As the number of cracks increase, there is bigger chance of harvesting more cassava (*habang dumadami ang bitak ng lupa, mas marami ang mahuhukay na ube*) Harvest about fishing in the river: The more water in the river, the less fish to catch; the less water, the more fish to catch (*habang dumadami ang tubig sa ilog, kumukonti ang huling isda at kapag kaunti lang ang tubig sa ilog, mas maraming nahuhuli*)
Conventional mathematics	Direct and inverse variation	Mathematics lesson on variation; types of variation; examples Valuing (*Pagpapahalaga*): Why is there a need to learn variation? (*Bakit kailangang matutunan ang variation*?); Where else do we see or experience variation? (*Saan saan eto nakikita o nararasan?*) Drill (*Pagsasanay*)
Activities from other cultures	Description of how other ethnolinguistic groups (especially in a heterogeneous classroom or setting) do the activity; practices, beliefs, knowledge and skills needed in the conduct of the activity	How do other ethnolinguistic groups harvest cassava; how do they fish? What are their beliefs and practices; what knowledge and skills are involved; what materials are used? Differences in ways and methods, materials, and beliefs (*pagkakaiba-iba ng pamamaran, gamit at paniniwala*); similarities (*pagkakapare-pareho*)

Guide Lesson 2

IP Competency: Community knowledge on stonewalling

DepEd Competency: Measurements, Slope, Area

Stone walling (*Menkabiti/Mentupeng*)		
Students' world activities in their culture	Description of the activity; practices, beliefs, knowledge and skills needed in the conduct of the activity	Activities related to the construction of stonewalls; materials; beliefs; profile and role of practitioners

(continued)

(continued)

Stone walling (*Menkabiti/Mentupeng*)		
Thinking mathematically	Qualitative, relational and spatial concepts in the activity	• Height of stonewall: How do decide on the height? How do we measure? • Inclination: How do we decide on the angle of inclination? How do we measure? • Area of rice paddy (*payeo*): How do we measure area? Why is area of the *payeo* important? What is the relationship of height of stonewall and area of *payeo*?
Conventional mathematics	Lengths and measurements, Slope, Perimeter and Area	Mathematics lesson on lengths and measurements, slope, perimeter and area; examples Valuing (*Pagpapahalaga*): Why is it important to learn about measurements, slope and area? Where else do we use length, inclination, area? Drill (*Pagsasanay*)
Activities from other cultures	Description of how other ethnolinguistic groups (especially in a heterogeneous classroom or setting) do the activity; practices, beliefs, knowledge and skills needed in the conduct of the activity.	How do other ethnolinguistic groups build stonewalls? What are their beliefs and practices; what knowledge and skills are involved; what materials are used? Differences in ways and methods, materials, and beliefs (*pagkakaiba-iba ng pamamaran, gamit at paniniwala*); similarities (*pagkakapare-pareho*)

Guide Lesson 3

IP Competency: Beliefs about Full Moon

DepEd Competency: Circle

Full moon (*Ongar Di Bulan*)		
Students' world activities in their culture	Description of the activity; practices, beliefs, knowledge and skills needed in the conduct of the activity	What activities are allowed during a full moon (*ongar di bulan*)? What activities are not allowed? What other beliefs are there about a full moon?
Thinking mathematically	Qualitative, relational and spatial concepts in the activity	What is the shape of a full moon? What other objects are similar to the shape of a full moon? In mathematics, the shape of a full

(continued)

(continued)

Full moon (*Ongar Di Bulan*)		
		moon is called a circle. Do we have a name for a circle in our language? How do we graph a circle? Introduction of the formula of a circle
Conventional mathematics	Circle	Mathematics lesson on circles; formula of a circle; examples Valuing (*Pagpapahalaga*): Why is it important to learn about circles? Drill (*Pagsasanay*)
Activities from other cultures	Description of how other ethnolinguistic groups (especially in a heterogeneous classroom or setting) do the activity; practices, beliefs, knowledge and skills needed in the conduct of the activity	In other ethnolinguistic groups, what activities are allowed during a full moon (*ongar di bulan*)? What activities are not allowed? What other beliefs are there about a full moon? Differences in beliefs and practices (*pagkakaiba-iba ng paniniwala at kaugalian*); similarities (*pagkakapare-pareho*)

The guide lessons above are examples of how the framework could help teachers organize and develop culturally relevant mathematics lessons out of a cultural practice or activity from the students' world or context. A template could be developed based on the framework, reflecting the four components: (1) Students' world (activities in their culture); (2) Thinking mathematically; (3) Conventional mathematics; and (4) Activities from other cultures. This template could be used as a starting point to develop more detailed mathematics lessons focusing on the connections between cultural practice, mathematical thinking, and conventional mathematics.

In PAMANA KA, the lessons developed by mathematics teachers reflect these components of the framework, with additional elements like Valuing (*Pagpapahalaga*) and Drill (*Pagsasanay*). The template is also guiding the GNHS teachers in their initial effort to develop mathematics lessons in the context of their being a special school for the conservation of the rice terraces.

8.6 Conclusion

What role can ethnomathematics play in IP education in the Philippines? My involvement in IP education efforts since 2011 has made me realize the difficulties faced by mathematics teachers in developing cross-cultural or culturally relevant mathematics lessons. My experiences with teachers in PAMANA KA, as well as my initial engagement with GNHS teachers, suggest that ethnomathematics may be used as a theory for culturally relevant mathematics education within IP education initiatives in the Philippines.

My discussions about the theory of ethnomathematics and my use of a modified framework for an ethnomathematical curriculum model (Adam 2004) show potential in providing direction and guidance to such efforts. Gohang mathematics teachers are developing lessons in the context of their rice terracing practice following this framework. After our discussions, PAMANA KA teachers and administration found the theory of ethnomathematics to be relevant to and supportive of their efforts at mathematics education in their school. Sr. Aristea Bautista, the PAMANA KA administrator expressed it succinctly after learning about ethnomathematics: *there's a theory behind what we are doing*!

As a final note, developing culturally relevant mathematics education for Indigenous students can only succeed and become effective if done within the context of a broader IP education program. The Department of Education is doing this precisely with its aggressive push for IP education within the context of the recently initiated K-12 program. There is a challenge to train more Filipino teachers in the theory of ethnomathematics for them to help advance culturally relevant mathematics education within the context of a comprehensive Indigenous peoples' education program.

References

Abadiano, B. (2011). *Proposed strategic directions on Indigenous peoples' education for the Department of Education*. Pasig: Department of Education, Republic of the Philippines.

Adam, S. (2004). *Ethnomathematics in the Maldivian curriculum: Trialling an implementation*. Auckland: The University of Auckland.

Alangui, W. (2010). *Stone walls and water flows: Interrogating cultural practice and mathematics*. Auckland: The University of Auckland.

Barton, W. D. (1996). *Ethnomathematics: Exploring cultural diversity in mathematics*. Auckland: The University of Auckland.

Department of Education. (2011). *Cordillera Administrative Region*. Retrieved from: http://www.nro.cordillera.gov.ph/en/rdp-2011-2016/323-social-development/all. Accessed on December 10, 2016.

Department of Education. Indigenous Peoples' Education Office, Gohang National High School, and the Schools Division of Ifugao (2014). Establishment of a Special Secondary School for the Conservation of the Ifugao Rice Terraces: Project Terminal Report submitted to the UNESCO National Commission.

De Villa, M. T., Alangui, W. V., Jaro-Amor, I., & Martin, F. (2013). *Developing an IP curriculum framework*. Philippines: Report submitted to the IP Education Office, Department of Education and Culture, Republic of the Philippines.

Donald, D., Glanfield, F., & Sterenberg, G. (2011). Culturally relational education in and with an Indigenous community aboriginal student performance in mathematics. *Education, 17*(3), 72–83.

ECIP. (n.d.). *Indigenous peoples education*: From Alienation To Rootedness. Philippines: A consolidated report by the Episcopal Commission on Indigenous Peoples.

Haldane, S., Lafond, G., & Krause, C. (2011). *Nurturing the Learning Spirit of First Nation Students*. Australia: Report of the National Panel on First Nation Elementary and Secondary Education for Students on Reserve.

Jaro-Amor, I. (2012). *PAMANA KA receives 2012 outstanding literacy program award*. IPsEO. Retrieved in: http://deped-ipseo.tk/pamana-ka-receives-2012-outstanding-literacy-program-award/. Accessed May 6, 2013.

Knijnik, G. (1997). An ethnomathematical approach in mathematical education: a Matter of political power. In A. Powell & M. Frankenstein (Eds.), *Ethnomathematics: Challenging Eurocentrism in mathematics education* (pp. 403–410). Albany: State University of New York Press.

Longboat, K. D. (2012). *Soul of sovereignty: The impact of culturally responsive education on the academic achievement of first nations students*. Canada: A report prepared for the Assembly of First Nations, Education, Jurisdiction, and Governance Sector.

McConaghy, C. (2000). *Rethinking indigenous education: Culturalism, colonialism and the politics of knowing*. Queensland: Post Pressed.

McEwan, P., & Trowbridge, M. (2007). The achievement of indigenous students in Guatemalan primary schools. *International Journal of Educational Development, 27*, 61–76.

PAMANA KA Report. (2012). San Jose: Unpublished report, PAMANA KA.

Salkind, N. J., & Rasmussen, K. (2008). *Encyclopedia of educational psychology* (Vol. 1). Thousand Oaks, CA: Sage Publications Inc.

Smith, L. T. (1999). *Decolonizing methodologies: Research and Indigenous peoples*. London & New York: Zed Books Ltd.

UNDP. (2010). Indigenous Peoples in the Philippines. UNDP Fast Facts: Lagom. http://www.ph.undp.org/content/dam/philippines/docs/Governance/fastFacts6%20-%20Indigenous%20Peoples%20in%20the%20Philippines%20rev%201.5.pdf. Accessed on November 15, 2016.

Chapter 9
The Role of Culture and Ecology in Visuospatial Reasoning: The Power of Ethnomathematics

Kay Owens

Abstract Reasoning is critical in mathematics but research has shown that people reason not only with numbers and algebra but also visuospatially (also called spatial thinking and visual reasoning). This kind of reasoning occurs with representations but also in everyday living. In other words, ecology and culture impact on mathematical reasoning and learning. To investigate this premise, Papua New Guinean Indigenous cultures and their mathematical activities were appropriate as they are rich in visuospatial reasoning. There is evidence of the strength of this reasoning and that it may or may not be associated with words. The challenge is how to enlist this cultural strength in schools.

Keywords Visuospatial reasoning · Culture · Ecology and mathematics · Mathematical activities · Ethnomathematics in schools

9.1 Introduction

Mathematics education has its fair share of controversy. This chapter discusses two of these. The first is whether visualisation can play a role in mathematics and be evaluated. The second is whether mathematics is universal. In 1992, the International Group for the Psychology of Mathematics raised the former. Visual imagery and visualisation were hot topics. There were discussions around whether visualisation was internal or external (Goldin 1998).

Needless-to-say, both were regarded as feasible and important but how could you assess visualisation? What role does it play in mathematics? Imagistic processing, that is visualising, played a critical role in problem solving along with symbols, language, affect, and heuristics, all interacting (Goldin 1987). Spatial abilities (Eliot 1987; Tartre 1990) and visual imagery psychology (Shepard 1975)

K. Owens (✉)
Charles Sturt University, Bathurst, NSW, Australia
e-mail: kowens@csu.edu.au

M. Rosa et al. (eds.), *Ethnomathematics and its Diverse Approaches for Mathematics Education*, ICME-13 Monographs,
DOI 10.1007/978-3-319-59220-6_9

were being drawn together, generally as a result of social considerations pertaining to education.

One early paper illustrated that proof can be established through visuospatial reasoning (Dreyfus 1991). Dreyfus argued that the validity of a proof is judged by experts who know the topic and that visual representation is an important part of deductive reasoning. He illustrated this with the problem of finding points with the sum of the distance to two intersecting lines being a constant given length. An illustration of this proof is given in Fig. 9.1.

He visually but logically explained the proof based on the diagram. This visual explanation can be dynamically illustrated either with a program like Geogebra or with a string that can slide on two perpendicular T bars that move along the given lines. Whether dynamic or static, the visual information is essentially used in a valid analytical argument. Since then the whole area of dynamic geometry software has been readily accepted as a visual means of proof. Figure 9.2 gives the example for the formula of a triangle as ½ × base × perpendicular height in which vertices at C are on the line *e* and perpendicular to the side AB; hence the perpendicular height is the same in all triangles with the common base of AB.

If a figure does not *mess up* when pulled, then properties of the shapes have been used to construct the shape. It is also a way, through measurements, to show that changes in shapes do not necessarily affect other properties.

In 1992, the importance of visualising was being described by adults and children in solving problems such as those involving:

(a) Pentominoes: how many shapes can be made with five squares joined by their sides, describe them in terms of symmetries and their possibility of being a net for an open cube,
(b) Tangrams: to make squares and explore the area of different combination of shapes in terms of area of the smallest triangle,
(c) Match stick designs and modifications: requiring embedding and disembedding shapes, and

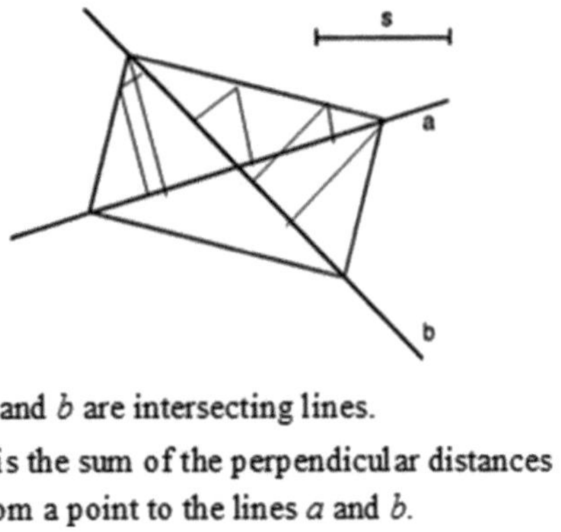

Fig. 9.1 The sum of the lengths of the perpendicular lines from each point to the lines *a* and *b* are equal. *Source* Based on Dreyfus (1991)

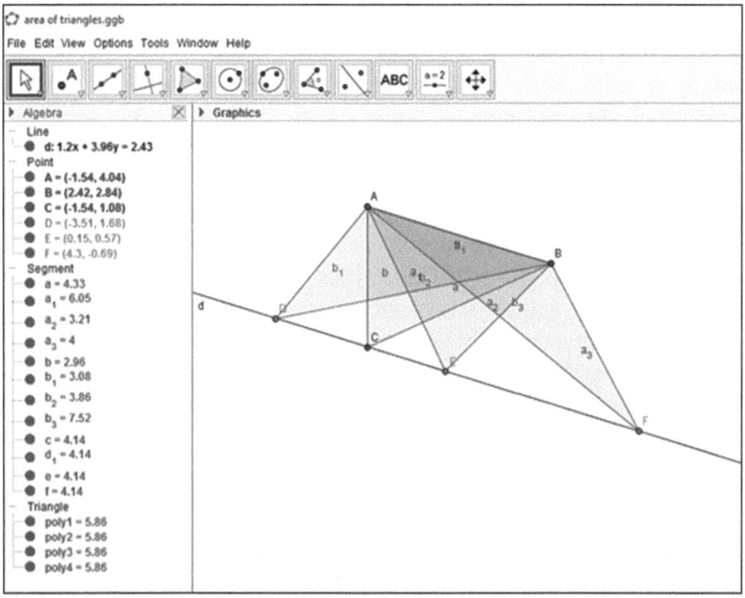

Fig. 9.2 Dynamic geometry image illustrating the area of a triangle formula (from *Geogebra*). *Source* Personal file

Fig. 9.3 Modifying pentomino *shapes* to create new ones using visuospatial reasoning. *Source* Personal file

(d) Pattern blocks: the size of their angles and those of tangrams, making enlargements.

At times they were able to capture their thinking in words but at other times there were systematic movements with the materials and both similarities and differences in the ways that people solved the problems (Owens 1992a, 1992b, 1993). It was, as Skemp (Campbell-Jones 1996) said, “out there on the table”. During the pentomino problem one child said “in my mind, I pictured my hand moving the pieces around the shape” illustrating how she was mentally moving the last square around the line of four squares (see Fig. 9.3).

A close analysis of the videotapes of other children’s work on the pentominoes suggested a similar tactic was being used, children were modifying one shape to form another (Owens 2015). First they often had to decide what is a shape? At first some were unsure of the question as they could not make triangles, rectangles or squares—that was the limit of their idea of a *shape* due to labels being the only emphasis in schools. Some initially considered that only familiar, symmetrical shapes for which they had a name would satisfy the word *shape*. Gradually they realised that they needed to consider other shapes and at that stage, they began modifying shapes in their head momentarily before trying something on the table.

Visuospatial reasoning was involving children in different kinds of visual imagery: holistic/pictorial, dynamic, action, pattern, and procedural visuospatial reasoning (Owens 2015, pp. 40–59). These types of imagery were first suggested by Presmeg (1986) from a high school study of children doing algebra. In brief, holistic/pictorial mental imagery are images generally of one or a few discrete representations of a drawing or object. Dynamic imagery is like the computer changing a shape or like transforming a shape such as from 2D to 3D shape or by rotating, turning, folding, or reflecting in one way or another. Action imagery involves bodily movements and may be thought about or embodied in various ways. Pattern imagery is where the pattern of the arrangement of the image parts is fairly dominant. If the imagery is well established then procedural imagery is simply following the steps or movements directing the changes to the imagery (Owens 2004; Owens et al. 2003). It is an efficient step in that the processing has been internalised.

This is similar to the idea of reification of the concept (Presmeg 2006) or the advanced structuring level of Pirie and Kieren's (1991) model. This model showed that understanding had a core of primitive knowing and image making, image having, and property noticing in order to formalize a concept with the support of imagery. Thus one important aspect of visuospatial reasoning is the diversity of types of visual imagery that help problem solving in a diversity of ways. Rivera (2011) has also emphasised the diversity and nature of visuospatial reasoning used in school learning indicating its importance in the agenda of schools.

Earlier training studies (Owens 2015, Chapter 2) suggested that visuospatial reasoning could be engaged through training, and even short training if pictures are not present in one's background to understand pictures (Bishop 1983). Recently there has been a renewed interest in visuospatial reasoning such as its malleability (Sinclair et al. 2016; Uttal et al. 2013). Thus teaching can improve this capability.

The second controversy was about the universality of mathematics. This chapter contends that people develop mathematics and, as Dreyfus (1991) suggested, mathematical visual reasoning is accepted as proof by informed people. To explore this further, everyday use of visuospatial reasoning of a differentiated population was investigated. Papua New Guinea has such a population and so an appropriate population in which to explore, with PNG colleagues, visuospatial reasoning which appeared to be important in everyday mathematical decision-making.

9.2 Research Project

The purpose of this project was to draw together many years of research to answer the question of how visuospatial reasoning is used in cultural mathematical activities in Papua New Guinea. A review of the literature of mathematical activities of PNG cultures and similar cultures in the Pacific in particular alerted me to understanding how visuospatial reasoning would be used in PNG. In particular, visuospatial reasoning was inextricably intertwined with cultural beliefs and practices (Owens 2015).

The research involved both ethnographic studies and questionnaire studies, and interrogation of 250 written reports of cultural mathematical practices by tertiary education students who were familiar with their cultural practices or undertook their own ethnographic studies. The questionnaires particularly asked about measurement practices and language using open-ended questions. These were collated and results from the same language group were compared and collated. Respondents often talked about time and location as well as area, length and volume.

Data from neighbouring language groups were also compared and focus groups were held to check the key ways in which measurement was being used. The use of visuospatial reasoning was part of these discussions although we talked about this topic as "how are people thinking to make those decisions?" and asked clarifying questions such as "What do you mean by measuring a garden area 'by leg'?".

In addition, we (Wilfred Kaleva, myself, and other interested colleagues with whom I was travelling) carried out interviews out of their places at the university or in towns. These were semi-structured interviews of people whom we knew had participated regularly in cultural activities, taking care to interview both coastal and highlands people.

However, most of the data came from ethnographic studies in villages. To get to villages in PNG, one often has to fly and/or walk, and sometimes take muddy roads in the backs of trucks or a dinghy or canoe for several hours. Villages generally have no power and no reticulated water; people have food gardens and sometimes cash crops, and use the bush and/or sea, river or swamp for food. Although we talk about all PNGians as Melanesians, with languages that are Austronesian Oceanic or Non-Austronesian (from multiple Families and Isolates), it should be noted that there are 850 languages and hence cultures in PNG.

Overall these cultural practices vary considerably although there are similarities and often cross-cultural marriages among neighbouring language groups. I was able to draw on 15 years of living in PNG and many village visits, yarning (talking, discussing) and videoing, during that time and subsequently. The particular ethnographic studies for the measurement project were carried out with educators who spoke the language or neighbouring languages and who had previously undertaken ethnomathematical studies. In particular, I want to acknowledge Charly Muke, Rex Matang, and Serongke Sondo. We also speak the ubiquitous lingua franca Tok Pisin.

All visits were negotiated with the community, generally in advance but if that was not possible due to transport difficulties, on arrival. We stayed in village huts. We observed and took videos of people going about their daily activities and often asked questions to see how they were thinking or why they were doing particular things. We particularly observed and listened to Elders, generally in a group, who might have demonstrated some of their activities if they were not in the process of building a house or making a canoe or another object. We sat and yarned for hours. We joined in activities where we had the skills. We walked along tracks with villagers, visiting gardens and other significant places.

From this rich data, I explain in this chapter what is meant by visuospatial reasoning in cultural contexts and how significant it is to mathematical activities and mathematics. It is an important mathematical thinking process that is under-utilised in schools and under-recognised in mathematics education especially in assessments that use paper-and-pencil tests.

9.3 Explaining Visuospatial Reasoning

The combined describing word *visuospatial* indicates that visual and also spatial sensing, perception and imagery are involved. The spatial aspects include the kinaesthetic and embodied sensing, perception and imagery that comes particularly with movement. Visuospatial reasoning is also about spatial relationships in and between objects, figures and positions. It involves spatial abilities such as visuospatial manipulations, alternative perspectives, and re-seeing in different ways including holistically, by completion, and in parts (Tartre 1990). Some spatial abilities such as efficient image rotation, image integration, adding detail, and image scanning but not image generation time as well as visual memory appear to assist visuospatial reasoning (Poltrock and Agnoli 1986).

Visuospatial reasoning involves generating images and generating representations, analysing and modifying both in an empirical way, predicting spatial change, and change in other fields represented spatially. It is about reasoning logically and visuospatially, making judgements, sometimes instantaneously or imperceptibly or by gazing (Mason 2008), but most importantly through applying knowledge and prior experience. Lohman et al. (1987) said “spatial ability may not consist so much in the ability to transform an image as in the ability to create the type of abstract, relation-preserving structure on which these sorts of transformations may be most easily and successfully performed” (p. 274).

Knowledge and experience are two areas that are most affected by culture and ecology whether influenced by technology, home, school, foundational cultural practices, activity, modelling, language or observation. For example, Jodie (a pseudonym) for whom English was a second language had realised that when the teacher said “bigger angle” that she was not referring to a sharper angle but to the larger angle thought of as the arms moving further apart (marked by thumb and finger) of the shapes. She later explained this as the arms being “more spread out”.

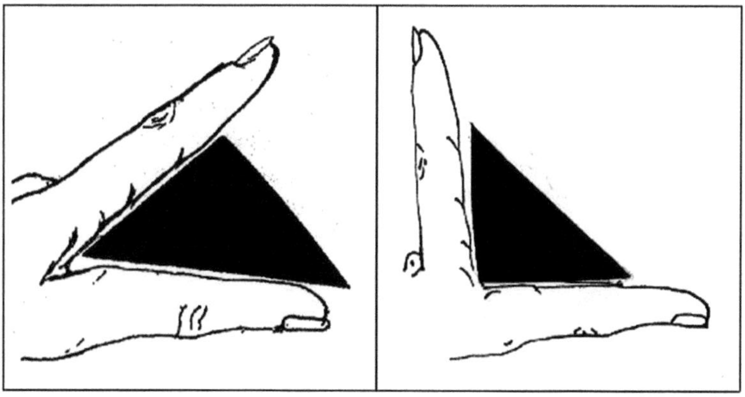

Fig. 9.4 A child's visuospatial reasoning about the size of an angle. *Source* Personal file

Her experience with the materials and background learning English at school influenced her visuospatial reasoning (Fig. 9.4).

The use of the fingers being spread apart as part of the description of angle size was engaging an embodiment of angle for visuospatial reasoning.

9.3.1 Spatial and Embodied Reasoning

Embodied knowledge is important, for example, for those who are used to change gears and turn a corner in a car, remember a phone number by touching the numbers on the phone. It is also known by those who feel the motion of water on the hull of a canoe, or the pull of a string from a flying kite or a swimming fish, or even the order of movements to create a string figure ('cat's cradle'). These pursuits are embedded in cultural contexts as illustrated by Fig. 9.5.

Sensing the feel of the swell of the sea may be learned by lying in the hull as well as by paddling and being out on the canoe feeling the wind and noting the impact on the sail also helps generate embodied visuospatial reasoning. Thus selecting the angle of a paddle, setting the position of the outrigger of a canoe, knowing the distance between places by the amount of time experienced by the body in moving between the places, assessing angles and slopes by gesturing with the hand, stretching out arms or parts of arms to assess lengths, will all be spatial decision-making times about objects in space, supported visually.

Comparisons made from one time to another are much easier when spatial, embodied reasoning is practiced (Owens 2015). In some traditional Indigenous communities people assess slope of the troughs made from the sago palm bark or the volume of water needed for sago processing (Fig. 9.6).

Fig. 9.5 Embodied cultural activities: sailing and paddling a canoe and making string figures. *Source* Personal file

Fig. 9.6 A PNG coastal activity: Sago processing; a PNG highland's approach to making a *mumu*. *Source* Personal file

The right amount of steam to cook karuka nuts in a dirt *mumu* (oven) is also estimated by visuospatial reasoning as they consider the size of the mumu and the amount of heat from the stones and the amount of steam escaping as they pour the water through a hole onto the hot stones (Fig. 9.6).

9.3.2 Experience and Estimation

Visuospatial reasoning is not unknown in western, especially farming, cultures. Repeating activities over time leads to greater efficiency and estimation in a form of informed scientific trail-and-error approach. Farmers might consider the amount of dryness and growth of plants to determine the number of cattle or sheep or length of time for them to graze on the paddock.

In three-dimensional play, children spatially sense how their block towers are balancing—examples are given in Fig. 9.7 (the sketch is from Ness and Farenga 2007). They use visuospatial reasoning through rotating and joining objects to create their play ideas. Children learn through play how to move their arm and fingers to hit a marble. Children also move in a confined space to solve spatial problems (Lábadi et al. 2012). Children use geometric ideas such as corners and distances to solve spatial problems. Children do not necessarily require language to explore these spaces and make decisions (Learmonth et al. 2008).

Many cultural activities in Papua New Guinea entail mathematical visuospatial reasoning to estimate and balance objects (Owens 2015). For example, when a group of men are building a house from bush materials, they will make many judgments such as the slope of the roof, the area of grass for thatching with kunai, the selection of trees for 'rafters', or the number and size of limbom palms to split for the sago-roofing panels or for floors or walls.

Figure 9.8 shows men modelling the angle of the roof to ensure it is steep enough for the rain to run off but not too steep because then the kunai thatching will fall off. It also has to be appropriate for an internal wall the height of a man's hand above his head and with side walls as high as his shoulder.

Fig. 9.7 Block play. *Source* Personal file

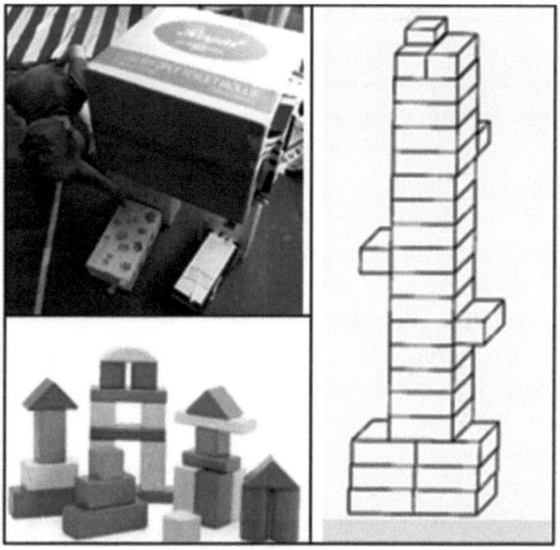

Fig. 9.8 Visuospatial reasoning for house building in Papua New Guinea. *Source* Personal file

In houses on the coast, sago-leaf roofing requires estimates of the size of sago trees to be cut down to supply the roofing along with the number of tulip trees whose inner bark is used to make rope. In addition, men will select bamboo lengths to split for floors or pitpit lengths for weaving for walls (see Fig. 9.10).

Visuospatial reasoning may begin with visualising lengths in the vertical that will become sloping or horizontal lengths in house building. In other cases there is a connection between lengths and areas or volumes. For other cases such as the use of palms, the overall size of the tree indicates whether it will be sufficient or not for the manufacture of roof panels or walls. In these cases, it seems there are ways of visualising these sizes associated with lengths. This is understandable if, for example, garden widths are of a fixed width or shapes are consistent (e.g. rectangular, square or trapezoidal as I have noted in different villages).

However, some highlands people have also added the length and width for a garden area but the decision on gardens is linked to more important matters such as quality of the soil, distance from the village, relationship of the people in the family or clan to the distribution of land, etc.

In terms of measuring the volume or mass of pigs which are important exchange and feast items, people might regard the count as more significant or they may use one or more length measures in their discussions of size (Fig. 9.9).

In most areas, two small pigs are just deemed as equal to a large one or the pigs are lined up in size so marking them off and noting the number and size for equality. Nevertheless, some cultural groups keep a track of the lengths of exchange pigs on a long rope which they add to as exchanges occur.

Some people's sweet potato (*kaukau)* mounds for exchange may be carefully piled into a cone shape and compared using a stick to measure height and rope for the circumference of the base of the cone. Again the complex relationships associated with the feast or exchange are more important than the actual size comparison. Hence various systems have been developed to provide rough estimates of equality.

Fig. 9.9 Lining up pigs for important exchanges in Papua New Guinea highlands. *Source* Google Images

Fig. 9.10 Weaving of walls in different ways in a PNG coastal village, a woven basket with lid from pandanus leaves by coastal Keapara Elder in PNG, and a basket and small purse from Timor Leste. *Source* Personal files

Figure 9.10 shows an example created by a Keapara elder. Her mental imagery of the pattern and how it will be created in terms of the width of the pandanus which is split into narrow sections to create the coloured finished pattern.

This process involves considerable estimation as the initial length depended on the size of a cereal box around which the pattern was created. It has an inner lining of wider strips similar to the base. In Timor Leste, women will also judge sizes to make a tight fitting lid on a small purse or on the basket. The hexagonal design is also carefully created for the particular basket size.

9.3.3 Estimations and Decisions Reflecting Ratio

Experience of creating with bush materials has led to some people having special skills in estimating the result of a raw product being used. Rather than thinking of the number of additional posts for a house, it is seen as half as much again—ratio thinking. This leads to how much more roofing, walls, floor, and posts; how much help will be needed; and how many gardens for a feast to recognise assistance or status. Thinking visuospatially in ratios does not require measurements and calculations.

However, not all amounts of the various items for the house are increased by half as much again. For example, the back and front walls, floor and roof might be half as much again but the side walls will not. For a nine post house, half as much again requires three more posts. If the roof slopes from the sides as well as the front and back, this will impact on the total amount of roofing required. Thus the sections of the roof are considered separately.

If the side slope was to half way along the length between posts then the overall increase can be set as the same amount for the side roofs and front and back to that point but the middle section is doubled as shown in Fig. 9.11.

In the case of a roof going up to one central point as in a curved ended rectangular house or round house as built in the highlands, the building's size will have an effect on roof and rafter lengths, reflecting trigonometry ratios. Figure 9.12 shows examples of men modelling these kinds of houses. Since the model was smaller than the example and since these houses could vary for family units or larger men's houses, certain aspects need to be considered. In practice, lengths can be shorter or longer to reflect these ratios.

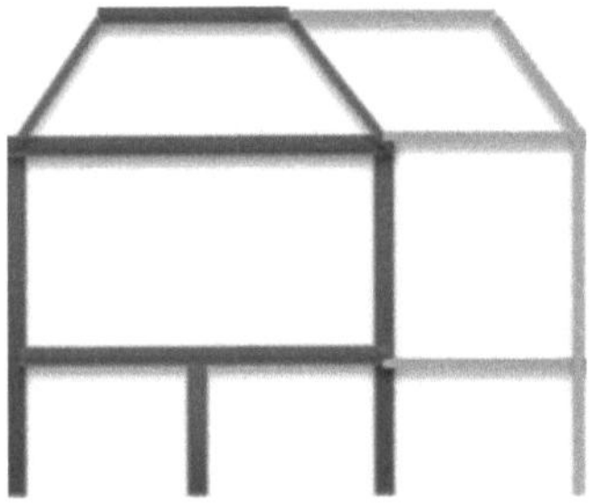

Fig. 9.11 Front view of a three × three post house extended in length by half to be a four × three post house. *Source* Personal file

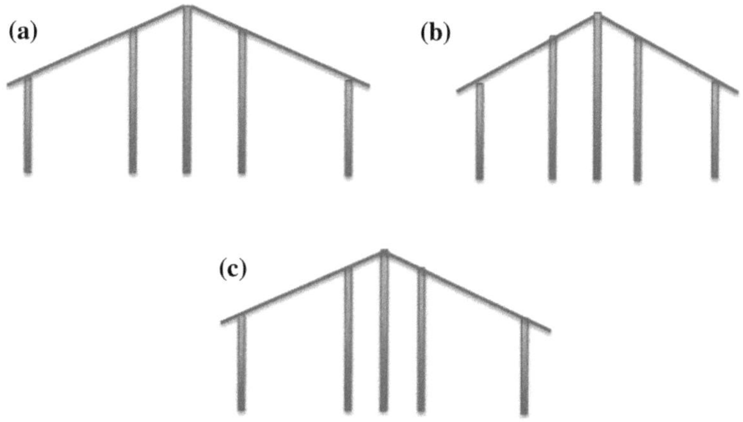

Fig. 9.12 **a** A highlands house front and **b** The model house with same lengths but larger angle on the roof or **c** The smaller middle section. *Source* Personal files

Fig. 9.13 **a** Measuring half way with an available stick, took the middle mark too far and **b** Tessellating area units are seen but not recognised as a measuring unit until pointed out. *Source* Personal files

Figure 9.12a illustrates the full house and then the model but the men kept the wall height equal to their shoulder height and the internal wall height to the raised arm height. They discussed whether the roof would now be too steep and whether they would have to compromise with the interior wall and overall height. They could see that the angle (unmeasured) would make the roof steeper.

The compromise for the same angle was to reduce the floor spacing from thirds for the outer sleeping areas and the middle talking areas so the middle section was slightly smaller as shown in the third house design. It was also possible for the width of the house to change allowing more people to sleep in the two ends. Each end space marked by the internal walls was divided into two and in Fig. 9.13a, the men are finding this halfway mark (see section on Geometry Knowledge below).

A further aspect of ratio was linked to the volume of a round house. If these houses are larger as in the background of Fig. 9.8a, then the ratio of the amount of materials needed for the walls, the rafters, and kunai for roof together with labour is associated with the size of the radius and circumference of the base circle. In fact, so strong is the aspect of ratio and labour that a person also has to consider the rate associated with the size of the food gardens needed to provide a feast for the helpers.

However, another significant aspect of the base of a house and its volume is that a large house also needs more warmth. A small house is not so cold with a little fire and one person, said one Elder. Similar ratio decisions associated with thanking labourers is associated with the size of a *mumu* pit for in ground cooking. Its shape might be partly determined by the number of pigs required but also for other foods like sweet potatoes.

9.4 Mathematical Conceptual Knowledge

While visuospatial reasoning is evident in processes such as mathematical problem solving, it is also evident in conceptual understanding. In particular, it is evident in geometry and measurement knowledge and in a sense of number. Relationships between numbers is particularly strengthened through visuospatial reasoning, a basis of Les Steffe's early work on early counting (Steffe 1991).

9.4.1 Geomtry Knowledge and Visuospatial Reasoning

In Fig. 9.13 to decide the centre of a line joining two points, a man took a long stick and marked from both points where the end of the stick fell. This made it easier to estimate the middle point even though the chosen length was more than half. The ends of the stick were recognised as equidistant from the end points and the middle point as shown in Fig. 9.14.

In house building, different groups of people know that for a rectangle, not only the opposite walls are equal but the corners must be right angles. Sometimes, people and sticks are placed in a line along the direction of the two arms of the right angle and the person at the corner has a good sense of the straightness of the line of people and whether it a right angle. They do it *by eye* was a common response to this visuospatial approach.

Fig. 9.14 A stick (*horizontal*) is used from both ends (*tall sticks*) to mark new points (*short sticks*) nearer the middle. *Source* Personal files

Others used the fact that not only the opposite sides have to be equal but the diagonals are also equal. For example, one man explained that, when he was a boy, he was in the way of his father measuring the diagonals with a rope. He immediately visualised that if the diagonals were different lengths, a parallelogram would be made. This lived experience with associated visuospatial reasoning was strong whereas in school, shapes are just labelled.

For a house on posts, the posts will be marked in one row equally spaced, then equal lengths will be taken to place the next row of posts. Checks are made by eye and with equal length sticks and ropes. Folding the rope for the full length of the house into three will position the centre row of posts. Equal lengths were known to form squares so each post was on the vertex of a square. When planting cash crops, two sticks are used to get equilateral triangles.

Only two sticks are needed as the first row is placed equally spaced and then the sticks are used for the other two sides of the equilateral triangle. Short sticks are put in the ground to check all the holes will be in the right position like a map. The pattern of tessellated triangles is checked out with straight oblique lines. "It is a beautiful pattern of triangles" said one villager.

Where round houses are common (Fig. 9.8a), people know a lot about circle geometry and volumes of cylinders and cones. They know about segments, equal and unequal lengths across circles, and different angles subtended by different chords of a circle. People know how much will spread out on a platform that is a segment of a circle, how much space will be available for sleeping, how much for drying nuts and storing belongings on the rafters at the top of the cylindrical part of house.

In one student's project he illustrated and described the use of two ropes to make an isosceles right-angled triangle. The shorter rope had two knots at the ends and the other three with one knot bisecting the rope with two knots at the ends. The first rope was the length of the hypotenuse if the other rope formed the other two sides with the knot at the vertex. Without knowing Pythagoras' theorem, this had become the practice to form right angles in his village. Another student's project noted how walking parallel to each other from one path to another when the paths were not at right angles and the two people walked away from the path but not perpendicularly, that a trapezium was formed. Despite the sides formed by the walking not being drawn, the student could visualise the overall shape.

9.4.2 Visuospatial Reasoning with Numbers

Visuospatial displays are often associated with the size of number in PNG as this is part of reciprocal exchanges. For example, Paraide (2010) showed some of the display of "bride price" exchange at a marriage. Interestingly, these bundles and rings are Fig. 9.15.

This sense of large numbers is significant in terms of estimates and values. People immediately recognise the amount in the displays while the displays

Fig. 9.15 Paraide's (2010) study of shell money in fathoms displayed in bundles and rings of 100 and 150 fathoms. *Source* Paraide (2010)

themselves add value and respect to the exchange. In addition, generosity is evident when the fathom length is generously larger than it should or used to be. The groupings involved in each display are also significant. For example, they can be in 10s or 100s. In addition, there are displays of bunches of bananas and baskets of food like sweet potato, taro and green leaves.

Counting is often accompanied by gestures. For example, people will often use the right hand to bend down the fingers of the left hand from the little finger to the thumb as they count 1, 2, 3, 4, 5 or the same of pairs or tens. Body-part tally is another visuospatial display in which parts of the body are used to signify numbers in order from the little finger up the arm and across the head and down the other side (Owens et al. 2017; Saxe 2012). The particular parts varies from language to language (Dwyer and Minnegal 2016).

Visuospatial reasoning is evident, for example, by pointing to 10 on their body when they tally with body parts especially when decimal counting and currency arrived in the village. How far a larger number is past the ten is quickly associated with the appropriate body part and it can also be a combination of smaller numbers with practice. Saxe (2012) also noted that the complete group for the Oksapmin had changed from 26 to either 20 or in a few cases 30, still represented by a body part as part of the body-part counting system.

In other languages, whole complete groups or groups of 4, 5, 8, 10, 12, or 20 may be accompanied by an action or object grouping. For example, in Hagen, people count to four on one hand, up to eight on the other and then slap the fists together for the complete group of 8. If they want to then count to 10 they will bend down the thumbs or in Gawigel dialect wipe these on their lips for the complete group of 10 in a combination of two counting systems with different cycles, the cycles of 8 dominating at least to $8 \times 8 \times 8$. In Waghi, people point to different body parts to represent different hundreds. When counting items, men will often stride down the lines of pigs or gifts gesturing to the items as they count. In Hagen and other areas, one person will to ten while another man keeps track of the tens.

9.4.3 Visuospatial Reasoning with Measurements

While a unit of volume or area is not specified, represented, nor generally used in these many different language communities, people do see and make area units in woven objects or garden plots. For example in Fig. 9.12b, the drains in the garden show area squares. These are recognised quickly to indicate the size of the garden through visualising and comparing with other gardens made with similar drainage systems and squares at another time or place. In another community, the gardens may be long rectangular shapes but often they are also marked off with a tankard bush or banana for every group of plants (5 or 10 or 20 depending on the plants, garden or culture). The number of plants in a garden is quickly estimated by the space they take up.

For example, people will know how many plants in a nursery by just looking or in the cash crop plot. They can estimate how many coffee bean kilograms they will make from a tree, or balls of string from a particular size tulip tree. Similarly, women estimate how wide to make a *bilum* (continuous string bag—see images in Owens 2015) for the number of balls of string they have. When sailing, men know how far they are from land by the birds they see as some birds will fly further from land than others but the winds and swell will be taken into account in estimating the amount of time it might take to reach the land. The time is not measured in hours but in the change in position of the sun. Similarly, distances are imagined in terms of known distances especially when walking.

Interesting measures or comparisons are used for making traps. The fence to direct an animal inside needs to be of a specific height and length and at a slight angle. It has to be positioned on an animal track recognised again visuospatially. Similarly if a noose trap is to be made the parts and weights need to be well devised. While part of this may be through trial and error, nevertheless the actual design requires a strong visual image. The same can be said of the making of a canoe. The prow has a certain shape but no drawings and often no measurements are made. They are made and modified according to need and available trees. Even the selection of trees from the supply on one's land requires carefully remembering of the types, sizes, shapes, and positions of many trees and estimates of their future growth and use. People feel to get the thickness of the hull correct. They estimate the positioning of the outrigger visuospatially but they will test this for balance and make adjustments to the ties.

9.4.4 Position

Grid maps are not available in the bush or ocean settings of Papua New Guinea. Nevertheless, people can find their way through the jungle or across the seas. As in many island nations, the stars, swells, birds and the position of the sun during the day assist the navigators on the sea. There are multiple ways of telling position include time taken to walk, slopes of the land, particularities of plants and natural features. Reference lines such as a river or coast are used and various places on these are demarcated so that one can find a place considered in terms of distance from the river on the left or right looking downstream. Often places are denoted as a specific area, for example, quadrants of a valley or areas with different vegetation. The source and flow of the river are noted. Tracks that are less frequently used are noted by slight foot marks or cut branches where an earlier hunter has passed through and cleared the track of overhanging shrub. Songs and stories of the man whose track it is are recited en route connecting activity with specific features of the landscape. From a young age, children can find places in the bush to gather food such as nuts, fruits and mushrooms. They reason about the features of the environment as they search for them having learnt where certain things are likely to grow.

9.5 Visuospatial Reasoning and Decision Making

One might ask how decisions are made if measurements especially with area units are not counted or calculated. Although the garden plots in the highlands may be square or rectangular, there are many that are trapezium shaped while on the coast, much is planted more opportunistically such as where an old tree root might help to hold the soil in the heavy rains. Visuospatial memory of different areas with multiple features such as quality of soil and distance from the village are taken into account.

Decisions are around sharing according to customary relationships and expectations and the people who are involved in the discussions may use length measures as part of the discussions, comparisons of lengths, or visual area comparisons. Similar comparisons as mentioned above under ratios are made for a range of decision-making requirements.

9.6 School Programs to Maintain Cultural Visuospatial Reasoning

Many participants thought that there was no connection between what was done at home and what was done at school. The languages were different, explaining school mathematics in home language or the lingua franca was not easy. If schools brought these common knowledges into the classroom with models, children's geometry knowledge would be far advanced of the linear trajectories of western school mathematics, often dominated by labelling shapes. The following examples illustrate this point:

- Two ropes used to form a right-angled isosceles triangle, one rope for the sum of equal sides so the midpoint becomes the vertex with the right-angle, and the other the hypotenuse. These ropes can then be used for right-angles in other situations.
- The physical walk to make parallel lines with a stick held between the two people. Other straight lines such as paths cutting across the parallel lines will form a trapezium (non-parallel paths) or parallelogram (parallel paths) or rectangle (parallel, orthogonal paths).
- The use of equal sticks or half rope lengths to form joined squares.
- The use of equal diagonals to ensure a rectangle rather than a parallelogram.

These are new introductions to these specific shapes based on well-established visuospatial reasoning from cultural practices rather than static proto-type images that might be misleading (e.g. all triangles are equilateral triangles with a horizontal base). Furthermore, these activities also emphasise properties that might not always be noted in school. This practical approach to school mathematics requires a model of teaching that incorporates culture.

Two programs in PNG harness these cultural ways of thinking mathematically: an elective at the University of Goroka on *Mathematics, Language and Culture* for secondary teachers (Owens 2014) and an inquiry approach in elementary schools (Owens et al. 2015). The inquiry model brings these village activities into the classroom to ensure that school mathematics is built on this background knowledge and visuospatial reasoning.

9.6.1 Teacher Education

The former project involved secondary teachers in preservice or inservice courses to establish the cultural mathematics in a particular activity of a community, generally their own, and to detail the activity and the mathematics involved in the activity and to link it to the syllabus for Grades 7–10 or 9–12 indicating how the mathematical thinking involved in the cultural activity would support school mathematics learning of various topics. It was clear from the reports how proud students were of their Elders and community members who shared the details of the activities with them indicating how this activity not only engaged them in self-regulated learning but also in establishing their mathematical thinking identity from their cultural identity (Owens 2014).

One teacher told the seasons by where the sun rose over the mountain range to the east of the village (PNG is just south of the equator). A secondary teacher used the curved vines that looped between the hand-rail suspension vine and the walking platform of a suspension bridge as a metaphor for a sine wave, an adequate image of the sine wave. Another student used a diagram of the framework for a wig-headdress to consider a parabolic curve using the width of his finger as a unit of measure, although again it was hard to fit the parabola to the curve of the wig.

This beginning grasp of the school concepts came with considerable cultural and mathematical pride and was a good image for a teacher to create in preparing for what was to be a school mathematical shape.

9.6.2 An Inquiry Model of Teaching in Elementary Schools

The inquiry model for the inservice elementary teachers was developed to be used over a week of five one-hour lessons. It specifically took the cultural mathematics as a starting point for engaging beginning school students. The outline of the inquiry process is based on Murdoch (1998). The purpose is for children to think and do mathematics through activities linked to cultural practices but extended to school expectations, to go further with the idea and connect it to other mathematical ideas. Children are expected to have a sense of belonging with the new ideas in culture and school through a good transition from known to new ideas.

The teachers were encouraged to plan for a week. They were first to notice key ideas such as what the new pattern and relationship were and how that leads to solve a problem. The teacher needed to identify what the children already knew as often children knew more from cultural activities than the teacher considered relevant. They also had to take account of the child feeling comfortable with school mathematics and how they can make it more appropriate by considering a place or cultural activity to introduce and develop a school mathematical idea.

They had to consider resources more broadly than just the chalk board at the front and a book for the children to copy off the board. They had to think about places to visit and people who could assist with the cultural mathematics. They needed to plan readily available materials for exploring, comparing, measuring, recording, and modelling; game cards, spinners, paper, etc. that might also be used in the classroom especially to generate conversation and reinforcement of new mathematical knowledge. The days given in the model below were adaptable to the situation.

Day 1
Tuning In

- Motivating
  - real world experience such as going outdoors to see something
  - telling a story
  - showing a video
  - Examples:

    house building Jiwaka;
    triangle gardens;
    hand clapping game
- Planning to find out
  - What questions to ask?

- What processes and equipment can be used?
- What we know and how might we extend this knowledge?

Finding Out

- Observing, noticing, comparing, measuring, discussing mathematical patterns

Day 2
Sorting Out

- Discussing, modelling, comparing, making a table, drawing a diagram, finding same and different

Day 3
Going Further (Thinking more deeply)

- Applying to other numbers or another situation, reading and discussing the maths book, using symbols, playing a game, solving an open problem

Day 4
Making Connections

- Summarising the mathematics and linking to other mathematics, whole class discussion or story writing
- Taking Action
- Share at home, solve a real problem, apply to a game

Day 5
Sharing, discussing, reflecting, evaluating

- Children explain the mathematics, write a maths story, write their own summary, say what new mathematics they have learnt
- Teacher reviews and decides where to next

Teachers also had to consider assessment. Some had learnt to record whether students had achieved some behavioural objectives such as adding two-digit numbers. They mostly recorded from observation of oral answers in class and in older children from written answers to questions involving symbolic operations recorded on the chalkboard.

However, they were not aware of diagnostic assessment or interview or questioning assessment. In order to reinforce the details around how young children learn to count, do arithmetic and learn about space and geometry and measurement, the teachers were taught how to ask diagnostic interview questions. The questions were similar to those found in overseas diagnostic assessments such as Count Me In Too (NSW Department of Education and Training 1998). To plan learning, teachers need to assess what their children know by observing ways children try things, what they say, how they problem solve, what they write, what they ask to make clear or to extend their exploring.

The teachers planned and tried out asking open-ended questions that allowed children to give more than one answer according to their understanding. Students were encouraged to visualise to operate on groups of numbers such as adding onto the larger number. Interestingly the extensive use of gestures and counting with fingers and toes both assisted visualisation but not necessarily assisted in getting students to stop counting by ones.

Interestingly, the project encouraged teachers to ask the interview questions in Tok Ples. This meant that teachers really had to consider what the questions meant in terms of their cultural ways of thinking. This task indicated how a number of teachers had not made the link between home mathematics and school mathematics. To follow the meaning through, there were discussions around not only the meaning of specific words but also the visuospatial reasoning by which the teachers understood the words and/or the mathematics.

For example, Wahgi language speakers north of the major east-west flowing river had used the term "source" for north as their small rivers started in what the English speakers were calling north, a direction they did not need in their local language. However, other dialect speakers south of the major river had tributaries starting in the south so for Wahgi speakers there was confusion. Nevertheless, it is clear how the visuospatial reasoning occurred for the speakers.

In another area, the teachers from the western coastal area of Central Province were discussing the idea of an area unit and how they might be able to speak of that notion. Interestingly, they began to refer to a "group" of area units such as a row of squares. It was the same word that they used for a composite group of ten, their counting system being a base 10 system. When they discussed addition they used the words for "joining together", which is commonly used across PNG in different languages.

Subtraction led to further discussion with the two meanings of difference and take away. They tended to use the idea of "take away" which they favoured in their discussion to "separate into two groups" but they also were happy to consider difference in terms of "compare and find out how much more one group was".

Similarly visuospatial embodiments and actions are used in western school mathematics classrooms. However, it was the first time some of these teachers had really worked through the true meaning of "addition" and "subtraction", transliteration words from the English addition and subtraction.

9.7 Conclusion

Given the number of everyday mathematical activities in which people participate in PNG, it is unsurprising that they reason with their gestures and bodily movements in a range of mathematical conceptual zones. However, it is only by identifying that visuospatial reasoning is being used that a link can be made to school mathematics. This rich cultural mathematical way of thinking needs to be harnessed for advancement of students succeeding in school mathematics.

This chapter shows that visuospatial reasoning was evident when spatial and embodied imagery was used to reason. Many times the pursuits are repeated for learning as contexts might require modifications. For example, the winds and swells vary so that learning to navigate and sail needs many visuospatial perceptions and reasons to be considered. Activities requiring spatial and embodied imagery could be encouraged in school mathematics by having students experience concepts in a physical way such as in outdoor mathematics or modelling cultural activities, walking trajectories for functions or illustrating concepts with gestures.

Typically angles appear to lend themselves to these approaches. For example, the consequences of a modification in an activity such as a ball rolling down a slope or bouncing off a wall require visuospatial reasoning by associating the visual trials and results with the language associated with the angles. Volume and mass activities would also strengthen understanding of these concepts, especially in distinguishing them from the dominance of length. Such activities encourage comparison and informal ratio ideas. For example, when one factor such as length is increased, the resultant comparative changes in volume and mass are estimated and compared.

To build model structures that work best when symmetry assists with balance and stresses encourages strong visuospatial reasoning. Actually participating in group activities to work collaboratively ensures estimations and comparisons are communicated in the reasoning process. A number of shape properties could be recognised. For example, the equal radii of circles, the lines of symmetry denoted by centre points of lines, and equal areas of opposite faces of a prism may be visualised and discussed. These kinds of physical activities require problem solving and hence the vehicle for using visuospatial reasoning.

Similarly the use of visual displays of numbers could strength the conceptualisation of numbers and could be set within a story or dramatisation. Such opportunities may strengthen multiplicative thinking and an understanding of large numbers. At the same time the cultural mathematics is valued and cultural beliefs and systems acknowledged and intertwined.

Mapping too is best associated with outside physical activities that could be representing a larger area of interest to the children such as their community, its roads and tracks, and landmarks. Nevertheless, the links from culture and environment to school mathematics need to be set into a mathematical inquiry going further from one setting into a new setting that is emphasising the structures of school mathematics, its representative methods, and its vocabulary.

These examples illustrate how visuospatial reasoning as advanced in this chapter recognises a virtually untapped resource made evident by ethnomathematics. An emphasis on visuospatial reasoning advances our general understanding of mathematics. This chapter illustrates the meaning of visuospatial reasoning and its ubiquitous but varied role in mathematical thinking. This kind of thinking needs to be recognised worldwide as a significant and legitimate way of thinking mathematically. It has analytic value as well as conceptual and memory value for mathematics and mathematics education.

If problem solving and visuospatial reasoning take precedents over rote-learning content, then teachers are not blinkered by school mathematics and are more able to incorporate cultural mathematics into the school classroom. The advantage of linking culture to school mathematics is preservation of culture, use of cultural identity to promote mathematical identity, and better school mathematics education (Owens 2014).

References

Bishop, A. (1983). Space and geometry. In R. Lesh & M. Landau (Eds.), *Acquisition of mathematics concepts and processes* (pp. 176–204). New York, NY: Academic Press.

Campbell-Jones, S. (Writer). (1996). *Horizon*: *Twice five plus the wings of a bird*. London, UK: BBC.

Dreyfus, T. (1991). On the status of visual reasoning in mathematics and mathematics education. In F. Furinghetti (Ed.), *Proc. 15th Conference of International Group for the Psychology of Mathematics Education* (Vol. 1, pp. 33–48). Italy: PME.

Dwyer, P., & Minnegal, M. (2016). Counting systems of the Strickland-Bosavi languages, Papua New Guinea. *Language and linguistics in Melanesia, 34*(1), 1–36.

Eliot, J. (1987). *Models of psychological space: Psychometric, developmental, and experimental approaches*. New York, NY: Springer-Verlag.

Goldin, G. (1987). (a) Levels of language in mathematical problem solving; (b) Cognitive representational systems for mathematical problem solving. In C. Janvier (Ed.), *Problems of representation in the teaching and learning of mathematics* (pp. 59–65, 125–145). Hillsdale, NJ: Lawrence Eribaum.

Goldin, G. (1998). The PME working group on representation. *Journal of Mathematical Behavior, 17*(2), 283–301.

Lábadi, B., Horváth, D. Á., & Palotai, R. (2012). Do infants encode feature and geometry information in a two-dimensional space? *Infant Behavior and Development, 35*(3), 408–416.

Learmonth, A. E., Newcombe, N. S., Sheridan, N., & Jones, M. (2008). Why size counts: Children's spatial reorientation in large and small enclosures. *Developmental Science, 11*(3), 414–426.

Lohman, D. F., Pellegrino, J. W., Alderton, D. L., & Regian, J. W. (1987). Dimensions and components of individual differences in spatial abilities. In S. H. Irvine & S. E. Newstead (Eds.), *Intelligence and cognition* (pp. 253–312). Dordrecht, The Netherlands: Nijhoff Publishers.

Mason, J. (2008). Being mathematical with and in front of learners: Attention, awareness, and attitude as sources of differences between teacher educators, teachers and learners. In T. Wood (Ed.), *International handbook of mathematics teacher education* (Vol. 4, pp. 31–56). Rotterdam, The Netherlands: Sense Publishers.

Murdoch, K. (1998). *Classroom connections: Strategies for integrated learning*. Armadale, Victoria, Australia: Eleanor Curtain.

Ness, D., & Farenga, S. (2007). *Knowledge under construction: The importance of play in developing children's spatial and geometric thinking*. Lanham, MD: Rowan & Littlefield.

NSW Department of Education and Training. (1998). *Count me in too*. Retrieved from http://www.curriculumsupport.education.nsw.gov.au/primary/mathematics/countmeintoo/index.htm

Owens, K. (1992a). Spatial mathematics: A group test for primary school students. In K. Stephens & J. Izard (Eds.), *Reshaping assessment practices: Assessment in the mathematical sciences under challenge*. Melbourne, Australia: Australian Council for Education Research.

Owens, K. (1992b). Spatial thinking takes shape through primary school experiences. In W. Geeslin, & K. Graham (Eds.), *Proc. 16th Conference of the International Group for the Psychology of Mathematics Education* (Vol. 2, pp. 202–209.). University of New Hampshire, Durham, NH: PME.

Owens, K. (1993). *Spatial thinking processes employed by primary school students engaged in mathematical problem solving* (PhD). Deakin University, Geelong, Victoria, Australia. Retrieved from http://dro.deakin.edu.au/eserv/DU:30023339/owens-spatialthinking-1993.pdf

Owens, K. (2004). Improving the teaching and learning of space mathematics. In B. Clarke, D. Clarke, G. Emanuelsson, B. Johansson, D. Lambdin, F. Lester, A. Wallby, & K. Wallby (Eds.), *International perspectives on learning and teaching mathematics* (pp. 569–584). Gothenburg, Sweden: Göteborg University National Center for Mathematics Education.

Owens, K. (2014). The impact of a teacher education culture-based project on identity as a mathematics learner. *Asia-Pacific Journal of Teacher Education, 42*(2), 186–207. doi:10.1080/1359866X.2014.892568

Owens, K. (2015). *Visuospatial reasoning: An ecocultural perspective for space, geometry and measurement education*. New York, NY: Springer.

Owens, K., Edmonds-Wathen, C., & Bino, V. (2015). Bringing ethnomathematics to elementary teachers in Papua New Guinea: A design-based research project. *Revista Latinoamericana de Etnomatematica, 8*(2), 32–52.

Owens, K., McPhail, D., & Reddacliff, C. (2003). Facilitating the teaching of space mathematics: An evaluation. In N. Pateman, B. Dougherty, & J. Zilliox (Eds.), *Proceedings of 27th Annual Conference of the International Group for the Psychology of Mathematics Education* (Vol. 1, pp. 339–345). Honolulu, Hawaii: International Group for the Psychology of Mathematics Education.

Owens, K., Lean, G., Paraide, P., & Muke, C. (2017). *The history of number: Perspective from Papua New Guinea and Oceania*. New York. NY: Springer.

Paraide, P. (2010). *Integrating Indigenous and Western mathematical knowledge in PNG early schooling* (PhD). Deakin University, Geelong, Victoria, Australia.

Pirie, S., & Kieren, T. (1991). Folding back: Dynamics in the growth of mathematical understanding. In F. Furinghetti (Ed.), *Proceedings of the 15th annual conference of the International Group for the Psychology of Mathematics Education* (Vol. 3, pp. 169–176). Assisi, Italy: Program Committee for the International Group for the Psychology of Mathematics Education.

Poltrock, S., & Agnoli, F. (1986). Are spatial visualization ability and visual imagery ability equivalent? In R. J. Sternberg (Ed.), *Advances in the psychology of human intelligence* (Vol. 3, pp. 255–296). Hillsdale, NJ: Lawrence Erlbaum.

Presmeg, N. (1986). Visualisation in high school mathematics. *For the Learning of Mathematics, 6* (3), 42–46.

Presmeg, N. (2006). Research on visualization in learning and teaching mathematics. In A. Gutiérrez & P. Boero (Eds.), *Handbook of research on the psychology of mathematics education* (pp. 205–304). Rotterdam, The Netherlands: Sense Publishers.

Rivera, F. (2011). *Towards a visually-oriented school mathematics classrooms: Research, theory, practice, and issues*. New York, NY: Springer.

Saxe, G. (2012). *Cultural development of mathematical ideas: Papua New Guinea studies*. New York: Cambridge University Press

Shepard, R. (1975). Form, formation, and transformation of internal representations. In R. Solso (Ed.), *Information processing and cognition: The Loyola Symposium* (pp. 87–122). Hillsdale, NJ: Lawrence Erlbaum.

Sinclair, N., Bartolini Bussi, M. G., de Villiers, M., Jones, K., Kortenkamp, U., et al. (2016). Recent research on geometry education: An ICME-13 survey team report. *ZDM* (pp. 1–29). doi:10.1007/s11858-016-0796-6

Steffe, L. (Ed.). (1991). *Epistemological foundations of mathematical experience*. New York, NY: Springer-Verlag.

Tartre, L. (1990). Spatial skills, gender, and mathematics. In E. L. Fennema, G. (Ed.), *Mathematics and gender*. New York, NY: Teachers College Press.

Uttal, D., et al. (2013). The malleability of spatial skills: A meta-analysis of training studies. *Psychological Bulletin, 139*(2), 352–402. doi:10.1037/a0028446

Chapter 10
Cultural and Mathematical Symmetry in Māori Meeting Houses (Wharenui)

Tony Trinick, Tamsin Meaney and Uenuku Fairhall

Abstract This chapter draws on the symbolism found in various artefacts in *wharenui* (meeting houses), with a particular focus on Rauru, a traditional Māori meeting house located in Hamburg, Germany, the location also of ICME 13, to illustrate an aspect of ethnomathematics: cultural and mathematical symmetry. This addresses a gap in Māori-medium education whereby much of the focus to date has been on revitalising the endangered Indigenous language of Aotearoa/New Zealand, *te reo Māori*. In contrast, reviving the associated cultural knowledge has been somewhat stymied for three key reasons: the dwindling number of elders with the knowledge, the tension associated with transposing traditional tribal knowledge to contemporary learning environments and resistance on the part of state agencies to acknowledging Indigenous knowledge. After 100 years of cultural and linguistic assimilation, reviving cultural knowledge is a big challenge for marginalised groups. Fortunately, aspects of both cultural and mathematical knowledge remain embedded in highly valued artefacts such as the meeting house, and are thus able to be reconnected in the contemporary Indigenous mathematics classroom. However, there is a need to better understand how the cultural significance can be connected to mathematical understandings in a way that gives value to both.

Keywords Ethnomathematics · Cultural & mathematical symmetry · Endangered indigenous languages · Indigenous language schooling · Mathematics classrooms

---

T. Trinick (✉)
The University of Auckland, Auckland, New Zealand
e-mail: t.trinick@auckland.ac.nz

T. Meaney
Bergen University College, Bergen, Norway
e-mail: Tamsin.Jillian.Meaney@hib.no

U. Fairhall
Te Kura o Te Koutu, Rotorua, New Zealand
e-mail: uenuku@koutu.school.nz

M. Rosa et al. (eds.), *Ethnomathematics and its Diverse Approaches for Mathematics Education*, ICME-13 Monographs,
DOI 10.1007/978-3-319-59220-6_10

10.1 Background to Māori-Medium Schooling

In this chapter, we explore the ethnomathematics connected to the symmetry found in a Māori meeting house called Rauru, relocated from New Zealand to Hamburg in Germany more than 100 years ago, and discuss the implications for the use of cultural artefacts in mathematics. Though a traditional meeting house from that period, much of its symmetrical nature is *reflected* in contemporary meeting houses, making it possible to transfer traditional understanding from this investigation to the more contemporary houses located throughout New Zealand and into mathematics classrooms.

This exploration is part of a wider Indigenous mathematics education project located in Aotearoa/New Zealand which responds to the challenge of reconnecting Indigenous knowledge to formal schooling. Such investigations are important because, although the Indigenous language of Aotearoa/New Zealand is now used successfully in Māori-medium schooling as part of an ongoing revitalisation project, much less work has been done on overcoming the challenges associated with revitalising Māori knowledge connected to mathematical ideas. One issue is to determine how to value the cultural and mathematical knowledge in contemporary settings such as schools, especially when these are isolated from traditional contexts, while simultaneously recognising that culture is not static but changes to meet contemporary challenges.

For Māori, the issue of access to Indigenous knowledge, how the knowledge is transmitted and where the knowledge is learnt and by whom was disrupted by colonisation in the nineteenth century. When the first missionaries and settlers arrived in Aotearoa/New Zealand, Māori had a robust system for educating their children in relevant and appropriate cultural contexts to ensure the survival of their communities (Riini and Riini 1993). After 1840, with the arrival of more and more European settlers and the establishment of the British colony, European forms of government were imposed alongside missionary schooling.

The hegemonic function of the missionary schools was to provide a formalised context to assimilate Māori communities into European beliefs, attitudes and practices, with the intent to "civilise" the Māori population (Simon 1998). The goal of assimilation was maintained by successive governments and their agencies over the next 100 years, resulting in a range of educational policies, both overt (English-language-only schooling policy) and covert (English-language-only workplaces), to privilege English as the sole language of education.

Consequently, mathematics was taught solely in the medium of English and was based exclusively on Western mathematical practices. For Māori students to progress through the education system to higher education, they had to achieve a variety of benchmarks (matriculation) at particular ages, such as transitioning from primary school to secondary, and then from junior secondary to senior secondary school in mathematics. For many native speakers of the Māori language, learning Western knowledge in an additional language was a challenge that they could not overcome and they exited the schooling system at a much earlier age than their European peers, generally into menial work with low pay (Simon 1998).

Bishop (1990) used the expression *cultural imperialism* in describing how "western mathematics' [has been] one of the most powerful weapons in the imposition of western culture" (p. 51). For colonised and marginalised Indigenous groups such as Māori, this view of mathematics and mathematics education resonates. Nevertheless, it may be the case that if schooling in Aotearoa/New Zealand had been allowed to continue to develop in the Indigenous language, from the colonial era into modern times, then some form of mathematics, related to what is currently taught in schools, would have been present.

This is because, from the Māori perspective, obtaining European knowledge, in particular, the knowledge associated with technology and trade, could enhance their traditional ways of life (Spolsky 2005). It may also have been the case that multiple, hybrid forms of mathematics would have been present, depending on the needs of individual communities or *iwi*[1] (tribe). Lamentably, for over 100 years Māori were denied the opportunity to incrementally develop subjects such as mathematics in their Indigenous language based on their cultural perspectives for schooling and higher education.

The change in the status of *te reo Māori* (the Māori language) to a low-status language in Aotearoa/New Zealand, including through its exclusion from use in schools, contributed to the language shift to English in the Māori community, to such an extent that by the 1970s *te reo Māori* was considered an endangered language (Spolsky 2005). It was against this background of rapid and significant language loss that the Māori community initiated Māori-medium education, in particular, *kura kaupapa Māori* schools such as Te Koutu (Meaney et al. 2012).

Initially, *kura kaupapa Māori*, a grassroots initiative, were developed from outside the state system, not only to revitalise *te reo Māori* but also as a resistance movement against the assimilationist nature of New Zealand European schooling (Penetito 2010). Along with language revitalisation, cultural knowledge revitalisation was also an aim for *kura kaupapa Māori* from the early days of its development (Smith 2004). However, as we discuss later, the revitalisation of mathematical practices embedded in cultural knowledge has somewhat languished in comparison with language revitalisation.

By the 1990s, *kura kaupapa Māori* agreed to become state funded in order to minimise the financial drain on the community, which was often manifested in, among other things, students being located in sub-standard dwellings for their schools. However, becoming state funded proved to be a double-edged sword, in that *kura* were required to implement state-mandated curricula, such as *pāngarau* (mathematics) and assessment practices developed from essentially Eurocentric interests (McMurchy-Pilkington and Trinick 2008). Furthermore, the Ministry of

[1]Māori terms are italicised with the English translations following in brackets for their first mention.

Education insisted that the structure of the first Māori-medium mathematics curriculum "mirror" the hegemonic English-medium version (McMurchy-Pilkington and Trinick 2008).

The old colonial hierarchies of European versus non-European remain in place. A number of studies show how imperial power is still exercised well after colonialism in an interconnected matrix of power that includes hierarchies of political, epistemic, economic, spiritual, linguistic and racial forms of domination where the racial/ethnic hierarchy of the European continues to position Indigenous knowledge as the *other* or inferior knowledge (Grosfoguel 2002).

As a consequence of these lingering Eurocentric interests, a major opportunity was denied to the Māori-medium schooling community to interrogate the place of traditional mathematics practices in a contemporary mathematics curriculum. At this time, there was still a pool of native speakers with the relevant knowledge to participate in and lead these discussions. However, these restrictions in curricula development have eased over time, as a response to changes in the prevailing discourse led by Māori campaigning for a greater say in the education of Māori children. Albeit slowly, such changes have been integrated into government policy about enabling Māori to live as Māori in contemporary society (McMurchy-Pilkington et al. 2013).

While providing considerable state support to elaborate the Māori language in a systemised way (McMurchy-Pilkington et al. 2013), thus enabling the teaching of mathematics in the medium of Māori to senior secondary school levels, the goal of supporting Māori to live as Māori has been much more challenging in terms of revitalising traditional knowledge and culture. Opportunities to formally advance traditional Māori cultural knowledge or alternative ways of thinking about mathematics have been lost. Whereas support for developing Māori language could easily be connected to improving students' mathematical achievement in assessments, something that successive governments have been keen to promote, there is still a belief held by many in positions of power in New Zealand that promoting non-Western knowledge is not in the interest of the state, particularly economic interests.

As well, as time has gone on, the access to native speakers who have the relevant knowledge for cultural revival has changed. The broad pool of elders who provided significant input into language revitalisation in the 1980s is no longer with us (McMurchy-Pilkington et al. 2013). Thus, the younger generation of speakers, often L2 learners who have been attempting to revitalise Māori knowledge have had to defer to the written historical record. However, identifying and understanding traditional practices via the written record have complications. Paradoxically, preserving the *authenticity* of traditional knowledge has often resulted in cultural knowledge being seen as unchanging and unchangeable.

Despite these challenges, our contention is that revitalisation and maintenance of the language must be considered insufficient unless cultural knowledge is also revitalised and maintained. Cultural knowledge and language loss as well as their revitalisation are human rights issues, connected to issues of power relations around who gets to determine what should be lost or saved (May 2005). The loss of a language and culture reflects the exercise of power by the dominant over the disenfranchised, and is concretely experienced "in the concomitant destruction of intimacy, family and community" (Fishman 1991, p. 4). However, in our view, revitalisation cannot be achieved if both language and culture are revered as fossilised, untouchable museum pieces, rather than aspects of Māori life that have changed and will continue to change to meet new circumstances.

10.2 The Relationship Between Culture, Language and Mathematics

The relationship between language and culture is deeply rooted. Languages are the repositories of cultural knowledge about the world, built up over many thousands of years of observations and experience, and it is argued that this knowledge is of benefit to all humankind (Chrisp 2005; Hale 1992). Consequently, language loss can be viewed as an erosion or extinction of ideas, of ways of knowing and ways of talking about the world, and is a loss, not only for the community of speakers itself, but for human knowledge generally (Harrison 2007).

Fishman (1991) noted that, traditionally, the primary argument for language maintenance in sociolinguistic work is that culture and language "stand for each other" (p. 22). It is argued that languages are a fundamental part of a people's culture (Lemke 1990). They relate to local customs, beliefs, rituals and the whole display of personal behaviours (Crystal 2003). Fishman (1991) also presented the idea that most of the culture is in the language and is expressed in the language. He further added, "take language away from the culture, and the culture loses its literature, its songs, its wisdom, ways of expressing kinships relations and so on" (Fishman 1991, p. 72). Crystal (2003) linked language to the issue of identity: "if we want to make sense of a community's identity, we need to look at its language" (p. 39). Therefore, when a community loses its language, it often loses a great deal of its cultural identity.

As noted, while there have been significant gains in addressing Māori language loss, the same cannot be said about reviving Indigenous knowledge. Nationally and internationally, incorporating cultural elements connected to mathematics education or mathematics is often met with strong resistance (Barton 2008; Ernest 1991), particularly from those who view mathematical thought as culture free (see discussions of this in Bishop 1994; Gerdes 1988).

This was certainly the approach taken by European educators to justify the exclusion of Māori knowledge and culture from the classroom for over 100 years, Western mathematics is universal, Indigenous particular to the group only. As Ernest (1991) pointed out, traditionally in Western mathematics philosophy based on hegemonic European paradigms, mathematical knowledge has been understood as universal and absolute, with the structure and objects of mathematics existing outside of human invention. A number of mathematicians, such as Thomas (1996), strongly argued that the contextualising of mathematics, such as is the case of ethnomathematics, needed to be resisted so that it did not become a watered-down version of mathematics in regard to what it should be considered to be and to be able to do.

In contrast, a number of scholars, including Bishop (1988), have long argued that mathematics is a cultural product arising from participation in various activities such as locating, designing, measuring and so on. For example, concepts such as rotation are often considered culture free. Yet, when examining this concept in Western mathematics, rotation is always considered to happen in a clockwise direction, unless stated otherwise. This suggests that a convention has developed over time within a societal group and thus the cultural history of mathematics can be recognised. When ideas are decontextualised and abstracted, understandings about these activities can be applied in a range of situations and so can be considered universal (Bishop 1988). However, the cultural connections still need to be recognised.

As a result, ethnomathematics has evolved since the 1980s to express the relationship between mathematics and culture (D'Ambrosio 1999). Ethnomathematics research examines a diverse range of ideas, including numeric traditions and patterns, as well as education policy and pedagogy in mathematics education. "One of the goals of ethnomathematics is to contribute both to the understanding of culture and the understanding of mathematics, and mainly to lead to an appreciation of the connections between the two" (D'Ambrosio 1999, p. 146).

Nevertheless, concerns have been raised about programmes that only consider the cultural nature of mathematics and the mathematics in cultural artefacts. For example, several authors have questioned whether ethnomathematics challenges the colonial structures imposed by the cultural imperialism of mathematics, because of its reliance on a comparison with Western mathematics (Meaney 2002). Pais (2011) suggested that although learners may engage in a range of activities, it is not until these activities are recognised as mathematics that they *become* mathematics and thus come to be considered valuable. Labelling traditional activities as mathematics runs the risk that they will be seen as having no intrinsic value in their own right, except as potentially Indigenous examples of Western knowledge (Roberts 1997).

Contributing to this process, ethnomathematical practices tend to be described in the mathematics register of the language of instruction and/or the language of the researcher rather than the language or dialect of the cultural activity. While it is important to make research accessible and contestable through describing it in international languages such as English, as Stillman and Balatti (2000) warned, this process potentially "divorces the cultural practices from their context and trivialises and fragments them from their real meaning in context" (p. 325).

If ethnomathematics is to support the decolonising of cultural knowledge, then there is a need to recognise that cultural considerations are as important as mathematical ones. For this to happen there is a need to define culture. In his seminal book, Geertz (1975) stated:

> Culture is best seen not as complexes of concrete behaviour patterns—customs, usages, traditions, habit clusters—as has, by and large, been the case up till now, but as a set of control mechanisms—plans, recipes, rules, instructions (what computer engineers call "programs")—for the governing of behaviour (p. 44).

If this is the case, then there is a need to make students aware that in learning about ethnomathematics, they are engaging in learning and responding to mathematics as well as how to learn and respond to the traditional understandings embedded within the cultural activity. This includes understanding the implications in regard to the language and culture, such as that a cultural practice or activity can change over time to meet contextual challenges.

This chapter continues two of the trends within ethnomathematics research—the investigation of mathematics in a non-Western culture, namely Māori (Gerdes 1986; Zaslavsky 1979), and the political and pedagogical trend challenging the colonial structures that imposed and maintained Eurocentric ideas and values (Bishop 1990; D'Ambrosio 1985). In addition, this chapter explores how ethnomathematics can be incorporated into the curriculum and simultaneously used to revitalise Indigenous cultural knowledge, without revering this knowledge as a museum artefact, frozen in time, and without denying the usefulness of Western mathematics. Our aim is not to position one knowledge as superior or inferior in some way, but to draw on cultural knowledge(s) that best support(s) the aspirations of children and families participating in Māori-medium education.

This investigation builds on our previous work in this area. We have noted for some time that Indigenous families are likely to have valuable insights into cultural practices that could be incorporated into mathematics lessons (Meaney and Fairhall 2003). Previously, we have described activities that Uenuku has used in his school, such as the mathematics activities in the division of land among descendants (Meaney et al. 2008, 2012) and traditional understandings about location and direction (Trinick et al. 2015, 2016).

In these papers, we emphasised how mathematics added value to the activity, rather than that the activity was valuable only because of the mathematics connected to it. Discussing the symmetry in meeting houses extends this research further in that we consider in greater detail how cultural knowledge can be revitalised in appropriate ways by connecting it to mathematics.

10.3 Symmetry in Traditional Meeting Houses

Wharenui (meeting houses) are important components of the *marae* (collection of buildings and sacred grounds) where many rituals and ceremonies take place (Hāwera and Taylor 2014). Nowadays, the term *marae* evokes two related meanings. In the first place, *marae* is used to denote an open space, a clearing or plaza in front of a meeting house, reserved and used for Māori assembly, particularly ceremonies of welcome. In the second place, the concept of *marae* is used in the broader sense for the combination of the *marae* proper, the courtyard and a set of communal buildings which normally include a meeting house, a dining hall and so on. As Selby (2009) wrote, "The marae is also a powerful symbol of place, of home, of belonging, of tradition and provides a link between today, yesteryear and the future" (p. 7).

Wharenui and *marae* are seen as "going together" in more than one way (Metge 1976, p. 230). The complementary relationship between *marae* and *wharenui* is often expressed by analogy with the gods of war and peace. Traditionally, the *marae* was the area of *Tū-matauenga*, the god and father of war, whereas the *wharenui* was associated with *Rongo-ma-tane*, the ancestor of the *kūmara* (sweet potato) and the god of all other cultivated food as well as the god of peace.

A *wharenui* is usually the dominant feature of any *marae* complex. *Wharenui* are generally named after an ancestor, of either gender of a particular group, and their structure frequently represents the body of an important ancestor, often the eponymous ancestor (Fig. 10.1). *Wharenui* are predominately rectangular, with a gabled roof and a front veranda, often, but certainly not always, marked by embellishments of carving, curvilinear rafter patterns (*kōwhaiwhai*) painted in black, red and white with lattice-work panels (*tukutuku*) on the wall. Houses range in length from approximately 12 or 13 m to nearly 30 m.

In many tribal areas, the rituals of engagement with other visiting groups, tribes and so on, often take place on the front veranda or just in front of it. The *tāhuhu* (ridge beam) represents the backbone, the *heke* (rafters) the ribs and the *maihi* (barge boards) the arms. The front window is seen as the *matapihi* (eye), and the interior of the meeting house is the *poho* (chest) of the ancestor (Fig. 10.1 provides an outline of the inside of a *wharenui*). Inside, the meeting house is divided into two separate yet complementary domains: the *tarawhānui* (the *big* side) and the *tarawhāiti* (the *little* side). The *tarawhāiti* is generally reserved for the home people and *tarawhānui* for visitors (Fig. 10.1). Within the spatial orientation of meeting houses, special places of honour are allocated to distinguished guests or family

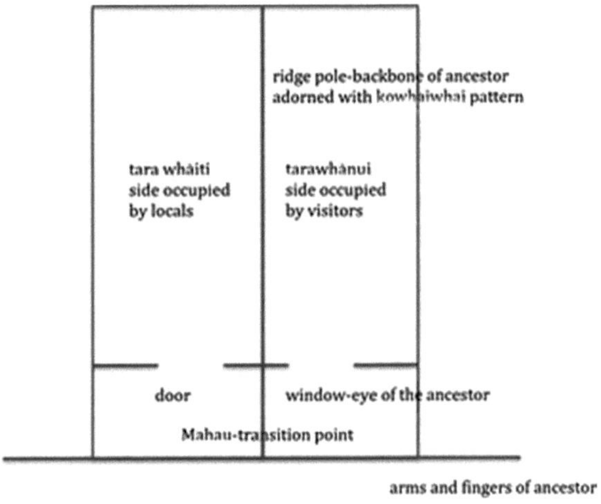

Fig. 10.1 Symbolic structure of meeting house. *Source* Personal file

elders. Often, this is by the front window. Both the size and the degree of ornamentation of a *wharenui* say something about the *mana* (prestige) of its owner group.

Wharenui are bilaterally symmetric and because they often represent ancestors they have reflection in the sagittal plane, which divides the body of the house (thus ancestor) vertically into left and right halves with the exception of the front door and window (Fig. 10.1). Bilateral symmetry, the one most commonly used in *wharenui* designs, is where an axis of symmetry divides a shape into equal halves (Booker et al. 2010). As is discussed in the following sections, *wharenui* are highly decorated with symmetrical patterns in different positions, used in a range of artefacts. One of the challenges of working with *wharenui* is that the parts of the *wharenui* where particular symmetrical patterns are located need to be the focus of mathematical discussions; at the same time, they should not be disconnected from the culture embedded within the *wharenui*.

Wharenui, built from about the middle of the nineteenth century, were redesigned from earlier versions, so that among other things they could hold the political community meetings for discussing colonisation of Māori land (Jackson 1972; McCarthy 2005). From a cultural perspective, the *wharenui* and the various symmetrical artefacts that adorn it generally represent a family's links to an ancestor (Jackson 1972; Salmond 1978).

Ironically, both the state agency responsible for schooling and schools themselves, even with large Māori populations, resisted the erection of *wharenui* on their premises for many decades. As a result of strong lobbying from Māori communities and Māori teachers, policy was eventually changed so that such decisions defaulted to schools if they were prepared to pay for them. Most Māori-medium schools, including Te Koutu, where Uenuku is principal, have meeting houses located on their premises. Therefore, students may be familiar with them as a cultural construct, although not necessarily as a mathematical one.

Many mathematics educators highlight the importance of teaching symmetry as it is embedded in reality, and this is considered to help students connect geometry with their real-life experiences (Leikin et al. 2000). Potentially, symmetry connects to both cultural and mathematical aspects, and this was the reason why *wharenui* were chosen as the focus for this chapter.

However, the use of symmetrical designs found in *wharenui* as contexts for teaching school mathematics has had a somewhat tumultuous history in Aotearoa/New Zealand, as have other concepts decontextualised from their Māori contexts (Anderson et al. 2005). In the 1980s, with the first endeavours and good intentions to make connections to Māori culture in mathematics, some of the symmetrical patterns within the *wharenui* found in the community rather than on school grounds were identified and used as examples of cultural mathematics (Knight 1984a, b; MacKenzie 1989). However, by the 1990s, the sole focus on patterns as mathematics objects was disparaged as insufficient to count as ethnomathematics (Barton 1993).

Barton (1993) stated, "in our search for cultural mathematics, it is not enough to find examples of mathematics in use—we need to study systems which describe patterns and make powerful generalisations that can be used in more than one practical application" (p. 60). Some of the earlier dissatisfaction with using the symmetry found in *wharenui* came from the abstraction, which enabled students in English-medium schools to gain mathematical understandings without visiting *wharenui* or needing any knowledge of the cultural significance of those patterns. Some felt that in this way an iconic cultural focus had been relegated to a desultory mathematical activity. In recent times, the symmetrical nature of some of the designs found in *wharenui* are again being incorporated into Māori-medium mathematics lessons, but this time in connection with cultural understandings (Manuel et al. 2015).

In the next section, we discuss the symmetrical and cultural knowledge connected to one *wharenui* and suggest some strategies for ensuring that it would add value to Māori students' understanding of their mathematics and culture.

10.4 The Wharenui (Meeting House)—Rauru

We draw on examples of symmetry in Rauru, a traditional meeting house built around 1900 and now located in the Hamburgisches Museum für Völkerkunde, Hamburg, Germany, where the International Congress of Mathematics Education (ICME 13) was held (information about it can be found at http://www.voelkerkundemuseum.com/247-1-Maori-Haus.html). A number of ancient *wharenui* are also found within museums within Aotearoa/New Zealand. In discussing the placement of *wharenui* within museums in Aotearoa/New Zealand, McCarthy (2005) discussed the relationship between the impact of a building (the *wharenui*) within a building (the museum) as a movement from one culture, Māori, into another, Western/Pākehā:

> The exhibited *wharenui* (Hotunui, Te Hau ki Turanga, or Mataatua) bring their own contextual and interior space into the museum. These spatialities test the conceptual perimeters of the museum as a building which determines relationships between inside and outside. These *wharenui* also exist as venerated artefacts which have been translated into and consumed as museum objects (p. 80).

Wharenui encased within museums have an impact on visitors which is different to their viewing of the smaller artefacts housed in isolated, glass cases. They also often have histories that are different to those of *wharenui* still in use on ancestral land. This is the case for Rauru, the meeting house, which is named after Rauru the son of Kuraimonoa and her husband Toi-te-huatahi (Fig. 10.2). Rauru is synonymous with the art of carving in Te Arawa, a Māori tribal group (Thomas et al. 2009). Te Koutu, the school of which Uenuku Fairhall is principal, is located on Te Arawa traditional land, and most staff and students belong to this tribe. As is

Fig. 10.2 Two of the authors at Rauru in Hamburgisches Museum für Völkerkunde. *Source* Personal file

discussed later, members of Te Arawa retain a strong connection to Rauru, even though it is situated on the other side of the world, in Germany.

The carvings, *kōwhaiwhai*, painted along the ridge poles and *tukutuku* (woven panels) represent Te Arawa ancestors, heroes and gods, including Tāne-Te-Pupuke, Kataore, Hinenui-Te-Pō and Māui (Thomas et al. 2009). Māui is represented in the narrative of fishing up the North Island of New Zealand (Fig. 10.7). *Kōwhaiwhai* refers to the traditional red, white and black coloured patterns most often found on the ridgepole or rafters in meeting houses such as Rauru.

Tukutuku panels sit between the posts (Figs. 10.2, 10.4 and 10.5) and are an integral part of the story of the meeting house. While the *pou* (posts) are carved by men, the *tukutuku* panels are woven by women (Jackson 1972). As Jackson (1972) stated, although there are distinctions of this kind present in the artefacts within a *wharenui*, "the house presents time past and present in a totality and a unity and it also effects a unity, through its symbolic design, among human events" (p. 64).

As noted earlier, the shape of meeting houses such as Rauru is an example of bilateral reflection (Witehira 2013). When we look at the meeting house front on, the figures and designs tend to be reflected on either side of the structure with a vertical axis of symmetry in the form of the *pou* (Fig. 10.3). Culturally, this focus on symmetry mirrors the pervasive duality in Māori culture as seen in origin stories and social exchanges (Hanson 1983). For example, the Rauru story is associated with the prototypical duality of celestial and terrestrial beings. Bilateral symmetry is the most common type of symmetry found in nature. An understanding of symmetry would also have had a functional application when building Māori meeting houses.

Fig. 10.3 The inside of Rauru, showing its symmetrical nature. *Source* Personal file

Although Rauru is a *traditional* meeting house, the carvers were experimenting with new techniques and designs while simultaneously preserving traditional stories (Thomas et al. 2009). The original carver, Tene Waitere, was one of the most renowned carvers of his time, producing amazing stylised figures, such as the one in Fig. 10.4.

However, family illness caused Tene Waitere to withdraw from building Rauru, before he[2] was completed. The carvers, who came in to finish the carving, Anaha te Rahui and Neke Kapua, instead of continuing the style of Tene Waitere, challenged the design parameters of their time, just as modern mathematical exploration challenges traditional mathematics. As can be seen in Fig. 10.7, their carvings are more realistic. It is interesting to note that since the time of Rauru's initial construction, many *wharenui* now incorporate realistic carvings which are considered just as traditional as those of Tene Waitere.

The possibility for experimentation was probably supported by the fact that Rauru was commissioned, not to be placed on a *marae*, but as a tourist attraction, to be situated near Whakarewarewa, the village near the main geothermal visitor site in Rotorua (Thomas et al. 2009).

[2]Rauru, like all *wharenui*, was given the same personal pronoun as the ancestor whose name he shares.

Fig. 10.4 One of the traditional stylised carvings of Tene Waitere. *Source* Personal file

Fig. 10.5 *Pūhoro*, glide reflection. *Source* Personal file

Fig. 10.6 *Pātiki*, vertical and horizontal translation. *Source* Personal file

10.4.1 Kōwhaiwhai Patterns

Kōwhaiwhai patterns express important cultural values such as unity, genealogy and family interconnectedness (Witehira 2013). The patterns painted on the ridgepole represent the tribal genealogy, power and spirits of the ancestors (Fig. 10.3). The patterns differ from tribe to tribe, many having *kōwhaiwhai* unique to their particular areas, defining the environment where the tribe exists. One of the patterns in Rauru is *pūhoro* (Fig. 10.5), which represents power and speed, and another is *pātiki* (flounder) (Fig. 10.6), which symbolises hospitality.

Kōwhaiwhai patterns involve combinations of transformations including reflection, rotation, translation (Fig. 10.6), enlargement, glide reflection (Fig. 10.5) and shears. Knight (1984a) identified seven different groups connected to the symmetry used in the freeze patterns of *kōwhaiwhai* and stated that these were present in traditional *wharenui* throughout Aotearoa/New Zealand:

1. Translational symmetry only;
2. Glide reflectional symmetry (together, of course, with translational symmetry);
3. Vertical reflectional symmetry;
4. Half-turn symmetry;
5. Half-turn and vertical symmetry (these two together ensure that there is glide reflectional symmetry too);
6. Horizontal reflectional symmetry (this, together with the translational symmetry, means that there is also glide reflectional symmetry); and
7. Horizontal and vertical reflectional symmetry (these ensure that there is also half-turn and glide reflectional symmetry) (p. 37).

In Figs. 10.2 and 10.3, it is possible to see translation, glide reflection, vertical reflection and half-turn with vertical reflection in the *kōwhaiwhai* in the rafters. The furthest to the right rafter in Fig. 10.2, next to the wall, appears to show a shear. A shear involves stretching a shape, in this case *pūhoro*, and distorting it on an angle (MacKenzie 1989).

Meaney et al. (2012) include several examples of patterns related to *kōwhaiwhai*, but they mostly examine them from a language and mathematical perspective. More recently, Manuel et al. (2015) took a class of Māori children to their local *wharenui* to investigate the symmetry in the *kōwhaiwhai* patterns. Before they began the mathematical investigation, an elder, *kaumatua*, described the cultural meaning behind the patterns. According to the researchers, "these children learned how tribal stories, reo Māori and transformation geometry were all connected and could be viewed in an integrated, meaningful way" (Manuel et al. 2015, p. 141).

10.4.2 *Tukutuku* Patterns

In contrast to the spirals, swirls and curved lines of the carvings and *kōwhaiwhai* paintwork, straight lines form the basis of *tukutuku* panels (Averill et al. 2009). The traditional *tukutuku* is a lattice-like frame made of vertical and horizontal rods, and flexible material is threaded through the rods to form the patterns and designs (Figs. 10.2, 10.3, 10.4 and 10.7).

They provide a visual representation of legends, in which *tukutuku*, like the *kōwhaiwhai* designs, are representational rather than figurative (Witehira 2013). The design of the *tukutuku* could represent an object or operation in the physical world as well as from nature (*tukutuku* on the left of Fig. 10.7 represent a fern frond and on the right side represent a fish).

Tukutuku designs consist of patterns which repeat and translate (see left-hand panel in Fig. 10.7), or which have reflection or rotational symmetry. Reflectional symmetry is seen in the top part of the left-hand panel in Fig. 10.2 as well as in the

Fig. 10.7 Māui flanked by *tukutuku* panels in Rauru meeting house. *Source* Personal file

right-hand panel of Fig. 10.7. The left-hand panel of Fig. 10.7 shows rotational symmetry. Whereas *kōwhaiwhai* patterns only have translational symmetry in one direction, *tukutuku* patterns are two-directional and so are known as wallpaper patterns (Knight 1984b).

Consequently, "they may have other symmetries in addition to their translational symmetry" (Knight 1984b, p. 80). Knight (1984b) identified 10 different examples of patterns based on different combinations of transformational geometry approaches in Māori art, which had been documented at the beginning of the twentieth century by Augustus Hamilton (1901).

Averill et al. (2009) described how they used *tukutuku* with preservice teachers both as a metaphor for the different mathematical topic covered in the course and as a practical activity whose final product was described in regard to the mathematics incorporated into it. The mathematics identified in the students' completed *tukutuku* included symmetry, although without specific descriptions of the types of symmetry involved (Anderson et al. 2005).

Using *tukutuku* as a metaphor for the whole course and engaging in the practical activity of making one seemed to make the preservice teachers understand better the responsibility and possibilities connected to incorporating Māori cultural activities into mathematics lessons:

> Reasons given by students for planning to use cultural activities such as tukutuku panel-making in their own teaching included cultural, pedagogical, and motivational aspects. They felt such activities would assist them to teach in a culturally responsive way (Averill et al. 2009, p. 171).

Therefore, it seems that if cultural understanding is connected to mathematical understanding, activities such as *kōwhaiwhai* and *tukutuku* could widen children's horizons in being able to transition (Meaney and Lange 2013) between the mathematics and the cultural learning. This is in contrast to focusing solely on mathematical understandings which might result in children restricting the value that they consider the cultural activity to have in its own right.

10.5 Cultural Symmetry in the Mathematics Classrooms

In Māori-medium contexts, when teaching symmetry, such as that found in *wharenui*, it is necessary to ensure that cultural knowledge is valued. If this does not occur, then the cultural context for the "real" mathematics learning is tokenistic and unlikely to be considered valuable by students. Nevertheless, just as Rauru should not be seen as referential to the mathematics, he should not be only considered as being reverential to the culture. Rauru shows that what it means to be a *wharenui* and how that meaning is expressed not only draws on traditional understandings, but is refigured as times and contexts change. Identifying mathematical understandings about symmetry in the different features of Rauru, like the incorporation of new carving techniques, can also be considered a renewal of the meanings attached to *wharenui*.

Rather than detracting from the cultural meanings already embedded within the designs, mathematical understandings can potentially provide another layer to the existing meanings. In this way, the mathematical understandings deepen and enrich the cultural meanings already present. Similarly, as Rauru is not simply a museum exhibit, locked behind glass doors, to be looked at with reverence but not to be touched, the existing cultural meanings can be enriched.

Members of Te Awara continue to value Rauru for his connections to the past, but celebrate his continued existence in the present by allowing new meanings to be developed. However, to support an enrichment rather than a colonisation of the existing cultural understandings, there is a need to engage in cultural symmetry in a respectful manner.

Consequently, we suggest a three-step approach. This three-step approach allows for mathematical understandings to be reflected into the cultural meanings already associated with cultural activities or products, such as *wharenui*, in a holistic manner. In this way, we anticipate that the valuing of the mathematical meanings will not colonise or distract from other understandings connected to cultural processes and artefacts such as *wharenui*.

The first step is for the cultural knowledge to be at the forefront of any learning, thus acknowledging the cultural dimension of mathematics. For example, if Rauru is to be discussed in mathematics classrooms, then it is important that the stories and cultural knowledge that are represented within the *wharenui* are discussed first and foremost. Only by knowing about the ancestors represented in the house and the legends to which they are connected can the cultural knowledge be valued. The use of *te reo Māori* in this discussion is important.

As Fishman (1991) argued, culture is in the language and is expressed in the language. He further added, "take language away from the culture, and the culture loses its literature, its songs, its wisdom, ways of expressing kinships relations and so on" (Fishman 1991, p. 72). It would also mean that the cultural nuances connected to symmetrical designs of *kōwhaiwhai* and *tukutuku* would be lost if they were discussed in the mathematics register of English only.

The second step is to identify the designs which have been used to create the different artefacts such as the *kōwhaiwhai* and *tukutuku* designs and what they symbolise. The designs used in Rauru are highly valued for their artistic qualities (Thomas et al. 2009) because although they are well-known designs, they were used in innovative ways, which makes them unique. Identification of the different design elements allows students to recognise their use in other Māori designs. To do this, students need to participate in the production of the patterns, not just see them as static artefacts that can be dissected for the knowledge they contain.

The most appropriate way to do this is for the students to be involved in the production or reconstruction of artefacts in an actual *wharenui*. However, it this is not possible, and then reproducing existing patterns (Fig. 10.8) or producing new patterns can be done with pencil and paper. Producing patterns provides possibilities for the students to learn some architectural terms as well as mathematics terms in order to discuss what they are doing and seeing and how they relate to the *wharenui* as a whole. The third and final stage is to discuss the designs in relationship to the symmetrical principles in order to understand how a pattern is repeated to produce the meanings connected back to the stories.

Recognition of the designs in the parts needs to be done in a constant interaction with recognition of the design of the whole *wharenui*. Discussions which include considerations of how meanings are represented in other *wharenui* and other cultural artefacts also allow for an understanding about how cultural meanings can change across both space and time. In this way, Māori culture is recognised as a living dynamic culture, not as a revered museum piece. For this to occur, the

Fig. 10.8 Student example of work explaining the symmetrical properties of kōwhaiwhai. *Source* Personal file

teaching of mathematics must add value to understanding the cultural knowledge embedded in the designs of the different artefacts, rather than detracting from that knowledge by purely focusing on the mathematics.

Cultural symmetry in mathematics classrooms is complex to achieve in that there are a number of different aspects that need to be considered simultaneously. Mathematical understandings are a form of cultural understandings but if they are merely presented as representative of Western mathematics, then the possibilities for using them to discuss Māori cultural artefacts and processes are likely to result in cultural imperialism: Māori culture is only considered valuable if mathematics can be connected to it.

Instead, giving equal balance to Māori cultural knowledge and mathematical cultural knowledge involves considering not just the language that is used, the mathematics register, for example, but also the purposes for employing cultural symmetry. As is the case with *wharenui*, adding value to learning needs to ensure that the learning is considered holistically and one small part does not become isolated from the rest.

10.6 Further Considerations

It is very convenient that *Rauru*, as the focus of this chapter, is presently located in Hamburg, where ICME 13 was held. Māori visiting Hamburg, from whence the *wharenui* originated, would consider it culturally unacceptable that the house be overlooked. It is as though *Rauru* were setting the agenda for the presentation, as much as that of such an important conference. Rauru the meeting house may be situated in Germany, but he is highly valued for the cultural knowledge he contains by members of Te Arawa tribe who travelled there in 2012 to celebrate his more than 100 years of being in Germany and the conservation of many of the features of this *wharenui*.

For the sake of brevity, we hope we are not contradicting ourselves by providing a very abbreviated narrative of *Rauru* the person. *Rauru* himself was human, the son of *Kuraimonoa* and her husband *Toi-te-huatahi*. However, a celestial being, Pūhaorangi, desired the mother and caused the infant Rauru to wet the sleeping mat he shared with his parents over several nights. Finally, his father left their house in disgust, whereupon Pūhaorangi descended from the sky, disguised himself as Toi-te-huatahi and entered the house to sleep with Kuraimonoa, who was unaware that it was not her husband who had returned.

Kuraimonoa conceived a child with Pūhaorangi, a son who was named Ohomairangi. The half-brothers' descendants would intermarry, giving rise to Te Arawa and many other tribes. And so the symmetry plays out, celestial/terrestrial, divine/human. Exploring the house, Rauru, could, and should be, much more than a desultory mathematical activity. It invites us to consider our place in *te ao tangata, the world of humans*, and *te ao tūroa, the world that is*: how we measure and make sense of our internal and external worlds. The discussion would continue well beyond the mathematics classroom but would also enrich what happens within it.

A final note is that Rauru's full name was Rauru-kī-tahi, *Rauru-of-the-single-utterance*. Now what could that utterance have been or possibly be?

References

Anderson, D., Averill, R., Easton, H., & Smith, D. (2005). Use of a cultural metaphor in pre-service mathematics teacher education. In P. Clarkson, A. Downton, D. Gronn, M. Horne, A. McDonough, R. Pierce, & A. Roche (Eds.), *Building connections: Theory, research and practice (Proceedings of the Annual Conference of the Mathematics Education Research Group of Australasia, held at RMIT, Melbourne, 7–9th July, 2005)* (Vol. 1, pp. 81–88). Melbourne, Australia: MERGA.

Averill, R., Anderson, D., Easton, H., Maro, P. T., Smith, D., & Hynds, A. (2009). Culturally responsive teaching of mathematics: Three models from linked studies. *Journal for Research in Mathematics Education, 40*(2), 157–186.

Barton, B. (1993). Ethnomathematics and its place in the classroom. In E. McKinley, P. Waiti, A. Begg, B. Bell, F. Biddulph, M. Carr, M. Carr, J. McChesney, & J. Young-Loveridge (Eds.), *SAMEpapers 1993* (pp. 47–65). Hamilton, New Zealand: Centre for Science and Mathematics Education Research, University of Waikato.

Barton, B. (2008). *The language of mathematics: telling mathematical tales*. New York, NY: Springer.

Bishop, A. J. (1988). *Mathematical enculturation: a cultural perspective on mathematics education*. Dordrecht, The Netherlands: Kluwer.

Bishop, A. J. (1990). Western mathematics: the secret weapon of cultural imperialism. *Race & Class, 32*(2), 51–65.

Bishop, A. J. (1994). Cultural conflicts in mathematics education: Developing a research agenda. *For the Learning of Mathematics, 14*(2), 15–18.

Booker, G., Bond, D., Sparrow, L., & Swan, P. (2010). *Teaching primary mathematics*. Sydney, Australia: Pearson.

Chrisp, S. (2005). Māori intergenerational language transmission. *International Journal of the Sociology of Language, 172*, 149–181.

Crystal, D. (2003). *Language death*. Cambridge, England: Cambridge University Press.

D'Ambrosio, U. (1985). Ethnomathematics and its place in the history and pedagogy of mathematics. *For the Learning of Mathematics, 5*(1), 44–48.

D'Ambrosio, U. (1999). Literacy, mathematics and technology: a trivium for today. *Mathematical Thinking and Learning, 1*(2), 131–153.

Ernest, P. (1991). *The philosophy of mathematics education*. London, England: Falmer Press.

Fishman, J. (1991). *Reversing language shift: Theoretical and empirical foundations of assistance to threatened languages*. Clevedon, England: Multilingual Matters.

Geertz, C. (1975). *The interpretation of cultures*. London, England: Hutchinson.

Gerdes, P. (1986). How to recognize hidden geometrical thinking? A contribution to the development of anthropological mathematics. *For the Learning of Mathematics, 6*(2), 10–17.

Gerdes, P. (1988). On culture, geometrical thinking and mathematics education. *Educational Studies in Mathematics, 19*(2), 137–162.

Grosfoguel, R. (2002). Colonial difference, geopolitics of knowledge and global coloniality in the modern/colonial capitalist world—system. *Review, 25*(3), 203–224.

Hale, K. (1992). On endangered languages and the safeguarding of diversity. *Language, 68*(1), 1–3.

Hamilton, A. (1901). *Maori art*. Wellington, New Zealand: New Zealand Institute.

Hanson, F. (1983). *Counterpoint in Māori culture*. London, England: Routledge & Paul.

Harrison, K. D. (2007). *When languages die: the extinction of the world's languages and the erosion of human knowledge*. Oxford, England: Oxford University Press.
Hāwera, N., & Taylor, M. (2014). Researcher-teacher collaboration in Māori medium education: Aspects of learning for a teacher and researchers in Aotearoa New Zealand when teaching mathematics. *AlterNative: An International Journal of Indigenous Peoples, 10*(2), 151–164.
Jackson, M. (1972). Aspects of symbolism and composition in Maori art. *Bijdragen tot de taal–, Land-en Volkenkunde* [Contributions to Linguistics, Land and Ethnology], *128*(1), 33–80.
Knight, G. H. (1984a). The geometry of Maori art-rafter patterns. *NZ Mathematics Magazine, 21*(2), 36–41.
Knight, G. (1984b). The geometry of Māori art—weaving patterns. *The New Zealand Mathematics Magazine, 21*(3), 80–87.
Leikin, R., Berman, A., & Zaslavsky, O. (2000). Learning through teaching: The case of symmetry. *Mathematics Education Research Journal, 12*(1), 18–36.
Lemke, J. (1990). *Talking science*. Norwood, NJ: Ablex.
MacKenzie, D. F. (1989). *Kōwhaiwhai: Geometry of Aotearoa*. Auckland, New Zealand: D. F. MacKenzie.
Manuel, H., Hāwera, N., & Taylor, M. (2015). Transformation geometry: Mate nekehanga, mate whakaata, mate hurihanga. In R. Averill (Ed.), *Mathematics and statistics in the middle years: evidence and practice* (pp. 131–145). Wellington, New Zealand: New Zealand Council of Educational Research.
May, S. (2005). Language rights: Moving the debate forward. *Journal of Sociolinguistics, 9*(3), 319–347.
McCarthy, C. (2005). Boundary arbitrations: Spatial complexities and tensions in recent New Zealand museum architecture. *New Zealand Sociology, 21*(1), 68–88.
McMurchy-Pilkington, C., & Trinick, T. (2008). Potential & possibilities. In V. Carpenter, J. Jesson, P. Roberts, & M. Stephenson (Eds.), *Ngā kaupapa here: Connections and contradictions in education* (pp. 133–144). Melbourne, Australia: Cengage Learning.
McMurchy-Pilkington, C., Trinick, T., & Meaney, T. (2013). Mathematics curriculum development and Indigenous language revitalisation: Contested spaces. *Mathematics Education Research Journal, 25*(3), 341–360.
Meaney, T. (2002). Symbiosis or cultural clash? Indigenous students learning mathematics. *Journal of Intercultural Studies, 23*(2), 167–187.
Meaney, T., & Fairhall, U. (2003). Tensions and possibilities: Indigenous parents doing mathematics education curriculum development. In L. Bragg, C. Campbell, G. Herbert, & J. Mousley (Eds.), *Mathematics Education Research: Innovation, networking, opportunities: Proceedings of the 26th Annual Conference of the Mathematics Education Research Group of Australasia* (pp. 507–514). Sydney, Australia: MERGA.
Meaney, T., Fairhall, U., & Trinick, T. (2008). The role of language in ethnomathematics. *Journal of Mathematics and Culture, 3*(1). Retrieved from http://nasgem.rpi.edu/index.php?siteid=37&pageid=543.
Meaney, T., & Lange, T. (2013). Learners in transition between contexts. In K. Clements, A. J. Bishop, C. Keitel, J. Kilpatrick, & F. Leung (Eds.), *Third international handbook of mathematics education* (pp. 169–202). New York, NY: Springer.
Meaney, T., Trinick, T., & Fairhall, U. (2012). *Collaborating to meet languages challenges in Indigenous mathematics classrooms*. Dordrecht, The Netherlands: Springer.
Metge, J. (1976). *The Maoris of New Zealand: Rautahi*. London, England: Routledge & Kegan Paul.
Pais, A. (2011). Criticism and contradictions of ethnomathematics. *Educational Studies in Mathematics, 76*(2), 209–230.
Penetito, W. (2010). *What's Māori about Māori education?*. Wellington, New Zealand: Victoria University Press.
Riini, M., & Riini, S. (1993). Historical perspectives of Māori and mathematics. In M. Ohia (Ed.), *Pāngarau: Māori mathematics and education* (pp. 16–20). Wellington, New Zealand: Ministry of Māori Development.

Roberts, T. (1997). Aboriginal maths: Can we use it in school? In N. Schott & H. Hollingworth (Eds.), Mathematics: Creating the future. *Proceedings of the 16th Biennial Conference of the Australian Association of Mathematics Teachers (AAMT)* (pp. 95–99). Adelaide, Australia: Australian Association of Mathematics Teachers.

Salmond, A. (1978). Te ao tawhito: a semantic approach to the traditional Maori cosmos. *The Journal of the Polynesian Society, 87*(1), 5–28.

Selby, R. (2009). Indigenous voices, indigenous symbols. In R. Selby (Ed.), *Indigenous voices, indigenous symbols* (pp. 4–9). Guovdageaidnu, Norway: Sàmi University College.

Simon, J. (1998). *Ngā kura Māori: The native schools system 1867–1969.* Auckland, New Zealand: Auckland University Press.

Smith, G. H. (2004). Mai i te maramatanga, ki te putanga mai o te tahuritanga: From conscientization to transformation. *Educational Perspectives, 37*(1), 46–52.

Spolsky, B. (2005). Māori lost and regained. In A. Bell, R. Harlow, & D. Starks (Eds.), *Languages of New Zealand* (pp. 67–85). Wellington, New Zealand: Victoria University Press.

Stillman, G., & Balatti, J. (2000). Contribution of ethnomathematics to mainstream mathematics classroom practices. In B. Atweh, H. Forgasz, & B. Nebres (Eds.), *Sociocultural research on mathematics education: An international perspective* (pp. 201–216). Mahwah, NJ: Lawrence Erlbaum.

Thomas, R. (1996). Proto-mathematics and/or real mathematics. *For the Learning of Mathematics, 16*(2), 11–18.

Thomas, N., Adams, M., Schuster, J., & Grant, L. (2009). *Rauru.* Dunedin, New Zealand: University of Otago Press.

Trinick, T., Meaney, T., & Fairhall, U. (2015). Reintroducing Māori ethnomathematical activities into the classroom: traditional Māori spatial orientation concepts. *Revista Latinoamericana de Etnomatemática, 8*(2), 415–431.

Trinick, T., Meaney, T., & Fairhall, U. (2016). The relationship between language, culture and ethnomathematics. *Journal of Mathematics and Culture, 10*(2), 175–191.

Witehira, J. (2013). *Tārai Kōrero Toi.* Doctoral thesis. Massey University, Palmerston North, New Zealand. Retrieved from http://mro.massey.ac.nz/handle/10179/5213

Zaslavsky, C. (1979). *Africa counts: number and pattern in African culture.* Brooklyn, NY: Lawrence Hill Books.

Chapter 11
An Ethnomodel of a Penobscot Lodge

Tod Shockey and John Bear Mitchell

Abstract When considering Pike's (Language in relation to a unified theory of the structure of human behavior. Mouton, Paris, 1967) linguistic work that introduced *emic* and *etic* through the ethnomathematics lens, new questions arise. With these new questions, emerge new perspectives. The building of a Penobscot hemispherical lodge, considered through the emic view, complimented with the etic view is the focus of this chapter. According to Albanese et al. (The evolution of ethnomathematics: two theoretical views and two approaches to education, 2016) this work overlaps what they reference as "mathematics in cultural practice" that is to say "recognizing mathematics, mathematics that is somehow similar to the academic one in cultural practices" and "different ways of knowing [because] different ways of knowing is based on the idea that there are different ways of conceptualizing quantities, relations and space" (p. 1). An ethnomodel of a Penobscot hemispherical lodge contributes to the ethnomathematics scholarship and supports the emerging research in ethnomodelling.

Keywords Ethnomathematics · Ethnomodel · Penobscot · Emic · Etic

11.1 Introduction

Rosa and Clark (2013) state that ethnomodelling is "a practical application of ethnomathematics, and which adds the cultural perspective to modelling concepts" (p. 78). A cultural perspective broadens views of modelling, which Bassanezi (2002, cited in Rosa and Clark 2013) recognizes as a potential pedagogical bridge for students in their learning of mathematics. "Inclusion of a diversity of ideas

T. Shockey (✉)
University of Toledo, Toledo, USA
e-mail: tod.shockey@utoledo.edu

J.B. Mitchell
University of Maine, Orono, USA
e-mail: John.mitchell@umit.maine.edu

M. Rosa et al. (eds.), *Ethnomathematics and its Diverse Approaches for Mathematics Education*, ICME-13 Monographs,
DOI 10.1007/978-3-319-59220-6_11

brought by students from other cultural groups can give confidence and dignity to students, while allowing them to see a variety of perspectives and provide a base for learning academic-Western mathematics" (Bassenezi 2002, cited in Rosa and Clark 2013, p. 78).

In 1990, when D'Ambrosio introduced *mathemematise*, this set the stage for emerging scholarship in ethnomodelling. According to Rosa and Orey (2006) "mathematisation is a process in which individuals from different cultural groups come up with different mathematical tools that help them organize, analyze, comprehend, understand, and solve specific problems located in the context of their real-life situation" (cited in Rosa and Orey 2013, p. 118). Rosa and Orey's (2013) "cultural perspective" is a different way to state what Pike (1967) was discussing in when he introduced the *emic* perspective, which is the insiders' view.

This described collaboration between Mitchell, Penobscot, and Shockey on the construction of a hemispherical lodge brings two perspectives to bear on the *practical application* of ethnomathematics. Mitchell is a member of the Penobscot Nation, a Native American population in the United States. Shockey is a mathematics educator. Mitchell is frequently invited to teach about his culture and the broader cultures of New England and the Maritimes. Wabanaki, *People of the Dawn*, is the referent to four tribes: Penobscot; Passamaquoddy; Micmac; and Maliseet. When Mitchell was invited to oversee the construction of a traditional village on the Great Salt Bay of Maine, Shockey invited himself to attend with two intentions: first to begin to understand Mitchell's Native pedagogy and second to focus attention on the mathematics embedded in the activities that Mitchell directed.

Previously Shockey and Silverman (2016) wrote of this experience, with an emphasis on the *etic view*. Shockey is trained as a mathematics educator and does not bring the important Native perspective of Mitchell. With the emergence of ethnomodelling in English, the two have revisited their previous work to bring the Native perspective to the forefront. Important implications emerge with the emic view, implications for connecting Western-mathematics to Indigenous views as well as implications for pedagogy.

The practical application of ethnomathematics, through the construction of a Penobscot lodge, an ethnomodel, and the mathematisation, through a western and cultural lens, of the lodge construction, is the focus of this chapter. The diversity of ideas brought together, support the ethnomathematics of this construction and offers pedagogical insights.

Ethnomathematics continues to rely on multiple scholarships to further understandings. Linguist Pike (1967) provided the emic and etic perspectives that have implications in the building of this summer lodge. Citing his work from 1947, Pike (1967) stated: "emic systems and emic units of these systems are in some sense to be discovered by the analyst, not created by him" (p. 64). The etic, according to Pike, "etic systems on the other hand, are assumed to be classifications created by the analyst" (p. 64). For our purposes, we borrow the units of analysis from Bishop (1991): counting; playing; locating; measuring; designing; and explaining.

The construction of this lodge was a new experience for Shockey. Shockey was not able to anticipate or oftentimes understand Mitchell's motivations from mere observation. Conversation between the two of them led to and opened insights through the emic and etic dialogues. From a western perspective, if these two perspectives are considered as a Venn diagram, there is overlap (intersection), but more importantly is the understandings and perspectives that do not lie in the intersection. The academic, western, circle of the Venn (Fig. 11.1) is shared understanding for Mitchell and Shockey, the learning occurred for Shockey as Mitchell described his emic view, which is based on his Penobscot worldview.

But, in any society as Edward Sapir said, "Forms and significances which seem obvious to an outsider will be denied outright by those who carry out patterns; outlines and implications that are perfectly clear to these may be absent to the eye of the onlooker" (cited in Kluckhohn 1949, p. 36). Shockey, in the role of the onlooker, was not understanding or seeing everything that was occurring. Kluckhohn (1949) again quoting Sapir:

> The fact of the matter is that the *real world* is to a large extent unconsciously built up on the language habits of the group (…). We see and hear other experiences as we do because the language habits of our community predispose certain choices of interpretation (cited in Kluckhohn 1949, p. 167).

Shockey's was predisposed to certain choices that will be highlighted below with the inclusion of etic in parenthesis (etic) to bring a focus for the readers on the influence of western academic influence. For clarity, Mitchell and Shockey are both English speakers and that is the language of interaction. Mitchell has the Pasamaquoddy language, but used English in his interactions. Shockey brings the language of mathematics, his etic view, to bear on the exchanges, with the intention of creating awareness of how connections may be made to classrooms.

While this chapter has a primary focus on views of a Native educator and a Western trained academic, it is important to acknowledge that there were many different cultural groups interacting in the construction of the lodge. When D'Ambrosio (1985) defined ethnomathematics, he included "children of a certain age bracket" (p. 45). We acknowledge the children, predominantly middle school

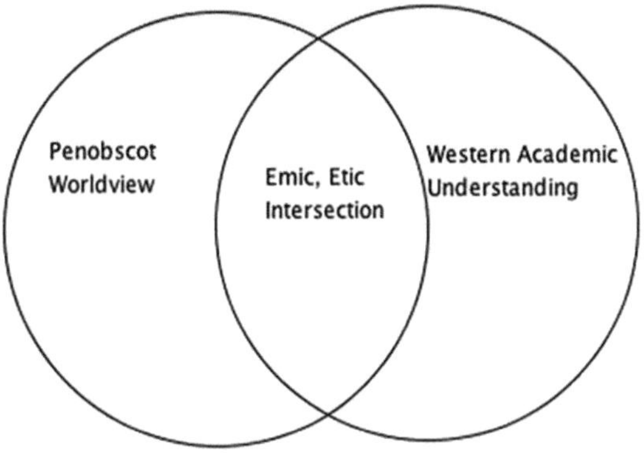

Fig. 11.1 Venn diagram comes from two worldviews. *Source* Personal file

age (in the United States these are children in grades six, seven, and eight) that come from varied backgrounds as described below. This acknowledgement is due to the emic perspective the children bring, but that is not a focus of this chapter.

11.2 Methodology

For two weeks in 2005 Shockey and Mitchell were on site at the Great Salt Bay in Maine (USA). It is worth stating; initially the pair had no idea as to what *mathematics* would emerge during the sessions. Shockey was unaware of what it meant to construct a hemispherical lodge and could only guess as to emerging mathematics relative to the type of lodge. Mitchell did not see this as an exercise in mathematics, nor was that his intent. As Jorgenson (1989) stated:

> The methodology of participant observation is appropriate for studies of almost every aspect of human existence. Through participant observation, it is possible to describe what goes on, who or what is involved, when and where things happen, how they occur and why (p. 12).

Shockey and Mitchell are revisiting this project from 2005 with the purpose of exploring the linguistic themes of etic and emic as put forth by Pike (1967). Neither of them were aware of ethnomodelling in 2005. The first mention of this experience by Mitchell resulted in Shockey imposing himself to be included as the mathematics educator. Jorgenson (1989) states that there are seven basic features of participant observation. His first defined feature: “a special interest in human meaning and interactions as viewed from the perspective of people who are insiders or members of particular situations and settings” (p. 13) best captures Shockey’s motivation.

Shockey was interested in the experience, interested in ethnomathematics as a lens to develop understanding of what was going to occur, and very motivated to observe a Native educator, educating. This qualitative study has no intent of generalization. During this time data was collected through videotaping and field notes. Videos were transcribed verbatim and images captured from the videos that are used throughout this manuscript to assist the reader’s understanding. Throughout the included transcripts we include *Remark* section to state our conjecture inferences of meaning. Emic and etic as defined by Pike (1967) are the categories used to analyze transcribed data.

11.3 Constructing the Lodge, Pedagogical Implications

The Penobscot hemispherical lodge that Mitchell oversaw the construction of was built by visiting groups of school children. For two weeks students from public, private, and home-schooled environments visited the site and *built* the lodge.

Students would arrive for two time periods each day, typically a group of students would arrive near 9:00 a.m. and return to their respective school for lunch at noon. A new group of students would arrive near noon and work until about 3:00 p.m. Home schooled children would frequently stay for the entire day. Verbatim transcripts are included below. The legends for the transcripts are as follows: JB—John Bear Mitchell speaking; S—student or students speaking; FN—field notes; events that fit Ancestral Engineering are noted as such; T—teacher speaking; and pedagogical remarks are contained in parenthesis.

Pedagogical remarks are not analyzed as etic or emic. Remarks are inserted in the transcripts that allow the inclusion of inferences and highlights elements of cultural perspectives, mathematisation, and pedagogical modelling. While we attempt to create an ethnomodel, it needs to be stated that many of the actions of Mitchell physically model what students will do as participants, this, we refer to as pedagogical modelling.

11.3.1 Preparing the Lodge Base and Setting Lodge Poles

This is the introduction to laying out the circular base to the lodge. The etic is placed in square brackets in italic.

JB: (…) so it's even on all sides, are there sides to a circle?
S: well not sides.
JB: you're right, so we want it even, so how does this make it even? I'm putting this peg in the ground, this is the center of my lodge right here [pounds the peg into the ground], how does that make my lodge even?

Remark: Here John Bear has a short stake he drives into the ground, this stake will serve as the center of the circle, the circumference of the circle will define where lodge poles are placed for the hemispherical lodge. John Bear's use of *even* is a remark about radius and all the lodge poles being equi-spaced about the center he defined with the stake. This is the first specific problem, mathematisation, recognized by Shockey and described with language of western mathematics, an etic perspective. The cultural perspective is that this lodge needs a circular base, with a very specific purpose to serve a family of four, described later.

JB: If I'm going to make this perfectly round, how much string am I going to use? Keeping in mind those are the size of our trees over there, you guys carried them down. Could I make a lodge that's as round as the circle?

Remark: John Bear is trying to get the students to recognize the size of the trees is a variable to consider since their size will determine how big of a lodge that can be built. Available resources, in this case the trees used as lodge poles, are part of the cultural perspective. The trees were harvested specifically for the construction of the lodge. Heights and diameters, etically speaking, taken into account to assure that the lodge would satisfy the mental image that Mitchell was working from. With respect to pedagogy, Mitchell uses a pedagogy of modelling actions for students. This is the first time Mitchell has shared his knowledge of lodge construction, so he frequently pauses to consider his articulation of the sequence of events and that these events are in the correct order.

S: Half of one of those trees

Remark: This student's remark has to do with the lodge radius (etic). Since lodge poles are opposite one another on the circle circumference and have to be bent inward and lashed together, the diameter of the lodge cannot exceed the lengths of opposite lodge pole lengths. This was a problem that Mitchell took into account with the harvesting of the trees, making sure these trees were long enough to construct a lodge that has habitable. (**FN**: Pedagogically, John Bear does not have a habit of repeating students; his repetition is usually of his own thoughts).

JB: Okay you're right, so I've already looked at the trees and the trees are approximately how long in feet?
S: Ten feet.
JB: Ten feet?
S: Seven or eight.
S: Six feet.
JB: Maybe about, I was thinking about eight feet, all about eight (he is getting student consensus on this length), the ones you guys brought down, probably about eight feet on average and there are some up there that are longer [*this makes sense since his remark is about average (etic)*] I know if I get an eight foot tree and I'm going to bend it in and make two trees come together by bending them in, I'm going to my lodge to about ten to twelve feet across [*diameter (etic)*].
JB: What's your name?
S: Thor.
JB: Thor, how far are we apart, stay right there, how far do you guys think we're standing apart from each other?
Thor: About nine feet.
JB: I would say about nine feet, so if I went one more foot, that's how long [diameter (etic)], that's how big that the lodge would be, which is huge.

Remark: At nine feet the area of the circular base would be 4.5^2 pi, increasing the diameter by a foot, increases the area to 25 pi. The area difference is approximately 15 square feet, or an increase of approximately 19%. With a 4.5 foot radius the area is about 63.62 square feet and a radius of 5 feet yields an area of about 78.54 square feet. This is defined as *huge*.

> I think if we come together a little bit more, what's going to happen, we're going to have a lodge that was only meant to sleep in, normally was only this high, like this high right about here [John Bear holds his hand above the ground about shoulder high, describing the height of a sleeping lodge].

> Like about this high cause we're not going to walk around in there, it's a sleeping lodge, a summer lodge. I mean there's very few tents that we set up nowadays we actually, I mean some of you might have them, you get a tent, you're in it, you're hanging out, you're walking around changing and all that without standing right up, but normally these were our sleeping lodges. Now if we got closer, what would happen to our dome? If we go closer what would happen to it, it gets what?

Remark: The cultural perspective is that the lodge is only for sleeping, so the height of the lodge does not have to accommodate walking about inside the lodge. The mathematisation occurring is a negotiation as to the how the lodge poles will determine the height of the lodge.

S: Smaller.
JB: It gets smaller, but does the roof go lower?
S: No.
JB: What happens?
S: It gets higher.
JB: It gets higher, I'd rather have one a little bit higher, about here (holds his hand up to his nose) so we're probably about eight feet, maybe six to eight feet, let's say eight feet and how tall am I?

Remarks: John Bear stretches out a piece of twine fingertip to fingertip, arms stretched approximating his height, this cultural, emic, perspective is understood by Mitchell.

S: (After John Bear says "I love you this much" to student laughter). Five foot ten, eleven.
JB: Well about five nine or ten, I want to be taller, guys want to be taller, I don't know why, doesn't make you see any better (humor is part of his pedagogy). So, let's say I went like this, it's like five and a half feet right, and then, I'm going to estimate a little bit more. Why am taking a little bit more?
S: For your knot [*this is good estimation on the students part, considering that length of the radius would be lost if "a little bit more" was not added*].
JB: For my knot, I'm going to tie it around my pole now. Okay so let's take a try because this is not like a serious math problem where (…) you know you got people that are actually watching you do this you know [*In the discussions leading up to this event, John Bear did not present that he was convinced that mathematics, western, would be used, at this point, he began to realize that mathematics was playing a role*]. It's going to hit or miss, so if I have this string here and I have Mari come stand on the end of it [*John Bear has established the radius which allows him to place the first lodge pole*]. (John Bear drives a steel bar into the ground to create a hole for placing the lodge pole in the ground).

Remark: The etic of radius used by Shockey, is Mitchell's emic of *even* used initially.

JB: Okay, so here I am, this is huge and this is probably about six feet, I don't want it to be this big

Remark: Six feet is a reference to radius, but now we have determined that the *little bit more* for the knot that was added to the five and a half feet of string, created a radius of about six feet. The cultural perspective that a lodge with a radius of six feet is huge has presented a problem to be solved.

> (…) because looking at the trees over there we got, because we didn't go out hunting for trees, we landed (history reference, building based on local materials) and used what we had, so this is huge, I don't think we want to do one this big.

Remark: Now the lodge floor has grown to 36 pi, approximately 113 square feet, an increase of about 35 square feet from the previous dimensions. This etic perspective is not part of Mitchell's problem solving. Here we witness Mitchell's mathematisation of the problem associated with created a lodge that is not "huge." By changing the length of the string he is using to make sure the lodge is *even* he is working toward a solution of this real-life situation as described by D'Ambrosio (1990).

JB: Okay, I like this a little better. This is going to give us a lodge that's about how far [*a question about diameter based on adjustments that began with John Bear's height*].
S: Eight feet.
JB: I would say about eight feet.

11.3.2 Discussion of the Circular Base Layout

The above-described interactions are the negotiations between John Bear and the students for laying out the circular base on the lodge. The discourse between them encourages mental estimation, introduced using body dimensions for approximating lengths, and reveal that John Bear is considering many of the variable that play a role in determining the size of the circle.

Particularly, he is concerned that they avoid too large a lodge, that the trees used for establishing the hemisphere will be long enough when arched over to lash together, and he wants to a have a lodge with a height such that a person could stand upright inside. He has brought the students along with his thinking based on the questions he has posed and revealed a mutual understanding that by decreasing the diameter the height of the lodge will be increased. This was a new experience for Shockey, he had no prior conceptions or pictures to consider, for what Mitchell was constructing.

The emic perspective for the floor layout reveals the importance of approximation, body dimensions for measurement, and the use of the word *huge* that is not quantified. The emic perspective becomes clearer as the circular base of this summer sleeping lodge that John Bear described would serve a family of four. In his mind, he was visualizing how the circular base would serve a family and as the lodge began to emerge, he articulated this purpose. The lodge's purpose beyond sleeping is that it would allow the family a place to retreat from inclement weather and be large enough that if the family had a visitor, there would be ample room for the guest to sleep as well. John Bear faced many practical problems associated with this base layout.

In the instances where he was repeating a students' remark, Shockey inferred that it allowed him mental time to determine if the dimensions associated the circular base, tree length, and height were going to work. From an etic perspective, there was a tremendous amount of spatial reasoning occurring on his part. Shockey's inference on spatial reasoning began with the considerations Mitchell was making for the layout of the circular base, spacing of the lodge poles on the circumference, and the consideration of a family of four and how they would be situated within the floor space of the lodge.

This emic perspective, we recognize as the mathematisation of solving the problem for the base of this lodge. The cultural perspectives of Mitchell begin the formation of his ethnomodel of this lodge.

11.3.2.1 Creating the Circumference

The second phase of the lodge layout dealt with laying out the circumference and spacing of the lodge poles. The etic is contained within square brackets in italic.

JB: Now we need to start getting poles and we will do one at a time. We'll put on pole in a hole right there and then we'll completely opposite [180°]. Can we use an odd number of poles, can we use thirteen poles?
S: No.
JB: What would happen?
S: (The students' chorus of response was not understandable).
JB: Yeah, it would be bad, in that we have an odd number of poles and we're trying to make a dome and you have two pole connecting.

Remark: Mathematically speaking, opposite poles, 180°, are arched toward the center of the circular base and lashed together, so there is a need for pairs. But the need for pairs is a cultural perspective.

JB: Where would be a better place to go? This way, right? (This question is about the placement of the second lodge pole, one strategy would be to estimate arc length and place lodge poles equi-spaced on the circumference. There is no response from the students as none of them have ever built a hemispherical lodge).

Remark: When John Bear states "this way, right?" he waves his arm in a perpendicular motion with respect to an imaginary segment joining the first two lodge poles. With this gesture he has created two perpendicular axes that intersect at the stake that serves as the center of the circular base. Figure 11.2 shows the field notes that depicted potential position of a lodge pole. This remark is an etic perspective by Shockey, through his use of perpendicular, axes, and circular center.

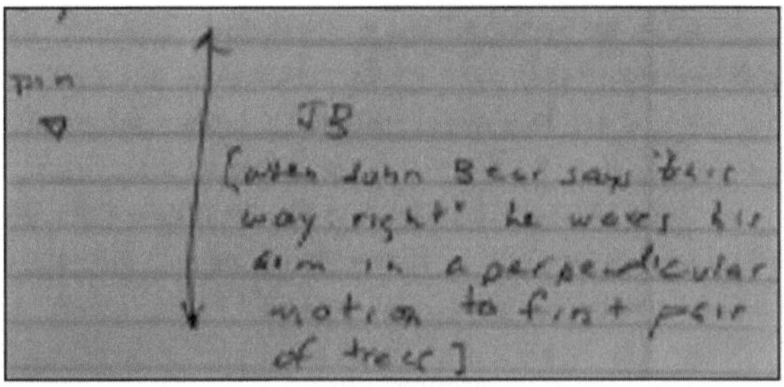

Fig. 11.2 Field notes depicted potential position of a lodge pole. *Source* Personal file

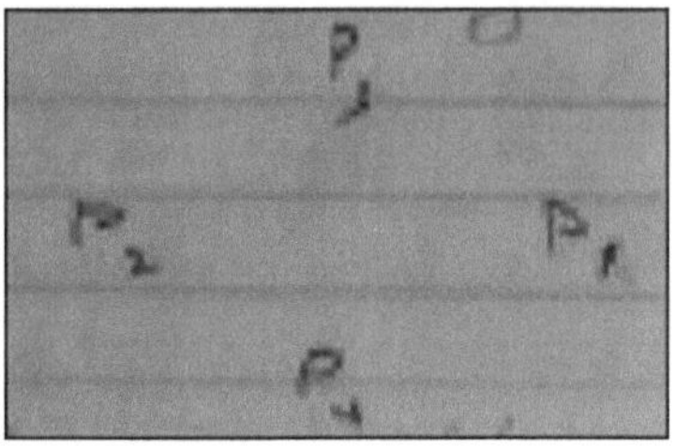

Fig. 11.3 Initial placement of first four lodge poles. *Source* Personal file

Figure 11.2 reads “When John Bear says “this way, right” he waves his arm in a perpendicular motion to first pair of trees.” This is a cultural perspective and can be considered through the lens of the four directions, north, south, east and west. Through his gesture he has created a mental image of the first four lodge poles (Fig. 11.3), this is important as revealed below, so as to make sure the lodge door is east facing.

This orthogonal arrangement, imagining the center pin not shown in Fig. 11.3, creates a mental image of four quadrants. This action of using *completely opposite* allows the first four poles to be equal spaced about the circumference, allowing for placement of the remaining poles to equal spaced as well. Culturally the four quadrants are necessary as the floor space is considered from the perspective of how a family of four would use the space.

JB: And then, so we have our four poles, we have our four poles standing (FN: this repetition allows time for John to think about the next event in the sequence of laying out the lodge poles), and then we probably want to go in between again, we’re going to estimate the distance between poles, so it’s like we’re cutting a pieces of pie (Fig. 11.4).

Remark: Fig. 11.4 reads: “When I cut pie I consider central angles, what about JB?” John Bear’s analogy to cutting pie, represented etically in Fig. 11.4 through Shockey’s use of central angles, could be about central angle, bisecting angles or arc length from a western perspective. His creation through language of this mental image gives the students an insight into what is going to occur next and how he is visualizing the process. Through this cultural process of “go in between” John Bear is placing four more lodge poles, with his mental attention to making sure it is “even.”

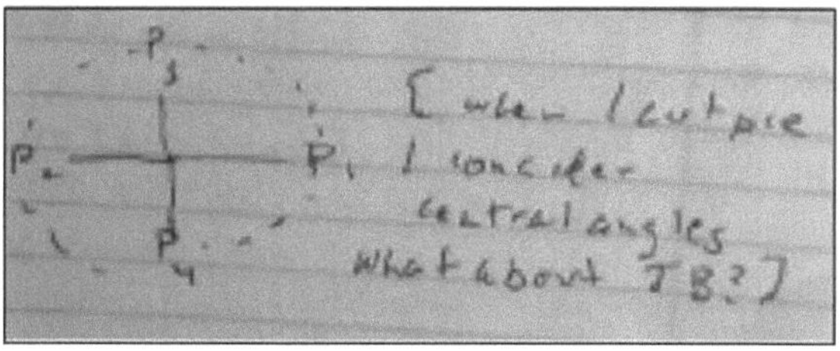

Fig. 11.4 Four quadrants of the lodge floor. *Source* Personal file

JB: We've got to go really even [*making sure that opposite lodge poles lash together, can be considered from a symmetry perspective*] and we have another issue here that we have to look at, that's where is the door going to go? Traditionally and still today when we're building one of these kinds of lodges we always go to the east, we always face our door to the east. That way when the sun rose over our head that's a ceremonial direction, that's the direction we come from (John Bear points to himself) as Wabanaki, the Dawn People, but it's also good to have the door to the east because we didn't have any windows right, so what happens when the sun comes up on a cold morning?

Remark: The placement of the door with an easterly opening is an important cultural perspective as stated by Mitchell above. Bishop's (1991) designing plays a role in how this doorway is going to occur.

S: [The chorus of response is not understandable].
JB: Okay, it's colder right before the sun come up, when the sun starts coming up, the sun throws off this stuff and makes us feel wet (he begins fanning himself, pulling his shirt as if he's sweating).
S: Heat.
JB: Heat, that you that's basic science and when you open your door the heat will come into your house right? And it will warm it up if you have door facing east. [*Ancestral Engineering*]. So, our doors going to face the east, where's the east? Just point to it.
S: (There is pointing and discussion, the sun is directly overhead when this question is posed, so the students have to negotiate from their collective understanding, they finally reach consensus and point toward a pick-up truck).
JB: Yeah, it's right about where that truck is, east is right there so it came down from somewhere over here.

Remark: This fits with Bishop's (1991) location but it also fits with the above discussion of the placement of the lodge poles being opposite one another or 180°, the sunrise and sunset are opposites.

> We're going to have a door facing about where that truck is, so is this a good place to have a pole? Put our door in the middle of a pole? It wouldn't be good, so we want our door facing east, well we're going to have a pole here (John Bear is physically locating the spot for students to see) and we're going to have a door here, so our door could go right here between these two poles, is this east?

S's: No.
JB: Not exactly, but is it close enough?
S's: Yeah.
JB: Yeah, it will do right? So, we're going to put our door here so this is a good place.

Remark: While the initial four poles are placed as shown above, from an etic perspective as described above, the remaining four poles would be placed either by central angle or arc length, making sure each pair is opposite. As an outsider looking in, what is "obvious" is not: Kluckhohn (1949) was clear on his view of the outsider looking in when he quoted Sapir: "But, in any society as Edward Sapir said, "Forms and significances which seem obvious to an outsider will be denied outright by those who carry out patterns; outlines and implications that are perfectly clear to those may be absent to the eye of the onlooker"" (p. 36). As the outsider looking in, Shockey had no consideration of the door placement and how that might impact the placement of the final four lodge poles.

JB: I think we're doing an eight pole lodge, one, two, three, four, five, six, seven, eight [*he is counting while pointing to the circumference*]. I made a mistake when I was thinking to myself earlier, twelve pole lodges are ceremonial lodges, if we did a twelve pole lodge you're going to be able to crawl on the thing it's going to be so solid, eight pole lodge is going to be solid as well, so I need eight poles, could somebody go, could we have eight people go choose, may we get teachers to help? [*At the lodge pole pile, students are trying to dig through the pile to choose eight, while John Bear is culling out the poles that don't look bendable*]. Okay without having to say it, you guys paired these trees up pretty good, now we're not pairing up from the base, we're pairing up from the top, right?

Remark: This is the first mention of this being an eight-pole lodge. Figure 11.4 and Mitchell's language of "go in between" hinted at this above. Mitchell's cultural perspective allowed him to design and locate this eight-pole lodge mentally and to pedagogically model its physical manifestation.

S's: Yes.
JB: So, what we're going to do is, I have a number of adults here, probably three of us that can do this, we have to sharpen the bottom of these poles. And so what we need to do is take a look at the bases and the base of this one is pretty bent, right? But, when we're building this lodge are we going to put the poles straight like this? And then bend them this way [toward the center of the circular floor], what happens if we put a pole here and we tried to bend it?

Remark: Here John Bear models his question by holding the pole perpendicular to the ground, pedagogical modelling. Embedded in Mitchell's question "are we going to put the poles straight like this?" is his cultural understanding, ancestral engineering, of why this type of placement can be harmful.

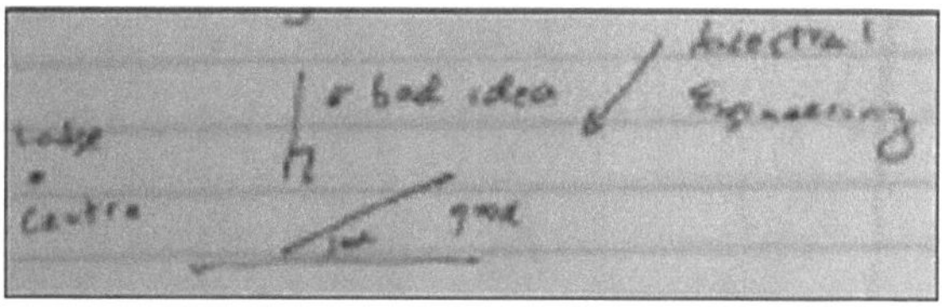

Fig. 11.5 Acute angle placement of lodge pole. *Source* Personal file

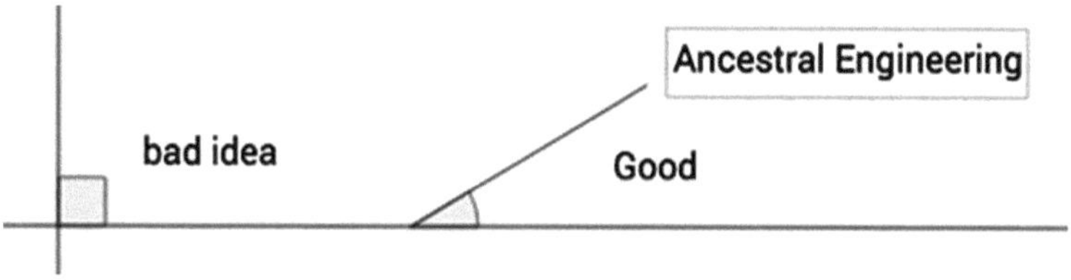

Fig. 11.6 An etic pictorial representation. *Source* Personal file

S: It's going to break.
JB: It's going to break, because it's standing straight up and we're just pulling it over right? Something else is going to happen to it, it will pop out [Ancestral Engineering] So I want to get an angle on here that's not straight [*straight is the insider view for perpendicular*] What we're going to do is we're going to come in at an angle with the poles, okay? Figure 11.5 is from the field notes.

Remark: While the description by John Bear is clear to himself, as an onlooker, Shockey had not idea of the need for an acute angular placement of the lodge pole. John Bear has a five foot steel bar that he uses to create holes about the circumference where the sharpened base of the lodge poles will be inserted. This is an instance of the pole position being "perfectly clear" but "absent to the eye of the onlooker" (Kluckhohn 1949, p. 36). A clearer representation of Fig. 11.5 is shown in Fig. 11.6.

The notation used in Fig. 11.6, representing perpendicular with the square representation, introduces a non-linguistic etic representation for Mitchell's cultural perspective of a bad ancestral engineering design.

JB: So, basically we're going to put the pole in the ground, we're going down, just try to make the hole at about this angle so that's where our tree going to stick in, so if this tree that I am holding, watch your head over there, and I stick it in the ground, is this a good angle?
S: Yeah.
JB: This one will work, it just so happens the one I, another one I cut was real squiggly and it just doesn't work, so what we need is someone to hold this up and someone to sharpen it.

Remark: John Bear is modelling holding the pole with the base against a log, someone else will then use an axe to sharpen the resting base of the pole. A teacher volunteers to hold the tree allowing John Bear to demonstrate how to use the axe to sharpen the base of the tree, making it easier to place in the hole on the circumference. To this point in the construction, John Bear has laid out the radius for the circular lodge base and has begun giving instructions and demonstrating how to place the lodge poles. Before he places the first lodge pole, he says to the students "so basically we know our concept, right?" This statement was met with agreement by the students.

JB: What I'm going to do, I'm going to do one of these by myself and let volunteers do the rest. (Pedagogically, John Bear emphasized modelling what was to occur next in the construction of the lodge). I just want the hole to get kind of wide, but not too wide (he places the first lodge pole), okay that's probably good, now we want to go opposite this [*this can be visualized as rotating the second lodge pole 180° around the circular bases center from the first lodge pole*] our strings down here somewhere [the string represents the radius] you want to make sure it goes at an angle (John Bear has already modeled how to create the hole for the lodge pole, this remark is a reminder of the needed angle. John Bear is not actively engaged in placing the second pole, he is standing opposite holding the first lodge pole). Want to go down about a foot and a half (this is how deep the lodge poles are placed in the ground) that means we need to sharpen another pole. (As for sharpening the poles, while this has been modeled, it is not something the students are allowed to do, due to the axe being very sharp. John Bear continues to sharpen the poles so as not put anyone in harms way).
T: At this angle, then jam that thing in there (The first adult that John Bear modeled using the steel bar to create an angled hole in the ground for placing a lodge pole, in this statement is now modelling for a student of how to use the steel bar for that purpose. For this part of placing the lodge poles, creating a hole in the earth about a foot and a half deep at an acute angle, John Bear is now done as others have taken over that responsibility).

Remark: At this point in time, the lodge poles are all placed about the circumference of the circular base. What follows are the instructions for how to bend opposite lodge poles together so these can be lashed together. Figure 11.7 shows how the poles are placed prior to being lashed together.

JB: I'm left handed so I've got to stand with my back to you (John Bear has his back to everyone in the Fig. 11.6 and is preparing to model how to bend the lodge pole toward the center of the circular base). I'm going to bend it this way, but what we're going to do, I'm going to stand on the base [*Ancestral Engineering*], we're going to need someone in the middle who is tall, and your job is going to be to grab the tree, but not pull it, just hold it. And then what we'll do is let somebody eyeball it when they (the lodge poles) come together, they're probably going to bend way over each other, that's okay, you're going to hold them at the middle, at the peak where they touch and we're going to tie them. This will be the quick first tie and

Fig. 11.7 Demonstrating how bend lodge pole toward center. *Source* Personal file

then we'll cut the branches where they come down too far, okay? So this is what I'm going to do, I'm standing here right (there is a student at the opposite lodge pole, John Bear is talking her through and modelling the process for bending the lodge pole. In the Fig. 11.6, the student is wearing a short sleeve green t-shirt). Notice my feet are in the lodge (Referring to being inside the circle, which is the position he wants the student opposite him to also be in). I kind of walk (John Bear's hands are "walking" up the pole, pulling it toward the center to initiate an arch. With his feet placed firmly against the base of the lodge pole, he assures that the lodge pole will stay fixed in the earth) very slowly, don't pull.

Remark: From the field notes, an interesting, intrusive event occurred at this time. The student opposite John Bear was following how to move her hands so that she was "walking" her hands up the lodge pole. John Bear was speaking directly to her when he said "don't pull." The students' impatient teacher physically pushed her out of the way and took over the process of bending the lodge pole toward the center. From the field notes "the teacher's actions are not following what John Bear modeled, he has managed to stand the pole vertical (he did not place his feet against the base of the lodge pole) without regard for the student attempting to walk her hands up the lodge pole". The rudeness of this teacher is noteworthy for two reasons. First, his actions demonstrated that he did not value the effort of the student and second, he imposed his own strategy for bending the lodge pole which resulted in the lodge pole not being in the ground at an angle, but perpendicular to the surface of the earth. Later in the day John Bear and Shockey pondered how frequently teachers get impatient with students, resulting in the teacher not allowing the student to participate in the learning moment.

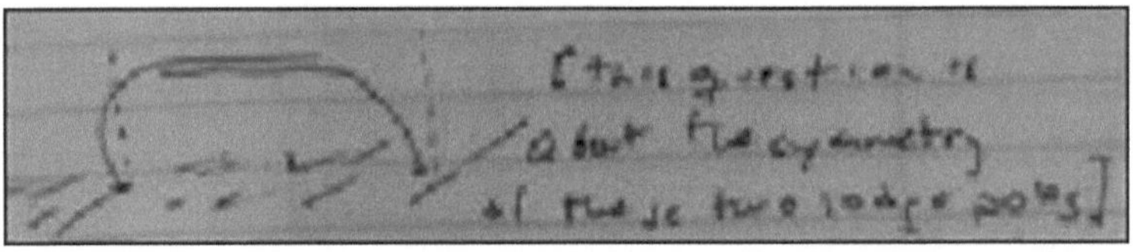

Fig. 11.8 Field note illustration of symmetry of opposite lodge poles. *Source* Personal file

JB: Come down, now are these pretty even [*"Pretty even" is about symmetry*].

Figure 11.8 shows a field note illustration of symmetry of opposite poles of the Penobscot lodge.

Remark: Fig. 11.8 reads: "This question is about the symmetry of these two lodge poles." Culturally, that is not what is happening. Here is an element of the ethnomodel for this lodge, the placement of opposite lodge poles and how these will be lashed together over the center of the lodge floor center.

S: Not bad (FN: the poles can be adjusted before being tied, so a viewer away from the lodge makes the symmetry determination).

JB: Let's tie this, go ahead you've got the string, kind of wrap it in the middle. Now this lodge is huge, we can stand up in it. (With the lashing of the first two lodge poles John Bear realizes that he standing under these with plenty of space between the top of his head and this first arch). [FN: previously the estimated lodge pole lengths were guessed to be about nine feet, these lodge poles are probably twelve to fifteen feet in length, maybe longer]. And I've actually made these so big that we've had to use ladders in the back of a pickup truck to reach the middle.

Remark: The last statement made by John Bear about using a ladder in a pick up truck references a lodge construction he had done in the past. The lodge was so large that the point where the lodge poles met once bent toward the middle, were so high off the ground that to be able to reach them and lash them together he needed a ladder placed in the bed of a truck (Figs. 11.9 and 11.10).

Remark: This process of bending lodge poles continues until all eight poles are secured forming the dome. The next phase in this construction is the placement of poles around the perimeter of the lodge poles, and these poles are parallel to the ground. Placement of these poles horizontal to the ground is another aspect of the ethnomodel. This is what Rosa and Clark (2013) state as a "practical application of ethnomathematics." By considering the constituent parts described in this construction, we are bringing the cultural perspective forward in the solving of specific problems. Elements of Bishop's (1991) six, designing, building, locating, explaining, counting are all present in the development of this ethnomodel.

JB: You'll see the true shape of the lodge. What you're going to do, you're going to take string and you're just going to tie it around here like an X. So you probably want to get a piece of string that's about four feet long and you want to wrap it around and wrap it around the other way and just tie it off. [FN: the instructions are

Fig. 11.9 Six lodge poles lodged together. *Source* Personal file

Fig. 11.10 Students tying lodge poles together. *Source* Personal file

vague enough that students have to make decisions]. We're going to do our first row about here [FN: about knee height] and we're going to do another one about a foot and half and in the smaller space we're going to start putting sticks across. (These side poles add tensile strength to the lodge. The "smaller space" is a reference to the moving toward the top of the dome, the lodge poles converge on the top and distance between them shortens. This highlights another example of ancestral engineering.)

Remark: From the field notes: "The students are busy putting the *rails* on the lodge, taller students are working on the top, shorter students are working closer to the ground, lots of eager hands. John Bear was off to the side talking with an adult, watching the students. His role was that of helper, now as he has modeled what the students needed to do, he is an observer. A student had a question; John Bear stopped his talking to help". These notes are shared to reinforce how John Bear's pedagogy engaged the students and how after he modeled a situation; he would get out of the way and allow the students to work. During this process, students were making decisions, negotiating meaning, and on the occasion when a question was posed to John Bear, it typically meant that the students had reached an impasse and were uncertain of to do next.

JB: Okay buddy, how high is our door going to be? We don't want it too high, no too high (Here a student is placing a horizontal stick that will serve as the top of the entry way). Okay, so what you're going to do is take a piece of string, okay swing around here, want to come out on the other side. (Here John Bear is giving verbal instructions for the student as he wraps the string around the horizontal stick to secure the two ends to lodge poles) (Figs. 11.11 and 11.12).

Remark: From the field notes: "Without roles being assigned students have just taken over, some are bringing poles to be lashed, some are distributing strings, adults are cutting strings, students are holding poles while other lash, just in a general sense of cooperation". This is important at this point the students *own* the project. They have come to understand what needs to happen next and have embraced the construction. In the episode described above of the placement and

Fig. 11.11 First layer of covering on hemispherical lodge. *Source* Personal file

Fig. 11.12 Final layer of covering on hemispherical lodge. *Source* Personal file

lashing of the horizontal poles, the video shows a girl stopping the placement of the horizontal poles as she realizes they are not parallel. The first horizontal pole was placed about a foot and a half from the ground, but when the horizontal pole above it was placed, the gap was not even, not parallel. This student imposed a unit of measure to assure that these poles would be parallel, the distance between her elbow and closed fist. She walked about the lodge, would place her arm to make sure poles were parallel and other students would lash the horizontal pole in place, it was brilliant.

11.4 Practical Applications

The practical application of ethnomathematics and the development of an ethnomodel of this lodge can be a conduit for children to engage in western mathematics. The youngster that used her arm to make sure parallel placement of the tensile strengthening poles occurred is an example of how a student may engage.

Considering the Ancestral Engineering (Corrine Mt. Pleasant Jette, personal communication) perspective that Mitchell brought to this construction, it is rich in relationships. Included in these relationships are Bishop's (1991) activities. Lodges were typically built based on location using available materials. The time of year determined what materials were available. These cultural perspectives for time of year and what materials are available are elements of the ethnomodel for this lodge.

Each of the vertices is connected in the Fig. 11.13 as we feel these are not independent events, these activities of Bishop (1991) are related. We go so far as to suggest that the accomplishing these activities would potentially be different based on the time of year. For example, this is a summer lodge being described; a winter

Fig. 11.13 Bishop's (1991) six activities expanded to include calendar. *Source* Personal file

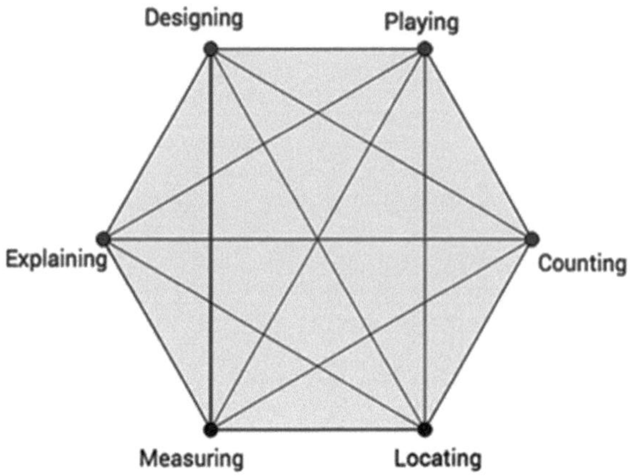

Fig. 11.14 Great Salt Bay in background. *Source* Personal file

lodge would begin with circular base that was a *cylinder* as it would be dug into the earth that would provide insulation and opportunities for heating during the winter.

Calendar as considered in this application was an important consideration; the types of materials that were available in the autumn and the time of this construction were different than early summer materials. For this construction, built on the Great Salt Bay, the consideration for a hemispherical lodge was important, as it would withstand strong winds. The Great Salt Bay (Fig. 11.14) is a tidal area; considerations for wind as well as tide were taken into consideration with respect to the type of lodge as well as the location of the lodge to avoid flooding.

The circular base of the lodge was purposeful. Mitchell approximated his height with outstretched arms, which served as a "radius" for the lodge floor. Mitchell cut a piece of twine to represent this radius, attached the twine to a center post allowing him to construct a circle as he walked around the center post as he held the twine. As Mitchell was establishing the perimeter of the base, he was also considering the purpose of the floor space. The floor space served as sleeping space as well as working space inside the lodge. Children, adults, and guests would sleep in different regions of the lodge and if there were no guests then there was interior workspace for preparing food. Cooking was not done inside the lodge, the workspace would allow getting out of inclement weather.

Another practical application of using body height to approximate the floor space dealt with the interior height of the lodge. Mitchell understood from his perspective and experience that a summer lodge with the stated radius would also provide an interior height so that he would not have to stoop while insider. This consideration of interior height is important for the inhabitants of the dwelling to be comfortable.

To this point in *designing and building*, Mitchell was considering the practical elements of the lodge. Another consideration was the length of the poles that would be placed about the boundary of the circular base. In his mind he was *calculating* the length of the needed poles, such that opposite poles would be lashed together over the center of the hemisphere. The lodge poles had to be long enough to satisfy three conditions, the portion of the pole that would buried in the ground, the length of the pole that while arched would be long enough to cover the lodge radius and the third portion has to be of suitable length such that when opposite poles were lashed together there was enough overlap.

Tall grass was available, which would be weaved into mats that would serve as one layer of the lodge covering. The lodge could not be larger than half the length of the lodge poles, but had to be large enough to be functional. The placement of eight lodge poles about the circumference was needed, but the placement had to allow for the placement of the lodge entrance to face the East.

As Mitchell shared with the students, "but, we always go to the east, we always face our door to the east and that way the sun rose over our head. Now two different things, first of all that's a ceremonial direction, that's the direction we came from as Wabanaki, the Dawn People. But it's also good to have your door to the east because we didn't have any windows right, so what happens when the sun comes up on a cold morning?".

The lodge had a circular base with eight lodge poles equi-spaced about the circumference. The etic perspective can focus on the determination of the placement of the lodge poles on the circumference, while the emic perspective brings much broader considerations. Considerations for height, for utilization of floor space, for size of materials, etc. all play important roles in Mitchell's spatial image of the final lodge. With the placement of a stake in the ground to represent the center of the lodge and a string to determine the base radius (emic), Mitchell supports the response of the students with respect to making the lodge *even*. Students immediately recognized that the lodge poles and the lodge radius are related.

When a lodge pole is placed in the ground, it is then bent toward the lodge center and tied to an opposing lodge pole that is also bent toward the center. Symmetry (etic) of the placement of the poles is important, but as important is how the length of the poles determines the area of the circular base of the lodge (emic). The circular base serves three distinct purposes: two areas for sleeping, children and adults; and a workspace. The lodge was a summer lodge that provided shelter from summer heat and summer rain.

Mitchell's pedagogy related to the initial placement of the lodge poles and the east-facing door revealed a Native pedagogy based on modelling. This is modelling that allows students to *see* what they will be doing. Modelling this and the discussion about the importance of the sun brings forward a cultural perspective. As noted above, the placement of the lodge door to the east allowed the heat of the sun to heat the lodge in the morning. The placement of the lodge poles into the earth revealed learned engineering that Mitchell modeled for the students.

Lodge poles were stuck in the ground at an acute angle, with the pole facing away from the lodge center (emic). Had the pole been placed straight into the ground, the likelihood that it would spring out of the ground causing harm was much greater than with the acute angle placement. Once this placement of the lodge pole was modeled, Mitchell would step aside and allow the students, and their teachers, to take over the construction. If assistance was needed, the learners understood that it was appropriate to pose questions. These elements are what we view as ethnomodelling, bringing the cultural perspectives to the practical application of ethnomathematics.

Once the eight lodge poles were placed and the placement of the door was negotiated, Mitchell intervened to provide pedagogical modelling of how to secure two, opposite lodge poles together over the center of the lodge. That is Mitchell physically lashed opposite lodge poles together allowing students to view the process. Once this was demonstrated Mitchell would again step aside and observe. Mitchell's pedagogy was also conversational. When he posed a question, he was very attentive to the students' responses.

As Mitchell directed the dialogue, sense making occurred, meanings were negotiated, and connections were made. For example, once the lodge poles were placed and secured to construct the dome frame additional poles were placed about the lodge, parallel to the ground, adding strength to the lodge. It became immediately apparent to one student that these poles were not parallel. She quickly determined that the unit of measure determined by the distance between her elbow and fist would ensure that these additional poles would be parallel.

11.5 Challenges

Gilsdorf (2012) in his book *Introduction to Cultural Mathematics*, makes clear that using Western mathematics to describe cultural phenomena is problematic.

> Writing about the topic of cultural mathematics for readers with backgrounds primarily in Western mathematics brings one to a dilemma: On one hand, using Western terminology and notation to describe mathematics of non-Western cultures is inherently inaccurate because people in such cultures would not think of the mathematical content in the same way as it is perceived in Western culture. On the other hand, if the goal is for people of Western backgrounds to understand how cultural activities can be understood as mathematics, then one must speak to readers in familiar mathematical terms (p. xii).

This is a dilemma for the authors as well. Shockey, a Western trained mathematics educator does not understand the Passamaquoddy language of Mitchell, thus the descriptions are in English, allowing Shockey to understand. Another challenge that could offer insight to this work could focus on the Passamaquoddy language and whether words exist for the western mathematical language used in this chapter. The cultural backgrounds of the participating children were not taken into account.

As a western trained mathematics educator, Shockey experienced the "because of the language habits of his community" (Kluckhohn 1949, p. 167), the community of mathematics educators. While a challenge, this is important to bring the etic forward. Certainly, Shockey and Mitchell have similar linguistic backgrounds (Whorf 1956), but only with respect to English. And as Whorf (1956) states, "We are thus introduced to a new principle of relativity, which holds that all observers are not led by the same physical evidence to the same picture of the universe" (p. 214).

How Mitchell viewed this and how Shockey viewed did not all reveal the "same picture". What Shockey viewed pedagogically was different that his academic preparation. The language use of Shockey, as shown throughout, was different than the language use of Mitchell. Is this complementary? It could be argued that Mitchell was also a mathematics educator, certainly through the framework of Bishop's (1991) six activities.

11.6 Concluding Remarks

The importance of how these youngsters were engaged, particular to the education of Native youth, is best stated by Cajete (1994):

> In Tribal education, knowledge gained from first-hand experience in the world is transmitted or explored through ritual, ceremony, art, and appropriate technology. Knowledge gained through these vehicles is then used in everyday living. Education, in this context, becomes education for life's sake. Education is, at its essence, learning about life through participation and relationship in community, including not only people, but plants, animals, and the whole of Nature (p. 26).

We concede the argument that the practicality of the lodge is questionable in modern society. The *mathematisation* that was captured and discussed we feel is valuable. The "first hand" experience of these students and what value it brought to them personally and educationally, we can never know. Although: during this two-week event, many students self revealed in confidence that they were Native. These students would mention a family member that lived on the reservation to learn if Mitchell knew them. This emerging "dignity and confidence" was emotional. We can only conjecture that the experience of working on the site directed by a Native Educator was respectful enough of each learner, such that the Native youth found the confidence to share who they were.

We feel that this adds to the ethnomodelling literature. Rosa and Orey (2013) state that ethnomodelling is "a practical application of ethnomathematics, and which adds the cultural perspective to modelling concepts" (p. 78). Bishop's (1991) six brings forward the ethnomathematics involved in the construction of this lodge. The many examples of mathematisation, through measuring, design, explaining, counting, and locating, highlighted throughout are important elements that bring Mitchell's cultural experience out.

The construction of his lodge is not intended to be generalizable, but the use of these activities of Bishop's (1991) may provide a lens for other to consider their "practical application of ethnomathematics." Throughout we emphasize the Penobscot perspective that Mitchell brings to this, particularly the pedagogical modelling he uses to engage his learners. We illuminate how important pedagogical modelling is, as does Cajete (1994) when he says "learning about life through participation". We acknowledge that the analysis using etic and emic may be problematic:

> So analysis from a western perspective breaks everything down to look at it. So you are breaking it down into its smallest pieces and then looking at those small pieces. And if we are saying that an Indigenous methodology includes all of these relationships, if you are breaking things down into their smallest pieces, you are destroying all of the relationships around it. So an Indigenous style of analysis has to look at all those relationships as a whole instead of breaking it down, cause it just won't work (Wilson 2008, p. 119).

We include Wilson's remarks above, acknowledging that while we did not engage an Indigenous methodology, we respect the importance of relationships.

While Mitchell is western trained as an educator, the teaching cliché of 'we teach the way we are taught' deserves elaborations. Had Mitchell taught the way the academy prepared him, the outcome, we believe would be very different. Fortunately, Mitchell taught the way his Elders taught him. He would model activities in the sequence that when finalized would yield the dwelling. This was not a linear trajectory. His teachings were mixed with stories and opportunities for discussion. Once he was convinced that students understood, he would stand by and allow them to work, to make sense, and to serve as a helper if the students needed help.

In 1928 Schlauch wrote:

> Any normal child is blessed with natural curiosity – that heritage of the evolutionary struggle during which not to comprehend the environment and its dangers meant death. Children take joy in mastering knowledge which they can see has some relation to the phenomena of their lives. It is only the mass of abstract material in a dull curriculum, unpedagogically presented, that finally kills the desire to learn (p. 28).

Let us all be mindful that a child's natural curiosity should be valued and supported.

References

Albanese, V., Adamuz-Povedano, N., & Bracho-Lopez, R. (2016). *The evolution of ethnomathematics: Two theoretical views and two approaches to education*. Paper Presented at TSG35: Role of Ethnomathematics in Mathematics Education, ICME13. Hamburg, Germany: University of Hamburg.

Bassenezi, R. C. (2002). *Ensino-aprendizagem com modelagem matemática [Teaching and learning with mathematical modelling]*. Sao Paulo, SP: Editora Contexto.

Bishop, A. J. (1991). *Mathematical enculturation*. Dordrecht, The Netherlands: Kluwer Academic Publishers.

Cajete, G. (1994). *Look to the mountain: An ecology of Indigenous education*. Skyland, NC: Kivaki Press.

D'Ambrosio, U. (1985). Ethnomathematics and its place in the history and pedagogy of mathematics. *For the Learning of Mathematics, 5*(1), 44–48.

D'Ambrosio, U. (1990). *Ethnomatematica [Ethnomathematics]*. São Paulo, SP: Editora Atica.

Gilsdorf, T. E. (2012). *Introduction to cultural mathematics: With case studies in the Otomies and Incas*. Hoboken, New Jersey, NJ: Wiley.

Jorgenson, D. L. (1989). *Participant observation: A methodology for human studies*. Thousand Oaks, CA: SAGE.

Kluckhohn, C. (1949). *Mirror for man*. New York: Whittlessy House, McGraw-Hill Book Company Inc.

Pike, K. L. (1967). *Language in relation to a unified theory of the structure of human behavior*. The Hague, Paris: Mouton.

Rosa, M., & Clark, D. (2013). Ethnomodelling as a methodology for ethnomathematics. In G. A. Stillman, K. Gabriele, W. Blum, & J. P. Brown (Eds.), *Teaching mathematical modelling: Connecting to research and practice* (pp. 77–88). Dordrecht: Springer.

Rosa, M., & Orey, D. (2006). Abordagens atuais do programa etnomatemática: Delinenando-se um caminho para a ação pedagógica [Current approaches in ethnomathematics as a program: Delineating a path toward pedagogical action]. *Bolema, 19*(26), 19–48.

Rosa, M., & Orey, D. (2013). Ethnomodelling as a research lens in ethnomathematics and modelling. In G. A. Stillman, G. Kaiser, W. Blum, & J. P. Brown (Eds.), *Teaching mathematical modelling: Connecting to research and practice* (pp. 117–130). Dordrecht: Springer.

Shockey, T. L., & Silverman, F. L. (2016). Ten years later: A look back on the history. *Journal of Mathematics and Culture, 10*(2), 12–36.

Schlauch, W. S. (1928). Mathematics as an interpretation of life. In J. R. Clark & W. D. Reeve (Eds.), *Selected topics in the teaching of mathematics* (pp. 24–34). New York, NY: Bureau of Publications Teachers College, Columbia University.

Whorf, B. L. (1956). Science and linguistics. In J. B. Carroll (Ed.), *Language, thought & reality: Selected writings of Benjamin Lee Whorf* (pp. 207–219). Cambridge, MA: MIT Press.

Wilson, S. (2008). *Research is ceremony: Indigenous research methodologies*. Black Point, Nova Scotia: Fernwood Publishing.

Part V
Philosophical Features of Ethnomathematics: A Theoretical Basis

It is important to raise the discussion regarding the theoretical basis for ethnomathematics as a program in order to enable the debate in regards to its epistemologies. There is a recognition that ethnomathematics as a science is located in the confluence zone between mathematics and cultural anthropology since this research highlights the presence of the mathematics developed by the unique contexts of the people from a variety of traditional cultures by illustrating how their mathematical ideas play a vital role in their diverse human endeavors. These perspectives show two of the numerous philosophical features of an ethnomathematics program.

Chapter 12
Ethnomathematics and Its Pedagogical Action in Mathematics Education

Ubiratan D'Ambrosio and Milton Rosa

Abstract Ethnomathematics school practices favor respect for, solidarity and cooperation with, the others. It is associated with the pursuit of peace. The main goal of ethnomathematics is building up a civilization free from truculence, arrogance, intolerance, discrimination, inequity, bigotry, and hatred of the others. These are basic questions that define philosophical and ideological postures, which are the roots of a holistic theory of knowledge, looking into the generation, the individual and social organization, and the institutionalization, transmission and diffusion of knowledge. This concept of ethnomathematics is primeval in recognizing the emergence of perceptions of space and time and the techniques of observing, comparing, classifying, ordering, measuring, quantifying and inferring that are different styles of abstract thinking in the school curricula. This is can be achieved by the application of the trivium curriculum, which is an innovative ethnomathematical approach that needs more investigations to address its pedagogical purposes.

Keywords Conceptual dimension · Ethnomathematics · Mathematics curriculum · Pedagogical dimension · Trivium

12.1 Introduction

Many students live in a world dominated by fear, uncertainty, and arrogance, as well as there is a general feeling that the planet is moving toward some form of catastrophic, either economic, political or environmental in nature (Rosa 2000). The most visible are the population and environment crises. Another, more subtle,

U. D'Ambrosio (✉)
Universidade Anhanguera, São Paulo, Brazil
e-mail: ubi@usp.br

M. Rosa
Universidade Federal de Ouro Preto, Ouro Preto, Brazil
e-mail: milton@cead.ufop.br

M. Rosa et al. (eds.), *Ethnomathematics and its Diverse Approaches for Mathematics Education*, ICME-13 Monographs,
DOI 10.1007/978-3-319-59220-6_12

but equally lethal crisis is the humankind's relations to its extensions, institutions, ideas, as well as the relationships between the many individuals and groups that inhabit the globe (Hall 1976; Toffler 1990). It seems that faith in humanity has been forgotten as part of the rationalistic model that characterizes much of contemporary society (D'Ambrosio 2011).

Many educators know that citizens must have specific knowledge and abilities to solve problems in an increasing complex and dynamic society. Teachers realize that students become motivated when they are involved in their own learning. According to Freire (2000), this is especially true when students are encouraged to deal directly with the issues of greatest concern to themselves and their communities.

Traditional school curriculum has neglected the contributions made by minority groups and non-dominant cultures. Given these conditions, educators need to prepare students to live and succeed in this uncertain world. While this crisis of confidence may contribute to the reason for many old skills giving ways to new technologies, it is the capacity to explore and discover, the freedom to succeed without fear of failure that combine to prepare and empower students to take change of their own future. Educators have the obligation to help students to uncover, understand, and comprehend the power of mathematics in their lives.

In this context, Rosa (2000) argues that traditional school curricular programs that focus on basic *skills* are more often than not a source of discouragement, anxiety, and repetition, especially for students who have been subjected to this form of mathematics for years on end. All students deserve access to curricular activities that allow them to see the usefulness, power, and excitement that mathematics can offer.

One of the most important objectives in teaching mathematics should be the development of mathematical capacity in all students. In addition to teach students techniques and tools to solve mathematical problems, they need to learn more than basic mathematical algorithms. All students need to extend their understanding to include how mathematics connects to other disciplines, to problems in society and the environment, and how diverse people around the world use it.

All students need to be encouraged to develop and apply higher-level critical and reflective thinking skills. They must learn mathematical concepts, they need to learn to question, to take risks, verbalize ideas, listen to others' ideas, and to analyze their own and others' ideas. This form of critical thinking must be practiced regularly within mathematical contexts.

A primary goal of learning mathematics is to develop powerful and critical students to be better citizens. The investigation and exploration of ethnomathematical relations can help to develop such students' talent. The teaching of mathematics must emphasize the development of ideas, topics, and themes relevant to the current and future needs of the students (Rosa 2000). The integration of ethnomathematics into the mathematics curriculum enables students to reason both sequentially and holistically to allow them to appreciate distinct relations among different forms of mathematics.

For these reasons, teaching mathematics today is a challenging job. It is necessary that teachers are prepared to develop students' mathematical abilities, critical attitudes, and empower them with mathematical expertise and confidence by including mathematical ideas from different cultures and assist students to acknowledge the contributions that individuals from distinct cultural groups have made to the development of mathematical knowledge.

Ethnomathematics is a pedagogical tool that helps teachers and students to understand both the influence that culture has on mathematics and how this influence results in diverse ways in which mathematics is used and communicated (D'Ambrosio 2006). In this regard, ethnomathematics should become central to a complete study of mathematics.

12.2 Ethnomathematics: Theory or Practice?

People frequently ask: *Is ethnomathematics a theory or a practice?* In order to answer this question, it is necessary to reply with another question: *Is there a practice without a theory supporting it? Is it justifiable to have a theory without a practice?* One of the main objectives of this chapter is to discuss the false the dichotomy between theory and practice.

The main reasons to bring ethnomathematics to schools are to:

1. Demystify school mathematics as a final, permanent, absolute, unique form of knowledge. There is a current misperception in societies, very damaging, that those who perform well in mathematics are more intelligent, indeed *superior* in relation to others. This erroneous impression given by traditional forms of teaching is easily extrapolated to religious, ideological, political, and racial creeds.
2. Illustrate intellectual achievement of various civilizations, cultures, peoples, professions, and genders. Western mathematics is absolutely integrated with conquest and colonialism that came to dominate the entire world. The acceptance, forced or voluntary, of western mathematics and western knowledge in general leads to the acceptance of behavior and values, of ideas like *the winner is the best, the losers are to be discarded.* More than any other form of knowledge, mathematics is identified with the winners. This is true in history, in professions, in everyday life, in families, and in schools. The only possibility of building up a planetary civilization depends on restoring the dignity of the losers and both winners and losers, moving together into the new. This requires respect for each other. Otherwise, the losers will direct their efforts to become winners and the winners will do the best to protect themselves from the losers, thus generating confrontation.

School ethnomathematics practices encourage respect, solidarity and cooperation with the other. It is thus associated with the pursuit of Peace. The main goal of ethnomathematics is to support the building up of a civilization free of truculence, arrogance, intolerance, discrimination, inequity, bigotry, and hatred.

These are basic questions that define philosophical and ideological postures. These postures are found in the roots of a holistic theory of knowledge, looking into the generation, the individual and social organization, and the institutionalization, transmission and diffusion of knowledge, as studied in the Program Ethnomathematics (D'Ambrosio 2015). Repeating what is written in many of D'Ambrosio's papers, his concept of ethnomathematics is primeval. It recognizes, in every corner of the planet, the different emergence of perceptions of space and time and the techniques of observing, comparing, classifying, ordering, measuring, quantifying and inferring and, as consequently, different styles of abstract thinking.

In each corner of the planet and at every time, individuals have developed strategies to satisfy the pulsions of survival and transcendence. These strategies are synthesized in three words: the *techné* ≈ *tics* [i.e., the ways, modes and styles, the arts and techniques] that people developed for *mathemá* [i.e., for explaining, learning and understanding, knowing and coping with] their *ethno* [i.e., their natural facts and phenomena and the social, cultural, mythical and imaginary environment]. This etymological exercise led D'Ambrosio to construct the concept of *tics* of *mathema* in distinct *ethnos*, or *ethno* + *mathema* + *tics*, hence, by rearranging the words, ethnomathematics.

Ethnomathematics as a research paradigm is much wider than traditional concepts of mathematics and ethnicity and any current sense of multiculturalism. *Ethno* is related to the members of distinct groups identified by cultural traditions, codes, symbols, myths, and specific ways of reasoning and inferring (D'Ambrosio 1985). Hence, ethnomathematics is the way that members of various cultural groups mathematize[1] their own reality because it examines how both mathematical ideas and practices are processed and used in daily activities. It is also described as the arts and techniques developed by individuals from diverse cultural and linguistic backgrounds to explain, to understand, and to cope with their own social, cultural, environmental, political, and economic environments (Rosa 2010).

There is no contradiction with this concept of ethnomathematics and the universally accepted concept of academic or scholarly mathematics. Indeed, a very synthetic view of ancient cultural history shows that the universally accepted concept called western mathematics is an elaboration of the specific way of the peoples of the *ethno* of the Mediterranean basin organized their *tics* of *mathemá* in that region, hence their own ethnomathematics. Since there were many civilizations

[1]Mathematization is a process in which members from different cultural groups come up with different mathematical tools that can help them to organize, analyze, comprehend, understand, and solve specific problems they face in the context of their real-life situation. These tools allow them to identify and describe a specific mathematical idea or practice in a general context by schematizing, formulating, and visualizing a problem in different ways, discovering relations, discovering regularities, and transferring a real world problem to a mathematical idea through mathematization (Rosa and Orey 2010).

in this broad region, they had many contacts and cultural encounters and their ethnomathematics, as every one, went through reformulations.

It is important to recall the encounters of civilizations around the Mediterranean, mainly those of Ancient Iraq, of Egypt, of Israel, of Greece, of Persia, of Rome, and many others. They were close enough to have mutual influence, through a dynamics of cultural encounters, and eventually gave origin to Greek philosophy, which was synthesized in the *Elements*, composed by Euclid of Alexandria, *c*300 a.C., and which inaugurated what became known as the *Euclidean Style*, by creating a specific kind of narrative and criteria of truth. Although no authentic complete copy is extant, the *Elements* became the canon of what is now called western mathematics.

For religious reasons, it was rejected by Christianity in the period known as the Early Middle Age, but in this period it was preserved and advanced in Hellenistic academies in Northern Africa, especially in Alexandria, and by Arabic and Islamic scholars. After the Crusades, the *Euclidean Style* was absorbed by Europe in the periods known as the Late Middle Age and Renaissance. This gave rise to the corpus of knowledge now called western, academic or school mathematics or, simply, mathematics.

This corpus of mathematics knowledge came to be organized as a discipline in the 15th century, and was successful as the theoretical foundation of modern science and technology and of the capitalist economic system. This discipline spread through Europe and after the great navigations, conquests and subsequent colonialism, to the entire world.

The colonial process established school systems in the occupied territories, which was continued after independence and which has prevailed to the present. In the schools, mathematics was and continues to be central in schools, in all levels, and also in the universities and in research. Mathematics became necessary for the commerce, for the production and for the economy systems, for technology and for sciences.

A major cause of social exclusion is the lack of competencies in mathematics. It results in difficulties in employment and in many common daily activities. It is the same as in illiteracy and innumeracy that excludes individuals from participation in society. It is the same as in learning to read and write, in learning to deal with basic quantities (arithmetic) and forms (geometry) that are essential in every cultural environment. They are essential as communicative instruments.

Good mathematical learning occurs with social communication and cultural interaction through dialogue and negotiation of meaning of the symbolic representations between teachers and students. To include and exclude differences and mathematical traditions brought to the school community by the students is a moral decision that governments and curriculum developers must consider.

In order to understand diverse mathematical cognitive strategies, it is necessary to envisage students in the context of their own sociocultural environment, which often includes a variety of traditions, behaviors, religions, and languages. This "approach helps the organization of the pedagogical action that occurs in classrooms through the use of the local aspects of these mathematical practices" (Rosa and Orey 2015a, p. 140).

Modifications in the classroom pedagogical practices are required because mathematical activities are not universal and they are unequally applicable across cultures. Thus, if the assumption that the origin, process, and manifestation of mathematical knowledge are not similar across cultures, then universal guidelines and strategies for the pedagogical work would appear to be inappropriate to be applied to all the members of distinct cultural groups (Rosa 2010).

Mathematics is central to school systems all over the world; it dominates and, sometimes is considered the most important subject. But in every society it has not eliminated the practices of the ethnomathematics of the many culturally identified groups. The ethnomathematics of the *invisible society*, of the *non-elite population*, which produces and provides for the basic needs of the people, and of the upper classes, is present and practiced. For example, craftsmen and retailers are responsible for small scale manufacturing, for producing and selling basic goods, like food, clothing, and other utilities for daily consumption.

These professionals deal with certain branches of the economy and they provide non-material needs such as religious and popular rituals and festivals, for popular medicine, for the arts and for sports and for the many other social and cultural activities. All these, craftsmen, retailers and other professionals, rely on people's knowledge and on the traditional wisdom, passed from generation to generation, and from their peers. It is impossible to deny that *official* competencies coexist with other people's competencies to deal with daily life.

To build a civilization that rejects inequity, arrogance, and bigotry, education must give special attention to the redemption of peoples that have been, for a long time, subordinated and must give priority to the empowerment of the excluded sectors of societies.

12.3 Conceptual Dimension of Program Ethnomathematics

The program ethnomathematics contributes to restoring cultural dignity and offers the intellectual tools for the exercise of citizenship. It enhances creativity, reinforces cultural self-respect, and offers a broad view of the mankind. This program offers the possibility of harmonious relations in human behavior and between humans and nature. Intrinsic to it is the ethics of diversity:

- Respect for the other (the different).
- Solidarity with the other.
- Cooperation with the other.

It is important to return to the question in the beginning of this chapter: is Ethnomathematics research or practice? Ethnomathematics arises from research, and this is the reason for calling it a Program. But, equally important, indeed what justifies this research, are the implications for curriculum innovation and development, for teaching and teacher education, for policy making, all focusing the effort to erase arrogance, inequity, and bigotry in society.

As discussed above, the theoretical approach of a program ethnomathematics recognizes the cultural dynamics found in the encounters between cultures, which result in the coexistence of both the *official* and people's competencies. All of this links the historical and epistemological dimensions of the program ethnomathematics, which brings new light into our understanding of how mathematical ideas are generated and how they evolved through the history of humankind. It is fundamental to recognize the contributions of various cultures and the importance of the dynamics of cultural encounters.

Culture, understood in its widest definition, includes aspects of art, history, languages, literature, medicine, music, philosophy, religion, science, and technology. It is characterized by shared knowledge systems, by compatible behavior and by acceptance of an assemblage of values. Research in ethnomathematics is, necessarily, transcultural and transdisciplinarian. The encounters between cultures are examined in its widest form to permit exploration of more indirect interactions and influences, and to permit an examination of subjects on a comparative basis.

At this moment, it is important to reinforce the concept that ethnomathematics should not be confused with ethnic-mathematics, as many researchers and educators mistakenly understand it. In this regard, it is necessary to stress the importance of the use of the denomination of the *Program Ethnomathematics* because ethnomathematics is not a final form of knowledge, but it is composed of a theoretical basis and a practice that is in permanent re-elaboration. Just like the various systems of knowledge, such as mathematics, religion, culinary, dress, sports and gaming, and several other practical and abstract manifestations of the human species in different contextual realities.

Ethnomathematics is permanently subjected to revision and reformulations. Although dealing primarily with space, time, classifying, comparing, which are practices proper to the human species, ethnomathematics examines the codes and techniques that we use to express and communicate our reflections on these practices, which are, undeniably, contextual. To express these ideas that are related to the assumptions of a research program, the neologism, *ethno* + *mathema* + *tics* was coined. In this context, D'Ambrosio (2007a) argues that:

> The resistance against Ethnomathematics may be the result of a damaging confusion of ethnomathematics with ethnic-mathematics. This is caused by a strong emphasis on ethnographic studies, sometimes not supported by theoretical foundations, which may lead to a folkloristic perception of ethnomathematics (p. IX).

According to this point of view, D'Ambrosio (2007b) argues that an etymological elaboration explains the meaning of ethnomathematics, which is a construct that uses the roots *ethno* that means the natural, social, cultural and imaginary environment, *mathema* that means explaining, learning, knowing, coping with, and *tics*, which is a simplified form of *techné*, meaning modes, styles, arts and techniques. Hence, ethnomathematics is a theoretical reflection on the *tics* of *mathema* in distinct *ethnos*. This etymological root caused much criticism since it does not reflect the etymology of academic mathematics.

For example, the term *mathema* is not related to mathematics, which is a neologism based on a Greek origin that was introduced worldwide in the 15th century during the colonization process through navigation. D'Ambrosio (2006) argues that the program ethnomathematics is not *mathema* + *tics* because it tries to understand and explain various knowledge systems such as mathematics, religion, culinary, and other practical and abstract manifestations of humanity in different and diverse cultural contexts.

12.4 Pedagogical Dimension of the Program Ethnomathematics

In general, education was classified as a commodity by the World Commerce Organization, which is a special agency of the United Nations Organization. Since then, Education has become increasingly subordinate to big corporations and the result is that contents and methodologies are out of context. This affects particularly Mathematics Education. The main concern is attaining pre-decided goals of proficiency, which favor uniformity and sameness and surely leads to the promotion of docile citizens and irresponsible forms of creativity. Tests are the best instruments to support this corporate aspect of education. Tests penalize creative and critical education, which leads to the intimidation of the new and to the reproduction of the current model of society.

Education, particularly mathematics education, must focus on the immediate questions facing the world, which includes both social and environmental threats. It is necessary that mathematics educators address these questions and the pursuit of peace in all four dimensions: individual peace, social peace, environmental peace and military peace (D'Ambrosio 2000). This is the relation of the program ethnomathematics with peace, ethics and citizenship.

It is important to insist that the program ethnomathematics is not ethnic mathematics, as some commentators interpret it. Of course, one has to work with different cultural environments and try to describe mathematical ideas and practices of other cultures, acting as an ethnographer. This is a style of doing ethnomathematics, which is absolutely necessary. Cultural environments include indigenous and also urban populations. In urban populations and in the periphery of major cities we must look into the practices of culturally identifies groups, through their professional practices, as retailers and laborers, as farmers, as artisans. The scenario is of individuals facing specific situations and problems in their daily life.

Individuals develop their own ad hoc practices to deal with the specific situations and to solve the problems they face, sometimes sharing their practices with their peers. When the ad hoc responses reveal that they are efficient in dealing with similar situations and solving similar problems, they are organized as methods, with specific jargon, and they are acquired as common practices by the culturally identified group. It is natural that practitioners ask: why does this method work? This is the moment a theory comes into the scenario. Individuals and groups reflect and theorize on their methods and ideas. This is the proper ground to germinate inventions, to propose the new. This is the most important element for the program ethnomathematics.

Synthesizing the arguments of the paragraph above, research and practice in ethnomathematics are the responses to three basic questions:

1. How are ad hoc practices and solution of problems developed into methods?
2. How are methods developed into theories?
3. How are theories developed into inventions?

The answers to these questions results in the learning process of methods and practices of individuals and their peers in cultural identified groups. This is an ethnographic approach to ethnomathematics. According to this context, the program ethnomathematics is a *creative insubordination* approach in Mathematics Education because it challenges the conventional view that Science, which includes Mathematics, is a uniquely modern Western phenomenon.

Science develops wherever there is a systematic search for the understanding of daily phenomena even though there is an absence of a systematic process or achieving certain results, which is a set of an ordered sequence of fixed steps in a recognized method. Indeed, the satisfaction of survival and transcendence that exists in every cultural group and is present during the developmental process of its members is resolved by the application of ad hoc solutions to deal with problems they face daily.

The improvement of these ad hoc practices leads to the development of methods as well as the search for comprehending and explaining hose these methods lead to the evolution of theories. Thus, the intellectual adventure of the humanity is synthesized in the expanding process of knowledge, which is given by:

ad hoc solutions* ⇨ *methods* ⇨ *theories* ⇨ *innovations

The program ethnomathematics develops this process further by trying to understand how procedures and practices, ethnographically studied, develop into methods, which are related to the studies on cognition, and how methods develop into theories, which are related to the studies on epistemology, and how theories develop into inventions, which are related to the studies on creativity.

An important issue in the Philosophy of Science is the acceptability of a theory, which acquires the reliability of a community recognized as elite specialists. Yet, it is difficult to deny that ad hoc solutions and methods to solve new problems that are both generated and socialized by the members of distinct cultural groups. In this case, these elites appropriate or expropriate the methods and become responsible for building up theories.

In this context, it is important to understand that ethnomathematics arises from research, which is the main reason to name it as a program. However, it is equally important to justify that the relevance of this research program is its implications for teaching, teacher education, policy making, and for the school curricula innovation and development, as well as its efforts to erase arrogance, inequity, and bigotry in society (D'Ambrosio 2007b).

This innovation in the mathematics curriculum is related to the elaboration of activities that motivates students to recognize that mathematical knowledge is part of their everyday life. Therefore, this curriculum enhances students' ability to make meaningful mathematical connections by deepening their understandings of all forms of mathematics (Rosa and Orey 2011).

In the current context, there is a necessity for innovative ethnomathematics-based programs to identify and seek teaching-learning practices directed towards its pedagogical action. The great challenge for researchers in ethnomathematics is to conduct studies and elaborate pedagogical practices that are in accordance with the philosophical-theoretical objectives of this program (Rosa and Orey 2015a).

As an example of innovation in the development of the school practices, the "trivium curriculum for mathematics proposed by D'Ambrosio (1999) is an important innovative ethnomathematics approach that needs more investigations in order to address pedagogical purposes" (Rosa and Orey 2016, p. 18). The proposal of this trivium is an answer to the criticism of the lack of balance between the quantitative and qualitative data in the mathematics curriculum.

12.5 Trivium Curriculum for Mathematics

The *dambrosian* proposal for the reconceptualization of the mathematics curriculum is based on the trivium, which is composed by three strands: literacy, matheracy, and technoracy. This curriculum provides, in a critical and reflective way, the communicative, analytical, and technological instruments that help students do develop necessary skills and abilities to perform daily activities in the twenty-first century.

Thus, Rosa and Orey (2015b) argue that this curriculum contributes to the development of school-based activities developed on an ethnomathematics perspective. This approach allows students to use communicative, analytical and technological instruments that are essential to the exercise of their rights and duties that are necessary to practice of citizenship as well as to the critical and reflective reading of daily phenomena that occur in their daily life.

12.5.1 Literacy as Communicative Instruments

Literacy is the capability that individuals possess to process information that are impregnated in their daily lives through the use of media, internet and mobile learning by means of reading, listening, speaking, writing, calculating, designing, and representing. This ability to process information is also performed by using signs, gestures, numbers, calculators, and computers (D'Ambrosio and D'Ambrosio 2013).

Mobile learning and internet enables educational possibilities, with constantly evolving information streams. However, the vastness possibilities of the internet and mobile learning can be a hindrance to those individuals who cannot effectively sift through to analyze and interpret the material presented and found online. Mobile learning integrates the use of mobile computing such as the application of small, portable, and wireless computing and communication with e-learning that facilitates the acquisition of information and communication through technological devices (Valdes and Valdes-Corbeil 2007).

Therefore, strong literacy skills and abilities are a key tool used when students discern and interpret information, enabling them to use them to its full potential in order to "support learning experiences that are collaborative, accessible, and integrated with the world beyond the classroom (Valdes and Valdes-Corbeil 2007, p. 54). These communicative instruments facilitate the integration of schools and communities through a cultural dynamics that allows them to exchange knowledge by processing information in this social interaction. This approach helps students to reduce cultural and communication barriers between schools and communities by using synchronous[2] and asynchronous[3] communication channels (Rosa and Orey 2016).

Currently, reading includes the competency of numeracy, which is the interpretation of graphs and tables, and other ways of informing individuals. Reading even includes the understanding of the condensed language of codes. These competencies also relates to getting information. The power of literacy relates to the individuals capacity to apply these skills to effectively connect, interpret and discern the intricacies of the world in which they live.

Since the role of the teachers is important to initiate and conduct this process, they need to know the sociocultural contexts of their students. In this regard, teachers guide the choice of a theme that is more meaningful to the students in order to drive a transformational action in the school community. By using these communicative instruments, students are able to analyze, understand, comprehend, process, and respond the stimuli offered by the modern phenomena such as inflation, urban growth, consumption, production sectors, elections, health, environmental, and educational policies.

---

[2]Synchronous communication literally means *at the same time* since it occurs in real time and can take place face-to-face. This kind of approach involves live communication either through sitting in a classroom, telephone conversations, instant messagings, online chattings, or teleconferencing via video/webconferences (Rosa and Orey 2016).

[3]Asynchronous communication literally means *not at the same time* since it is not immediately received or responded to by those involved. This kind of communication usually involves a set of weekly deadlines, but otherwise allows students to work at their own pace by using emails and message board forums which allow them to communicate on different schedules (Rosa and Orey 2016).

Rosa and Orey (2015b) stated that once the theme is chosen, teachers must prepare students for the field ethnographic work, which objective is to collect qualitative and quantitative data that must be analyzed and interpreted in order to allow them to formulate their questionings and conjectures to help them to solve problems they face in their school community.

12.5.2 Matheracy as Analytical Instruments

Matheracy is the capability individuals possess to infer, propose hypothesis, and drawing conclusions from data, which helps them to analyze and interpret signs and codes so they are able to propose and use models to find solutions to problems they face in their lives through the elaboration of abstract representations of the real world (Rosa and Orey 2015b).

It provides symbolic tools that help students to develop their own creativity in order to allow them to understand and solve problems and situations. It is a first step toward an intellectual posture, which is almost completely absent in our school systems. Regrettably, even conceding that problem solving, modeling, and projects can be seen in some mathematics classrooms, the main pedagogical importance is usually given to numeracy or the manipulation of numbers and operations (D'Ambrosio and D'Ambrosio 2013).

The elaboration of models helps students to read the world in order to acquire tools that allow them to propitiate in a clear and concise ways the search for solution to problems they face daily. Rosa and Orey (2015b) argue that the use of matheracy is affected by the analysis of the relation between variables that are considered essential to the understanding of the given phenomenon. In this approach, teachers need to act as mediators of this process by instrumentalizing students with the necessary tools that help them to analyze and interpret solutions to these problems in order to allow them to reflect about the issues involving society.

Thus, matheracy is the domain of skills, strategies, and abilities that empower students to be aware of the way in which they explain their beliefs, traditions, myths, symbols, and scientific and mathematical knowledge. According to D'Ambrosio (2008) states that representations given by these elements enable the expansion of reality by incorporating the sociofacts,[4] mentifacts,[5] and

[4]Sociofacts are structures and organizations of a culture that influences social behavior and the development of scientific and mathematical knowledge, which include families, governments, educational systems, sports organizations, religious groups, and any other grouping designed to develop specific sociocultural activities. Sociofacts define the social organization of the members of distinct cultural groups because they regulate how individuals' function in relation to the other members of a specific group (Rosa and Orey 2015b).

[5]Mentifacts refer to the shared ideas, values and beliefs such as religion, language, mathematics, sciences, viewpoints, law, and knowledge that are developed and diffused by the members of distinct cultural groups from generation to generation (Rosa and Orey 2015b). The main issue in

artifacts[6] regarding mathematical simulations and models. Rosa and Orey (2016) affirm that these competencies allow students to have access to a diverse set of signs, codes, symbols, and methods that are essential to the development of the decision-making process.

12.5.3 Technoracy as Technological Instruments

Technocracy is the critical and reflective capability that individuals possess to use and combine different technological instruments from the simplest to the most complex ones in order to help them to solve problems they encounter in everyday activities (Rosa and Orey 2016).

The operative aspects of the familiarity with technology, in most cases, is inaccessible to most individuals, yet basic ideas behind technological devices, their possibilities and dangers, and the morality supporting the use of technology, are essential issues to be raised among individuals. History shows that ethics and values are intimately related to the development of technological progress.

Mobile learning involves the use of mobile technology, either alone or in combination with other information and communication technology to enable learning anytime and anywhere. Learning can unfold in a variety of ways: individuals are able to use mobile devices to access educational resources, connect with others, and create content, both inside and outside classrooms. Mobile learning also encompasses efforts to support broad educational goals such as the effective administration of school systems and improved communication between schools and families.

Technoracy is an important feature of scientific and mathematical knowledge as well as its reification as technological artifacts manifest itself in technological tools that translate ways of dealing with natural, social, natural, cultural, political, and economic contexts. These environments allow the development and incorporation of technological tools used in specific sociocultural contexts (Rosa and Orey 2015b).

Technoracy uses different technological instruments that serve to mathematical purposes such as simple, graph, and scientific calculators, *softwares*, computational

(Footnote 5 continued)

the empowerment of individuals is the transition from the elaboration of mentifacts (theorizing about the events and phenomena that occur in everyday life) to the development of strategies and actions that are adequate to solve new problems and situations (D'Ambrosio 2006).

[6]Artifacts are the cultural objects, primarily material items and technologies created by members of distinct cultural groups. It is the technological subsystem composed of material objects as well as techniques of their use. Such objects are tools and other instruments that enable individuals to feed, clothe, house, defend, transport, and solve daily problems by using scientific and mathematical techniques and tools as well as informal and non-standardized mathematical knowledge found in other sociocultural contexts (Rosa and Orey 2015b).

programs, and simulators even though the manipulation of different types of equations that allow them to analyze, interpret, and evaluate models during the decision-making process in order to act and transform reality.

It is important to emphasize that a trivium mathematics curriculum intends to build a generalizable and applicable common body of mathematical knowledge by valuing locally, relevant, and specific mathematical knowledge. According to this point of view, Rosa and Orey (2015b) state that this curriculum seeks to make connections between local and academic knowledge, which can be jointly developed in a dialogical and interdisciplinary way.

It is important to highlight that the trivium curriculum addresses connections between the general aspects of academic mathematics (matheracy) and locally relevant mathematical practices developed by the members of distinct cultural groups (literacy) who use technological instruments (technoracy) to evaluate diverse ways to represent mathematical ideas, procedures, and practices as well as to assess the reasonableness of the results and their contextualizations.

12.6 Incorporating Ethnomathematics into the Mathematics Curriculum

A question that usually arises is about the objective of incorporating ethnomathematics in the curriculum. Would not be enough to teach the mathematics that will be useful to students, to pass in tests, to apply for jobs, or to improve their own professional practice? True, these are the immediate objectives expected by students, by parents and by society as a whole. It is important to do not deny these immediate expected objectives. But education cannot be resumed to practical immediate objectives. It is even more important to prepare students to face the new, to aim further than what they have now. They have to acquire high esteem of themselves.

There is a similarity with the offering of languages other than the mother language. Of course, it is important to learn and to use properly the language of the family and of the community. If we are proficient in only one language, we are less equipped to be successful in the modern World. No one denies this. But there is another factor, of cognitive nature, that applies also to mathematics. Studies in cognition show the advantages of knowing other languages.

There are evidences that the use of one language favors the utilization of another language. Bilingual speakers have to select the best from two competing options. This may be determined by the social context, but cognitive resources are activated. A very difficult question is which one to select in a determined circumstance. Hence, a possible explanation for the selection has to do with control systems recruited into linguistic processing.

According to this context, what do we expect of a good education? The answer is:

- To promote full citizenship by preparing the individual to be integrated and productive in society, through the transmission of values and the clarification of both their responsibilities and their rights in society.
- To promote the expression of the creativity of every individual, by encouraging people to achieve their potential at the highest level of their interest and ability, which leads to progress of the society.

To achieve full citizenship, no one can deny that the proficiency in mathematics is absolutely necessary. Like being illiterate, without proficiency in basic school mathematics the individual is excluded from participation in society. It will be difficult to have employment, to be a conscious consumer, to perform daily routines. But we have data that confirms the failure of school systems around the world in providing proficiency in basic school mathematics.

We have to question the meaning of basic school mathematics. There are no valid arguments that justify insisting on teaching geometry as organized 25 centuries ago, arithmetic as organized more than 10 centuries ago and algebra invented about 5 centuries ago, as it is done in most schools in the world. Teachers teach the way they were taught, ignoring progresses occurred during their professional career.

This is particularly serious in mathematics, which changed so much since the evolution of computers, calculators and informatics in various forms. The insistence in teaching geometry, arithmetic and algebra, and even calculus, as knowledge frozen for many centuries, memorizing techniques and availing the proficiency in tests, is unsustainable. The result is that students are bored and reject mathematics. As a consequence, students are not acquiring necessary competencies for full citizenship.

Of course, some basic geometry, arithmetic and algebra are important tools to solve real problems, but the access to these tools is possible, thanks to modern technological resources, without the need of memorizing techniques. The role of the teacher is to propose interesting real problems. If there is interest of the students, they will look for tools. Then, the teacher acts listening to the students and learning their approach to solve the problems, and in many cases their approach rely on ethnomathematics learned from their parents and from their communities.

The elaboration of a mathematics curriculum that is based on students' knowledge allows teachers to have more freedom and creativity to choose mathematical topics to be covered in the lessons. Through dialogue with the students, teachers can apply mathematical themes that help them to understand the content of the mathematics curriculum. Hence, teachers can engage students in the critical analysis of the dominant culture as well as of their own culture (Rosa and Orey 2011). This approach is extremely valuable and must be respected by the teacher and by fellow students.

At the same time, students may compare different approaches to deal with the problem, which reveals the existence of multiple ethnomathematics, which enriches the learning experience. Surely, the teacher may also reveal her/his own way of dealing with the problem, based on School Mathematics and the appropriate use of technological devices. The various approaches lead to possible new approaches, combining aspects of different ethnomathematics and of school mathematics. This is an important example of the dynamics of cultural encounters.

The most important aspect of the pedagogical dimension of the program ethnomathematics is the mutual exposition of different cultural approaches to face a situation or problem. This mutual exposition, with full mutual respect, is responsible to advances of knowledge throughout the evolution of the human species. This is the essence of the dynamics of cultural encounters.

Not only in mathematics, the dynamics of cultural encounters forms the great hope of new approaches concerning the assembly of environmental crises, particularly the fast exhaustion of water supplies, and to face the mounting health problems, and to confront the social and religious tensions around the world.

Regarding health care, the scientific developments since the Renaissance have led to a scientific and highly technical form of medicine, displacing traditional medicine of indigenous populations. The traditional practices survived, even if they became prohibited as criminal practices. Now, the World Health Organization has launched a project to support countries in developing proactive policies and implementing action plans that will strengthen the role Traditional Medicine keeping populations healthy (WHO 2013).

The eminent cultural historian Geoffrey Ernest Richard Lloyd refers to the encounter of Western medicine and traditional medicine saying that:

> (…) the possibilities of mismatch between what biomedicine [with a battery of tests to call on] pronounces to be the case and what individual patients feel, are unlikely ever to be completely removed. If so, alternative styles of medicine, with their more or less articulate elites to promote them, are likely to continue to bear witness to the complexities of our understanding of what it is to be truly well, and it would surely be foolhardy to suppose that biomedicine has nothing to learn from its rivals (Lloyd 2009, p. 92).

It is important to paraphrase what Geoffrey Ernest Richard Lloyd says by just replacing a few words and stating that the:

> (…) possibilities of mismatch between what Mathematics [with a battery of frozen knowledge and tests] pronounces to be an instrument to solve real life problems, are unlikely ever to be completely removed. If so, alternative styles of Mathematics, which are Ethnomathematics practiced by cultural identifies groups, are likely to continue to bear witness to the complexities of our understanding of what it is to face problems posed by real life, and it would surely be foolhardy to suppose that Academic and School Mathematics has nothing to learn from its rivals Ethnomathematics (D'Ambrosio and Rosa 2016, p. 7).

A good education cannot be limited to promote full citizenship, to succeed in employment, in being a conscious consumer and in performing well in daily routines. These are immediate, necessary goals, but not enough for a good education. A good education should also help students to attain their personal satisfaction, attaining higher objectives in life, according to their interest and abilities. This requires raising their self-esteem and creativity.

12.7 An Ethnomathematics Curriculum

Schools and classrooms cannot be isolated from the communities in which they are embedded. If schools are considered part of a community with defined cultural practices, then classrooms might be understood as environments that facilitate pedagogical practices in these learning environments.

When students come to school, they bring with them values, norms, and concepts that they have acquired in their sociocultural environment. Some of these ideas and procedural concepts are mathematical in nature. However, mathematical concepts developed in the school curriculum are presented in a way that may not be related to the students' cultural backgrounds.

It has been hypothesized that low attainment in mathematics could due to the lack of cultural consonance in the curriculum. Moreover, there is evidence from research that the inclusion of cultural aspects in the curriculum have long-term benefits for mathematics learners, which means that cultural aspects contribute to the recognition that mathematics is part of daily life of the students (Eglash et al. 2006; Rosa and Orey 2013).

The pedagogical work towards an ethnomathematics perspective allows for a broader analysis of the school context in which mathematical practices transcend the classroom environment because they embrace the sociocultural context of the students. Hence, pedagogical elements necessary to develop curricular activities are found in the school community. The field of ethnomathematics posits some possibilities for innovative educational initiatives that help to reach this goal (Rosa and Orey 2015b).

It is important to recognize that ethnomathematics is a research program that guides educational pedagogical practices. In this context, there is a need to examine the embeddedness of mathematics in culture, drawing from a body of literature that takes on the cultural nature of knowledge production into the mathematics curriculum (Rogoff 2003).

Mathematics as part of the school curriculum must reinforce and value the cultural knowledge of students rather than ignore or negate it. It is argued that this mathematics curriculum seek to change the way mathematics teachers construct their learning environments by producing teachers who are able to facilitate a mathematics learning environment grounded in real life experiences and to support students in the social construction of mathematics (Rosa and Orey 2007).

The trend towards ethnomathematical approaches to the mathematics curriculum reflects a comprehensive development in mathematics education. These approaches are intended to make school mathematics more relevant and meaningful to students and to promote the overall quality of education. It is necessary to investigate conceptions, traditions, and mathematical procedures developed by the members of a particular cultural group with the intention of incorporating these practices into the mathematics curriculum. In this direction, Rosa and Orey (2011) plead for a more culturally sensitive view of mathematics to be incorporated into the school curriculum.

This mathematical approach is presented as a cultural response to students' needs by making connections between their cultural background and mathematics, which supports the view that mathematics is conceived as a cultural product which has developed as a result of various activities. A learning environment using this type of curriculum would be full of examples that draw on the students' own experiences that are common in their own cultural contexts (Rosa and Orey 2008). Ethnomathematics uses these cultural experiences as vehicles to make mathematics learning more meaningful and to provide students with insights of mathematical knowledge embedded in their sociocultural environment. Students understand the nature of mathematics as they become aware of the presence of mathematical knowledge in their own culture.

In an ethnomathematical curriculum, it is necessary that teachers develop a different approach to mathematics instruction that empowers students to understand mathematical power more critically by considering the effects of culture on mathematical knowledge and work with their students to uncover the distorted and hidden history of the mathematical knowledge. According to Rosa (2010), this methodology is essential in developing the curricular practice of ethnomathematics and culturally relevant education: through the investigation of the cultural aspects of mathematics and an elaboration upon mathematics curricula that considers the contributions of people from other cultures, students' knowledge of mathematics becomes enabled and enriched.

12.8 Final Considerations

Modern Civilization is failing in supplying the humanity with the basic strategies for survival and, at the same time, it is denying cultural identities and social justice. Words like democracy, freedom and equity are political slogans suggesting affirmative action.

However, they have been used by social movements and political factions to draw applause and to recruit and align cadres, which are soon voided by internal contradictions and conflicts. There are no real benefits for the people. This opens a space for reactionary counter-action, which are well coordinated and silent, and gain space in maintaining the *status quo*. People see international corporations displacing public schools, all over the world co-opting educators for new programs and new methods of instruction, with the fallacious objective of getting better results in tests.

The proposals favor the economic interest of corporations, but fail to develop the dignity of the individual and the imperatives of justice, compassion and peaceful modes of resolving conflicts, aiming at a new future, in which the humanity can live together in grace and peace. It is necessary to recall the appeal of Bertrand Russell and Albert Einstein in the Pugwash Manifesto (1955):

There lies before us, if we choose continual progress in happiness, knowledge, and wisdom. Shall we, instead, choose death, because we cannot forget our quarrels? We appeal as human beings to human beings: Remember your humanity, and forget the rest. If you can do so, the way lies open to a new Paradise; if you cannot, there lies before you the risk of universal death (Russel and Einstein 1955).

The program ethnomathematics is an answer to this appeal. It is a theoretical framework that establishes the foundation for organizing practices and systems of explanations developed by the species throughout its evolution in order to survive and to transcend.

According to this discussion, a comprehensive view of mathematics curriculum is implicit in an ethnomathematical perspective. All students possess potential for understanding and communication through a variety of mathematical systems within distinct cultural contexts. This approach allows students to gain new perspectives on human potential and on the organization of the mathematics curriculum. Therefore, mathematics can only be truly learned and taught if it includes culture, natural language, and visual representations that are culturally relevant to learners and teachers alike.

An ethnomathematical perspective in the mathematics curriculum helps all participants to come to understand and appreciate alternative viewpoints, cultural diversity, natural language, mathematics, and visual representations which form a unique system for meaning-making. In this context, reorienting teaching and learning to include ethnomathematics can engage and excite students about learning and encourages them to see themselves as being able to do mathematics by validating their own cultural experiences, which serves as an essential component of understanding and celebrating the differences between diverse cultural groups.

Ethnomathematics is a system of knowledge that offers the possibility of a harmonious relation among humans and between humans and nature. It contributes to restoring cultural dignity and offers the intellectual tools for the exercise of citizenship. It enhances creativity, reinforces cultural self-respect and mutual respect, and offers a broad view of humanity.

References

D'Ambrosio, U. (1985). Ethnomathematics and its place in the history and pedagogy of mathematics. *For the Learning of Mathematics, 5*(1), 44–48.

D'Ambrosio, U. (1999). Literacy, matheracy, and technocracy: A trivium for today. *Mathematical Thinking and Learning, 1*(2), 131–153.

D'Ambrosio, U. (2000). Ethnomathematics: A step toward peace. *Chronicle of Higher Education, 12*(2), 16–18.

D'Ambrosio, U. (2006). *Ethnomathematics: Link between traditions and modernity*. Rotterdan, The Netherlands: Sense Publishers.

D'Ambrosio, U. (2007a). Preface. In F. Favilli (Ed.). *Ethnomathematics and mathematics education* (pp. V–X). Pisa, Italy: Tipografia Editrice Pisana.

D'Ambrosio, U. (2007b). Peace, social justice and ethnomathematics. *The Montana Mathematics Enthusiast, Monograph, 1,* 25–34.

D'Ambrosio, U. (2008). Educação numa era de transição [Education in a transitional era]. *Revista Matemática & Ciência, 1*(1), 8–18.
D'Ambrosio, U. (2011). Non-killing mathematics. In J. Evans Pim (Ed.), Engineering nonkilling: Scientific responsibility and the advancement of killing-free societies (pp. 121–148). Honolulu. HI: Center for Global Nonkilling.
D'Ambrosio, U. (2015). Mathematical modelling as a strategy for building-up systems of knowledge in different cultural environments. In G. A. Stillman, W. Blum, & M. S. Biembengut (Eds.), *Mathematical modelling in education research and practice: Cultural, social and cognitive influences* (pp. 35–44). New York, NY: Springer.
D'Ambrosio, U., & D'Ambrosio, B. S. (2013). The role of ethnomathematics in curricular leadership in mathematics education. *Journal of Mathematics Education at Teachers College, 4,* 19–25.
D'Ambrosio, U., & Rosa, M. (2016). *Ethnomathematics and its pedagogical action.* Paper presented at the 13th International Conference on Mathematics Education—ICME12. TSG35 Role of Ethnomathematics in Mathematics Education. Hamburg, Germany: University of Hamburg.
Eglash, R., Bennett, A., O'Donnell, C., Jennings, S., & Cintorino, M. (2006). Culturally situated designed tools: Ethnocomputing from field site to classroom. *American Anthropologist, 108*(2), 347–362.
Freire, P. (2000). *Pedagogy of the oppressed.* New York, NY: Continuum.
Hall, E. T. (1976). *Beyond culture.* New York, NY: Anchor Books.
Lloyd, G. E. R. (2009). *Disciplines in the making: Cross-cultural perspectives on elites, learning and innovation* (p. 2009). New York, NY: Oxford University Press.
Rogoff, B. (2003). *The cultural nature of human development.* New York, NY: Oxford University Press.
Rosa, M. (2000). From reality to mathematical modeling: A proposal for using ehtnomathematical knowledge. College of Education. Master Degree Thesis. Sacramento, CA: CSUS.
Rosa, M. (2010). The perceptions of high school leaders about English language learners (ELL): The case of Mathematics. College of Education. Doctorate Dissertation. Sacramento, CA: CSUS.
Rosa, M., & Orey, D. C. (2007). Cultural assertions and challenges towards pedagogical action of an ethnomathematics program. *For the Learning of Mathematics, 27*(1), 10–16.
Rosa, M., & Orey, D. C. (2008). Ethnomathematics and cultural representations: Teaching in highly diverse contexts. *Acta Scientiae - ULBRA, 10,* 27–46.
Rosa, M., & Orey, D. C. (2010). Ethnomodeling: an ethnomathematical holistic tool. *Academic Exchange Quarterly, 14*(3), 191–195.
Rosa, M., & Orey, D. C. (2011). Ethnomathematic: the cultural aspects of mathematics. *Revista Latinoamericana de Etnomatemática, 4*(2), 32–54.
Rosa, M., & Orey, D. C. (2013). Ethnomodelling as a methodology for ethnomathematics. In Gloria A. Stillman; Jill Brown. (Orgs.). *Teaching mathematical modelling*: connecting to research and practice. International perspectives on the teaching and learning of mathematical modelling (pp. 77–88). Dordrecht, The Netherlands: Springer Science + Business Media Dordrecht.
Rosa, M., & Orey, D. C. (2015a). Evidence of creative insubordination in the research of pedagogical action of ethnomathematics program. In: B. S. D'Ambrosio & C. E. Lopes (Eds.), *Creative insubordination in Brazilian mathematics education research* (pp. 131–146). Raleigh, NC: Lulu Press.
Rosa, M., & Orey, D. C. (2015b). A trivium curriculum for mathematics based on literacy, matheracy, and technoracy: An ethnomathematics perspective. *ZDM Mathematics Education, 47*(4), 587–598.
Rosa, M., & Orey, D. C. (2016). Developing mathematical modeling in virtual learning environments by applying critical and reflective dimensions. In Kristy Wallace. (Eds.). *Learning environments*: emerging theories, applications and future directions (pp. 1–20). New York, NY: Nova Science Publishers.

Russell, B., & Einstein, A. (1955). *The Russell-Einstein manifesto*. London, England: Pugwash Conferences on Science and World Affairs. Retrieved from http://pugwash.org/1955/07/09/statement-manifesto/.

Toffler, I. (1990). *Powershift*: knowledge, wealth, and violence at the edge of the 21st. century. Buffalo, NY: Bantam Books.

Valdes, J. R., & Valdes-l, M. E. (2007). Are you ready for mobile learning? *Educause Quarterly, 7*, 51–58.

WHO. (2013). *Traditional medicine strategy*: 2014–2023. Geneva, Switzerland: WHO Press, 2013. Retrieved from www.who.int.

Chapter 13
The Evolution of Ethnomathematics: Two Theoretical Views and Two Approaches to Education

Veronica Albanese, Natividad Adamuz-Povedano and Rafael Bracho-López

Abstract The Ethnomathematics program is in continuous evolution. At the beginning of its history, Ethnomathematics had the purpose of recognizing ideas and practices of different cultural groups but it subsequently evolved to embrace wider studies centered on the ways in which cultural and social context affects the process of generation, organization and communication of knowledge. We propose two views of Ethnomathematics that respond to this evolution: (1) mathematics of cultural practices and (2) different ways of thinking. The first view implies recognizing mathematics in the practices of cultural groups, and this research is performing in accordance to the categories and schemes of thinking of the researcher's culture. The second view implies discovering different ways of knowing quantity, space and relations, considering a broader concept of (ethno)mathematics where the categories and schemes of thinking of the studied cultural group are taken into consideration. Thus, we point out which implications for Mathematics Education each of these two views suggests. On the one hand, the idea of mathematics in cultural practices involves designing tasks that are contextualized in the cultural heritage. On the other hand, the idea of a different way of knowing can help to reflect on certain mathematical notions or on the nature of mathematical knowledge.

Keywords Evolution of ethnomathematics · Cultural practice · Mathematics education · Theoretical posture · Philosophical backgrounds

---

V. Albanese (✉) · R. Bracho-López
University of Granada, Granada, Spain
e-mail: vealbanese@ugr.es

R. Bracho-López
e-mail: rbracho@uco.es

N. Adamuz-Povedano · R. Bracho-López
University of Córdoba, Córdoba, Spain
e-mail: nadamuz@uco.es

M. Rosa et al. (eds.), *Ethnomathematics and its Diverse Approaches for Mathematics Education*, ICME-13 Monographs,
DOI 10.1007/978-3-319-59220-6_13

13.1 Introduction

In this chapter we present a reinterpretation of the evolution of the Ethnomathematics Program by identifying two theoretical views which depend on the focus of interest of the researchers: (1) mathematics of cultural practices and (2) different ways of thinking. The first view aims to recognize *known* mathematical concepts in cultural practices while the second view aims to discover new ways of thinking mathematically.

We will describe how these views are connected with the evolution of the theoretical bases and we will display how each view has different implications in educational issues. In our opinion, considering both views during research can help ethnomathematicians to overcome possible contradictions emerging from working with others' cultures.

13.1.1 The Evolution of the Ethnomathematical Program

At the end of the 19th century and at the beginning of the 20th century, anthropologists deepened their ethnographic research on indigenous peoples and their cultures and began to discover ways to conceptualize space and quantity existing in these cultures that were different from the mathematical ones spread by so-called *Western* culture.

Mathematicians, and most of all mathematics educators who deal with indigenous peoples, started to look at these ways of conceptualizing mathematics with interest, giving them different names: sociomathematics, spontaneous mathematics, informal mathematics, oppressed mathematics, non-standardized mathematics, popular mathematics, mathematics of know-how, oral mathematics, implicit mathematics, non-professional mathematics, contextualized mathematics, folk mathematics, and indigenous mathematics (Gerdes 2001; Sebastiani 1997).

D'Ambrosio (1985) proposed considering these ways of conceptualizing mathematics, different from the ones of Western culture, as ethnomathematics,[1] with the idea of including under this name the various previously used terminologies (with their nuances). Thus, Ethnomathematics, as a field of research, arose to catalyze and give voice to those researchers convinced of the need to assess the different ways of conceptualizing/doing mathematics which had developed within the various sociocultural contexts in response to the instincts of man to know how to survive and transcend (D'Ambrosio 2008).

[1]As many authors have done, we differentiate Ethnomathematics, singular and with capital letter, from ethnomathematics, plural with lowercase letter. Ethnomathematics is the research program, which is the field that studies ethnomathematics that are the different ways of conceptualizing mathematics of the cultural groups.

At the beginning of its history, Ethnomathematics had the purpose of recognizing ideas and practices of different cultural groups (Barton 1996a; D'Ambrosio 1985), but then, it evolved to embrace wider studies centered on the ways the cultural and social contexts affect the process of generation, organization and communication of knowledge (D'Ambrosio 2012). This shift of the focus of the program happened gradually.

When researchers observed mathematics in practice that were different from the expected, some believed it was just a different way to interpret or implement the same mathematical concepts, but when studies deepened, in some cases, it became evident that the mathematical concepts hardly fit in the mathematics known up until then. Thus, researchers' questions were mobilizing as far as how this knowledge, these ethnomathematics; arises, develops and is shared by the community.

In this chapter, we propose distinguishing between two different views or postures within the continuous and dynamic evolution of the Ethnomathematics program, connected with the two moments of the evolution just mentioned. The first view we propose considers Ethnomathematics as the study of *mathematics of cultural practices*, while the second one understands Ethnomathematics as the study of *different ways of knowing*.

In the first posture, the basic idea is that there are many concepts that can be recognized as mathematics in many practices of various cultures, but all of them are seen by the same scholar/academic mathematics point of view. Here, the idea that mathematics always has something in common prevails. Although, they may take different forms, what connects them is highlighted.

In contrast, in the second posture the basic thought is that valuing the ideas of other cultures implies the discovery of these other ways that cultural groups develop to survive and transcend (D'Ambrosio 2008). Here, the idea of the existence of different mathematics is stressed; the focus is more on the differences, strongly based on the philosophical foundation of a cultural relativism that states the contemporary existence of various kinds of mathematics (Barton 2012).

To sum up, on the one hand mathematics has to be *recognized* in the practices of cultural groups (D'Ambrosio 1985; Barton 1996a); on the other hand, new ways of knowing, and doing mathematics, have to be *discovered* in cultures, identifying Ethnomathematics as the research of the forms, ways (tics) of knowing, understanding and relating (mathema) to the environment (ethno) (D'Ambrosio 2008).

We observe that these two postures can appear in the theoretical or methodological proposal of the same authors, sometimes in different moments of their professional development, while other times they both appear in the same proposal as complementary views. Backing up our proposal with literature in the field, we go over the development of these two views in the history of the field, in the works of some significant authors and, finally, in our own research.

As a matter of fact, we determine that the view of Ethnomathematics as *mathematics of cultural practices* was historically the first one to emerge in the definitions of the field, and usually it has been the first one to be displayed in the work of the authors whose perspective would then embrace both postures. Our own research is an example of this, as we will explain later. Indirectly, D'Ambrosio

(2012) states the same idea when he says the program was conceived to recognize mathematical ideas and practices of cultural groups and then goes *further* to include a wider understanding of knowledge to make sense of the comparisons among the (ethno)mathematics of different cultural groups (views of *different ways of knowing*).

Likewise, each view implies a different approach to the issue of what role Ethnomathematics should have in mathematics education. On one hand, the idea of mathematics in cultural practices involves designing tasks that are contextualized in the cultural heritage. On the other hand, the idea of a different way of knowing can help us to reflect on certain mathematical notions or on the nature of mathematical knowledge.

With respect to the nature of mathematics, we present a systematization of the changes that the Ethnomathematics program introduces in the concept of mathematics, by means of the definition of three dimensions: the practical, the social, and the cultural. Therefore, we propose Wittgenstein's philosophy of language games as a philosophical foundation of the cultural relativism of the concept of mathematics (Barton 2012). We provide examples for both implications based again on our research. Finally, we highlight the necessity to integrate both of them in research as well as in educational actions.

13.2 The Two Views in the Ethnomathematics Theoretical Framework

In this section we look over the Ethnomathematical literature selecting the authors that allow us to delineate the differences between these two postures (Fig. 13.1).

In his first work, Barton (1996a) analyzes the definitions of Ethnomathematics published until then, and proposes considering Ethnomathematics as the study of concepts and practices, articulated and used by cultural groups, which according to the researchers are mathematics. This statement is similar to our first view of Ethnomathematics as *mathematics of cultural practices*, yet it contains a hint of something more enclosed in that *according to the researcher* that helps to clarify the idea. It means the researcher has the task of deciding and identifying mathematics based on the cultural baggage he/she has accumulated during his/her scholarly and academic career.

Initially, Barton (1996b) presented an analysis of the distinctive methodological features of Ethnomathematical research based on the author's review of and reflections on research carried out by the Ethnomathematical community up until that moment. These activities are: (a) the *descriptive activity*, i.e. the description stage in which an ethnographic study is performed on the practice, and the description is done from an anthropological perspective; (b) *archaeological-analytical activity*, involving the reconstruction of the history of the practice in search of the reasons and principles of the mathematical implications of the practice,

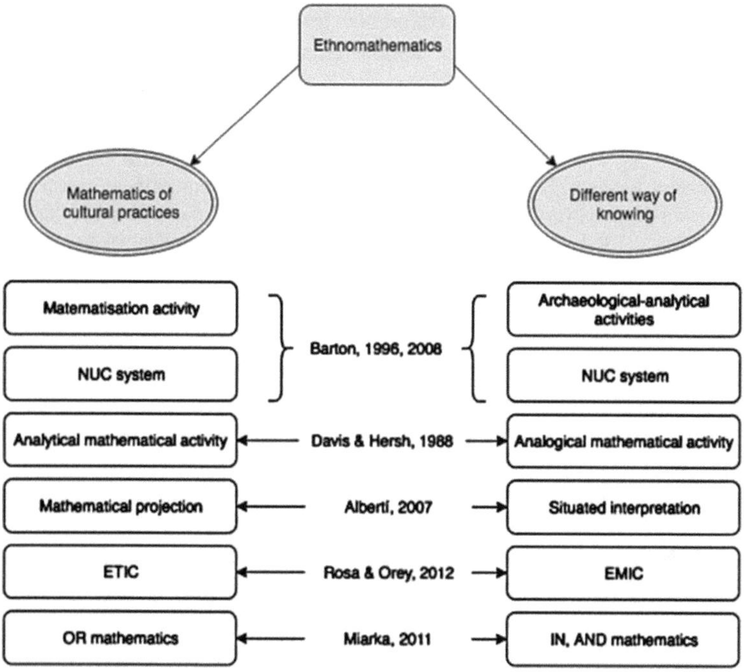

Fig. 13.1 The two postures in the work of Ethnomathematicians. *Source* Personal file

highlighting the mathematical aspects from the standpoint of the practitioners themselves; (c) *the mathematization activity*, where the researcher attempts to translate the cultural material found into a formal language for the development of new mathematical knowledge or better understanding of the cultural group from the perspective of the academic community; (d) *creative-synthetic activity* is a meta-mathematical reflection that involves taking into account a relativistic conception because it means accepting a conceptual-cultural diversity about what is mathematical.

In fact, among the four activities (Barton 1996b) we emphasize that the *mathematization activity* can be seen in the perspective of ethnomathematics as mathematics in cultural practice because, in other words, it consists of describing cultural practices from the point of view of the researcher, based on his/her cultural standpoint. Nevertheless, he admits that there are other activities that are close to the idea of Ethnomathematics as the study of different ways of knowing: it is the *archaeological-analytical activities* where the researcher describes practices from the point of view of the studied culture -and this is a task which is anything but easy.

Some years later, Barton (2008, 2012), strengthening the latter concept, proposed a wider understanding of ethnomathematics as *QRS systems*, which means systems to reveal the quantity, relation, and spatial aspects of human experience.

This new terminology brands the difference from the *NUC system*, the near universal conventional system with which Barton indicates scholarly/academic mathematics.

> Replac[e] the words 'mathematics' (or 'mathematical') with the phrase "(concerning) a system for dealing with quantitative, relational, or spatial aspects of human experience", or "QRS-system" for short. Thus, any system that helps us deal with quantity or measurement, or the relationships between things or ideas, or space, shapes or patterns, can be regarded as mathematics. (…) If I want to talk about the smaller, formal, conventional world of academic mathematics as it is exemplified in schools and universities all over the world, then I will use the words "near-universal, conventional mathematics", or "NUC-system" to refer to it (Barton 2008, p. 10).

This new definition of ethnomathematics as *QRS systems* stresses his interest in the search of *ways of knowing* and thinking about the real world that each culture develops. But, what are the differences between a *QRS system* and a *NUC system*? To clarify this, we consider the idea of an *analogical mathematical activity* in opposition to an *analytical mathematical activity* (Davis and Hersh 1988). In the latter the symbolic material predominates, it is considered difficult and tiring, a specific preparation is needed to deal with it, there are few people dedicated to this mathematics and their studies require constant verification by the mathematical community.

The *analogical mathematical activity* is easier and faster, uses few or no symbols, is available to almost everyone, and does not require excessive effort though it may be of some complexity when it comes to attaining an intuitive control of a complex system. On the other hand, its results are not always expressed in words, but in understanding, intuition or clinical eye. Albertí (2007) emphasizes the experimental, affordable nature of this *analogical activity* and stresses that it almost entirely lacks symbols with respect to the logical-deductive reasoning character, the symbolic equations of analytical mathematics. As ethnomathematical researchers, we seek analogical solutions that dominate cultural practices. An analogical solution with a reasoned justification is actually classifiable as a mathematical solution.

So, Barton (2004) proposes guidelines to consider that a practice belongs to a mathematical type, saying that the involved knowledge has to be in a systematic, formalized structure and has to deal with quantities, space and relations. This happens if those conducting the practice are able to discuss, formulate hypotheses and convince each other on issues related to the knowledge system when one is away from the practice. It does not necessarily include explanations in written formalization, but it is enough if there is any formal oral system.

Albertí (2007) developed a model to look for ethnomathematics in crafts, stating that we can study crafts looking at three moments or stages: *work in progress* is the process the artisan carries out to make the piece of craft, the *finished work* is the piece of craft, the product when it is ready to sell or use, and finally the *explained work* consists of the discourse of the artisan transforming into words the actions he/she realizes in his/her work. Albertí (2007) distinguishes *mathematical projection*, namely what the researcher can see of mathematics in the work in progress and in the finished work, from *situated mathematical interpretation*. The last one can be achieved only by contrasting if what the researcher sees as mathematics, is confirmed by the artisan's words in the explained work. In other words it is an analogical solution with a reasoned justification that we can consider mathematics from the point of view of the cultural group.

Connected to the latter observation, from a methodological perspective, using the terminology of Rosa and Orey (2012, 2013), our two postures are reflected in the dichotomy *etic* and *emic*, which were borrowed from the anthropologists. Taking an *emic* perspective means that research is done from the dynamics and relations of the studied cultural group so that the descriptions and analyses are carried out with the schemes and conceptual categories of the cultural group, respecting their perception of reality. The main purpose of the *emic* perspective is to respect different *ways of knowing*, or QRS systems, of other cultures, when this is the case, trying not to alter them by the researcher's point of view.

On the contrary, an *etic* perspective gives voice to the schemes and conceptual categories of the researcher's culture so that the descriptions follow patterns that are meaningful for the academic culture, but not necessarily for the studied cultural group. In fact, the researcher recognizes *mathematics in cultural practices* according to his/her own understanding of what mathematics is. The main purpose of the *etic* perspective is to build bridges between the cultures and interpret one in terms of the other. There is a risk of modifying the native point of view, but the intention is to establish communication between the two cultures, a translation to allow the approach, and although not a complete rapport, at least the possibility to interact.

Interviewing four of the main personalities of the Ethnomathematical community, Miarka (2011), Miarka and Viggiani (2012) finds out three lines of thinking that guide different Ethnomathematical research projects, summed up in the nature of the relation between Ethnomathematics and mathematics:

- Mathematics <u>in</u> Ethnomathematics: academic Mathematics is just one of the possible developments of knowledge; Ethnomathematics studies the other possibilities. Here, we find the posture of Ethnomathematics as different *ways of knowledge*, which is the emic perspective. According to Miarka, the interviewed authors who display this thinking are Barton and D'Ambrosio, and this is consistent with what has been said above.
- Mathematics <u>or</u> Ethnomathematics: mathematics is unique, so just one pattern exists that gets manifested in different forms depending on the culture. This expresses an idea of Ethnomathematics as *mathematics in cultural practices*, which is the etic perspective. According to Miarka, Sebastiani (1991) and Gerdes (1988, 1998) show this thinking.
- Mathematics <u>and</u> Ethnomathematics: they are different and incommensurable; even if there are similarities, there is no way to get from one to the other. Epistemologically, they are at the same level. According to Miarka, Knijnik (2012) shows this thinking. We propose interpreting this in the emic posture, even if this is still an open issue to discuss.

In our own research (Albanese 2014, 2015a, b), we have adopted both perspectives starting from the idea of Albertí (2007). We decided to follow his model; so we initially focused on the *mathematical projection* on craft practices of braiding from the point of view of the researcher, the *etic* perspective, and we got a model. We studied two braiding crafts, one from the region of Salta and the other one from the area of Buenos Aires. The first model achieved is based on the mathematical concept of graph (Albanese el al. 2014; Oliveras and Albanese 2012).

Then, we consider that this becomes a *situated mathematical interpretation* if we get evidence that the artisans manage the same model, i.e. the *emic* perspective. If this is not the case, we look for the model the artisans use. Specifically, we found out that the model with graphs was a *situated interpretation* for the craft of Salta, but this was not the case of the craft from Buenos Aires. So, we decided to deepen the research in this second fieldwork and we ended up with another representation of the craft, based on the notion of ethnomodel (Rosa and Orey 2012, 2013), where patterns and number relations become evident (Albanese and Perales 2014a; Albanese 2015c).

What we have just described is an example of how to use the theoretical framework presented as a tool to do research. Also, it is an example of how to intertwine the two postures. Since, as other authors, for example Rosa and Orey (2012, 2013) have already pointed out, we believe it is important, as far as it is possible, to consider both postures in every research project.

13.3 The Dialogue Between the Two Views

In this section, we provide some examples from the literature to show how the dialogue between the etic and the emic views is possible.

13.3.1 The Bricklayers' Rectangle

Bricklayers in rural areas of Mozambique usually build houses with rectangular floors. Since they do not have tools to design parallel lines (or, what is the same, right angles), they consider a concept of the geometric figure of the rectangle that does not need these geometric properties to build the figure (Gerdes 1998).

In particular, they use a stick and two ropes of equal measure; they tie the ropes to one other at the midpoint, bending them. Then, they tie two of the ends of the ropes to the stick and finally tighten the ropes. Thus, they find the vertices of the rectangle. This construction uses a concept of rectangle based on the diagonals: the rectangle is a quadrilateral whose diagonals are equal and bisect each other. In school mathematics, this is a property of the figure, but it becomes an operational definition for bricklayers, while the formal definition is usually that the rectangle is the quadrilateral with four right angles.

Therefore, considering an *emic* point of view, the bricklayers conceptualize the geometric figure of the rectangle in a different way with respect to scholarly mathematics textbooks. Still, in an *etic* point of view, we can recognize this new definition as a property and we can demonstrate that both definitions are equivalent. Thus, we may actually build a bridge between scholarly mathematics and ethnomathematics.

13.3.2 Kira-Kira Method

A guild of artisan-architects on an Indonesian island has developed special techniques to make home decorations. For example the kira-kira method is used to find the midpoint of a segment using only a stick and a pencil (Albertí 2007).

The method works as follows: a point is marked where half of the segment is estimated to be. Then, the stick is overlapped to the segment so that one endpoint of the stick coincides with one endpoint of the segment, and the midpoint is also estimated on the stick. Then the stick is moved so that its end now matches the mark of the midpoint in the segment, and half of the mistake made is gauged and marked on the stick. With this new mark the process is performed again with this new mark and it ends when there is no error perceived by the human eye.

It is worth noticing that greater accuracy in determining the midpoint is achieved with each repetition of the process, not to mention that the artisans-architects have a very good eye. If we stop to think, accustomed to reason as taught in school, most of us would have to measure the segment with a measuring instrument (probably a ruler), perform a division operation, and measure with the instrument again to find the midpoint.

All this needs some material such as cognitive resources and a good deal of time, which makes the kira-kira method much more convenient (and even more precise, if we take into account the numerous opportunities for errors provided during the performance of two measurements).

A researcher that assumes an *emic* point of view may observe that the method these artisans-architects use to find the midpoint of a segment is completely different from the one employed in school. Still, from an *etic* perspective, the researcher could be interested in studying the process in mathematical terms, and observes similarities with many of the interpolation algorithms used in computation to reduce errors.

13.3.3 Cabecars Counting

The indigenous people of Costa Rica have their own way of counting. The Cabecares people particularly, like others in the region, have different names for numbers according to the class of objects to which the number refers (Gavarrete

2012). For example, there is a word for 3 oranges (which are round) and another to say 3 bananas (which are elongated).

Likewise, objects that do not belong to the same class cannot be considered in the same group, so they cannot be added. Syntactically, these classes are called numeral classifiers. In the Cabecar language there are six classes: humans, elongated objects, rounded objects, flat objects, packages (bunches or bags), and parts or pieces. This way of counting is probably optimal in the context where it has been developed.

We have just shown an original *QRS system*, in particular to deal with quantities. A researcher with an emic point of view, observes that this is a way of knowing that could seem incommensurable with the *NUC system*. Yet, if the researcher thinks about algebraic rules, for example, *do not add or subtract x to/from y*, he/she can easily observes the same underlying idea. Moreover, at school we have often heard about the troubles of adding apples and oranges.

13.3.4 The Dancers' Diamond

The Chacarera dancers, Argentinian folk dancers, draw a diamond on the floor when performing some of the steps of the choreography. The dancers begin this series of equal steps on one of the edges of the figure, but the movement emphasizes the direction of the partner they are facing, unlike the lateral direction that is barely perceived. Then, the diamond is treated as a figure of four equal sides whose perpendicular bisecting diagonal should measure one more than the other[2] (Albanese and Perales 2014b).

If we recall the definition of the diamond in school textbooks, we are likely to find that a diamond is a figure with four equal sides and equal opposite angles. But, the notion of angle is very complex for a child, while length measurement is much more intuitive.

Again we have an example of a property of a figure used as a definition because of the easier notion on which it is based. As before, from an *emic* point of view, the researcher can underline that the definition is different from that found in a school text book but, from an *etic* perspective, the new definition is compatible with the other. Thus, it is possible to build bridges, demonstrating a figure that has one statement as a definition and has the other as a property.

---

[2]Usually in primary school a non-inclusive classification of the quadrilaterals is given. This means that, at this level, a square is not recognized as a particular case of a diamond.

13.3.5 Ocean Path Navigation

Polynesian rowers know the ocean so well; the currents, the waves, the winds; that they are able to always sail along the same paths, as if somehow they could visualize these paths in the water (Barton 2008); and as they know how to calculate the distance traveled, they can situate themselves along these paths. Modern navigation is based on the position provided by GPS. There is a grid reference on the Earth consisting of parallels and meridians that lets us determine our location with the coordinates and lets us calculate directions and distances.

From the emic point of view, the research makes evident that the Path-navigation is profoundly different from Position-navigation, but if we acquire an etic perspective we can easily find a metaphor to understand how Path-navigation works. Indeed, this is really similar to how a driver usually situates himself/herself during car trips. In this case, he/she travels along highways or freeways and the car odometer allows him/her to calculate the distances traveled and so provide the needed information to determine location along the path he/she decided to travel.

It is worth noting that in the cases described here, the methodological process has been the opposite compared to Albertí's proposal (2007). Indeed, as researchers, we have first taken an *emic* perspective to discover *different ways of knowing* and in second place we have adopted an *etic* perspective to find correspondence with some knowledge of *Western* mathematics. We perceive that, in this way, it could perhaps be easier to discover *different ways of knowing*, removing one's own cultural constraints at the beginning of the observation of the practices.

13.4 Two Approaches to Education

In the Ethnomathematical program, we consider Mathematics Education from a social and cultural perspective as a logical-mathematical enculturation (Bishop 1991). This is not a process of instruction with a passive transmission of knowledge, but it is a way to learn that involves direct experiences and inquiries as well as communitarian learning.

Now, regarding the possible implications of Ethnomathematics in mathematical curriculum, we start by taking into consideration what Adam et al. (2003) assert and, then, Pais (2011) summarizes:

> Ethnomathematics can appear in schools as an approach to mathematics; as a particular content distinct from the conventional mathematical concepts taught in schools; as a stage in the progression of mathematical thinking; and as awareness that classrooms are situated in a cultural context (Adam et al. cited in Pais 2011, p. 221).

Now, we try to analyze these possibilities from our two postures: *mathematics in cultural practices* or an *etic perspective*, and *ways of knowing* or an *emic perspective*.

The first option can be seen as the use of cultural practices to introduce mathematical concepts; this means that cultural practices are interpreted as examples or applications of scholarly-academic mathematics, mainly with the objective to contextualize tasks in the cultural context; this is an *etic* perspective.

The second option takes into account distinct contents from the conventional one, while the third implies the existence of different developments of mathematical thinking. Both options stand on the idea that different *ways of knowing* exist in mathematics, so they are discovered doing research from an *emic* perspective.

The fourth option talks about awareness. The awareness of the cultural context can be reached when both etic and emic perspectives are considered, valued, and accepted. So, we interpret that this last option matches the two postures.

It is worth exploring what each posture implies for education, answering the main challenge of the Ethnomathematics program at this point of its evolution: defining the role of Ethnomathematics in mathematics education (Gavarrete 2013).

The *etic* perspective, where Ethnomathematics is considered the recognition of mathematics in cultural practices, allows us to make comparisons and build bridges between different cultures, as we have already mentioned; so it involves research with the objective of tasks designed for compulsory education that look for meaningful contextualization. This research usually emphasizes scholarly mathematics as a social and cultural product developed to solve problems arising in daily life.

The *emic* perspective, where Ethnomathematics is the discovery of ways of knowing, involves the reflection of the epistemological nature of mathematics, sometimes with a noticeable anthropological interest. It is adopted in educational proposals in teacher education in order to generate reflection and enculturation through the experience of investigatory experiences.

The *emic* posture must also be taken into account when working in multicultural and intercultural environments as well as where the cultural groups preserve a strong identity, for example in Gavarrete (2012), in order to respect as much as possible their cultural knowledge.

Again, in our own research we try to apply both views, and their consequences in mathematics education. In fact, we design workshops for teacher education to promote both the tasks designed for primary education (Albanese and Perales 2014b, 2015b) and the reflection about the nature of mathematics (Albanese et al. 2016a). In the following paragraphs we present these ideas with some more details.

13.4.1 Mathematics Education… From an Etic Perspective

Mainly in an *etic* view, first Bishop and then Oliveras, proposed, respectively, the realization of projects (Bishop 1991) and Ethnomathematical Microprojects (Oliveras 1996) linking mathematics to cultural knowledge.

Bishop (1991) promotes the performance of work in small groups with elementary school children where mathematics is discovered in relation to any artefact or cultural practice. Similarly, work with integrated and cooperative Microprojects, based on Ethnomathematics proposed by Oliveras (1996) in the initial training of primary school teachers, combines the mathematical knowledge about a cultural sign, a tangible or intangible element of culture, with the mathematical potential that needs to be explored by the teacher before proposing mathematical activities around this cultural sign in the elementary classroom.

The structure of a Microproject is usually organized into two parts (Oliveras 1996; Gavarrete 2012; Albanese and Perales 2014b): the first consists of an ethnographic research on the cultural sign where the teacher first justifies the choice of the sign and then takes an ethnographic approach to expert knowledge of the sign, to carry out further analysis of the mathematical potential; the second, from a reflection on the research process and the results on the mathematical aspects found, one aspect of the cultural sign is identified to design a classroom activity that the teacher proposes to his/her primary school children.

This is one way in which Ethnomathematics can be introduced in the context of teacher education. Many authors look for relations between mathematics and some cultural knowledge or practices to design cultural contextualized tasks and activities, although without calling them projects or Microprojects.

In our research, we proposed the development of Microprojects in a primary teacher education course with teachers in training of Chaco, a region in the north of Argentina, choosing the folk dances of the region as the cultural sign (Albanese and Perales 2014b, 2015b). Even though the idea of designing tasks considering the mathematics of cultural signs is nearer to an *etic* approach, we took into consideration both the *etic* and *emic* views, at least in the first part of the project during the ethnographical study of the folk dances.

The results provided different ways of defining and conceptualizing some geometrical figures such as the diamond (discussed in a previous section) and the circle. Thereby, an important value of the Microproject was the reflection on definitions and properties of geometrical figures. In the second part of the project, taking an *etic* perspective, the future teachers built tasks to work on the relation between the definitions and properties previously found.

13.4.2 Mathematics Education… From an Emic Perspective

On the other hand, some authors state that Ethnomathematics should play a more *revolutionary* role in mathematics education, trying to change the educational system.

Pais (2011) raises a deeper involvement for education recognizing Ethnomathematics as a strong instrument of criticism and questioning of society and arguing that this should be reflected in a critique of the educational system, working toward a transformation of the present school organization. In this sense, for example, instead of focusing on bringing local knowledge to schools, he proposes problematisation, with the community of knowledge, of the skills needed in situ, with the idea of creating alternative educational settings outside the school (Pais and Mesquita 2013). He also identifies the main challenge of Ethnomathematics as being the creation of a competence to understand and change the world and the construction of a system that makes this an important competence for education (Pais 2013).

Also it is worth noting the existence of different positions on the relationship between Ethnomathematics and mathematics education manifested in the interview reports in Miarka's (2011) research. Barton and Gerdes agree that the potential of Ethnomathematics goes beyond the educational field and to restrict it to education means limiting this potential. Both authors perceive in Ethnomathematics the possibility of an expansion albeit with different shades: Gerdes pursues an extension of mathematics (which for him is a single science in expansion) as a field of knowledge while Barton enhances the expansion of the horizon of understanding mathematics as a concept, arguing that Ethnomathematics helps us understand better what mathematics is and broaden its conception (Barton 1996a, b).

With this last idea of Barton's we undertake our work but we approach it in relation to education. In fact, the same Barton alerts that educators (much more than professional mathematicians) are those with a very limited view of the concept of mathematics that is linked to the concept that is commonly taught in primary and secondary education curricula.

The contribution we provide from this reflection is in the following direction: to change education we have to begin from changing teachers' beliefs about mathematics. Adopting such a new vision of education, as proposed by Pais (2011), is closely linked to the awareness of the profound epistemological changes that an Ethnomathematics perspective brings to the concepts of mathematics, the same changes we have previously related and that Barton (in the interview in Miarka 2011) states are the deeper potential of Ethnomathematics, in terms of the expansion of the concept of mathematics. Understanding and accepting the magnitude of such changes is not easy for researchers (we will make this clearer in the following section, reviewing the critics) and must be even less so for teachers.

To follow these ideas, in Albanese's Ph.D. research (2014), we chose to study a cultural sign, namely a craft, to develop a workshop that had the purpose of influencing the concepts of teachers (both in training and in service) about the nature of mathematics (Albanese et al. 2016a). In this work, as a theoretical contribution, we systematized the changes Ethnomathematics introduces in the concept of mathematics. In fact, we proposed three dimensions:

1. the *practical dimension*: mathematics is a tool that Man develops to relate, understand, manage and eventually change his environment (…),
2. the *social dimension*: mathematics is a consensual construction of a set of rules and norms within a group of people that decide to share it (…),
3. the *cultural dimension*: there is a profound relationship between mathematics and culture so different forms of mathematics exist in different cultures (Albanese and Perales 2015a, p. 1540).

It is worth remembering that at the beginning of this chapter, we mentioned the shift of interest in Ethnomathematics focusing on how the cultural and social context affects the process of generation, organization and communication of knowledge. Now, these dimensions refer precisely to these processes.

The *practical dimension* points to the generation of knowledge, which is related to the need to understand and model the environment, and anticipate events that give man the ability to control his future by changing or adjusting his behavior to the environment.

The *social dimension* refers to the way of organizing and communicating knowledge, developing a system that is shared by the community to allow the exchange of information in an efficient form. Here mathematics is understood as a language, i.e. a means of communicating.

The *cultural dimension* arises as a consequence of the above, as needs and contexts are different, and ways to communicate about them may be different, then it must be possible to have different systems of knowledge, i.e. different Ethnomathematics.

Currently, our interest is in teachers and prospective teachers becoming aware of these arguments so that their work can be open and alert to cultural and social differences.

13.5 Philosophical Background

The authors who have recently tackled the issue of the cultural relativism have delved into Wittgenstein's philosophy of a language game. Each author, however, draws different conclusions.

Barton (2012[3]) highlights Wittgenstein's philosophy as a possible solution to the need of Ethnomathematics to justify the simultaneous existence of different mathematics. Wittgenstein argues that mathematics is not a description of reality or a useful science but rather a system whose propositions are rules that give meaning to the system, created as a *way of speaking*. Our interest as ethnomathematicians should focus on these rules, the way this system works and its elements in live language. This gives rise to the definition of *QRS systems*.

Barton (2012) also provides a justification for the similarities of the different *QRS-systems* and their usefulness as a tool for understanding reality: *QRS-systems* evolve in response to the environment in a way that corresponds to the context in which we live. Since the way we interact with the environment has common features, then *QRS systems* may resemble each other.

The same Barton (2012) and also D'Ambrosio (2012) propose the dynamics of cultural encounters as a possible reason for the similarities of different mathematics,

[3]First published in 1999 as: Barton, B. (1999). Ethnomathematics and philosophy. *ZDM, 31*(2), 54–58.

i.e. when two different systems come into contact with one another, the elements of one system integrate into the other, and sometimes they merge and a new system is born from the combination of the original systems.

Vilela (2010) expresses some similar ideas to those of Barton (2012). She searches for a philosophical basis for Ethnomathematics that includes and explains the coexistence of different concepts of rationality and proposes Wittgenstein's philosophy as an interesting starting point. She highlights two aspects of Wittgenstein's philosophy that support Ethnomathematics. The first is that meaning lies within linguistic practices, hence language is part of the context in which it develops. According to Wittgenstein's theory of language games, mathematics is a system of rules and procedures, and as the structure of the language allows reality to be structured, then mathematical language organizes experience, determines meaning and shapes reality.

The second aspect Vilela (2010) identifies refers to our second issue: although practice determines different meanings, they are similar. The explanation is linked to the structure of the language system dictated by *grammar*, which is a way to see and read the world; grammar is the lens that detects the regularity of the ways of life of the community. In other words, the attempt to interpret a reflection of reality (or similar realities) makes these systems look alike.

The above rules constitute *grammar*, which is the constructed and invented rationality that allows language to be articulated within lifestyles and establishes which logic should be accepted. All this builds a relativistic vision that suggests seeing Ethnomathematics as languages closely related to the culture of the context in which they develop.

Knijnik (2012) performed an in-depth study of some of Wittgenstein's notions to develop a theoretical basis for Ethnomathematics. She emphasizes that the meaning of words is determined by their use in practice, so language games and rules depend on the ways in which language is used. These language games should be understood within the life forms (context, culture) in which they have developed.

Knijnik (2012) therefore appeals to the construct of *family resemblance* to justify that we should consider all these languages and practices as mathematics because they are similar, but her discourse presents no evidence of the reasons for the existence of these similarities. But it is noteworthy that she draws very different conclusions as she refers to the incommensurability of the different mathematics because once separated from their *life form*, they lose meaning. This implies that there is no possibility of building bridges between different mathematics, or at least this is what she seems to assert, though we don't agree with this last statement.

However, the issue of the role of Wittgenstein's philosophy of a language game has been broadly discussed during the presentation of another one of our papers (Albanese et al. 2016b) from a few ethnomathematicians that presented there and during a paper presented in the ICME5 (François 2016).

13.6 Answering Some Critics

Concerning the issue of the role of Ethnomathematics in mathematics education, it is interesting to remember Pais' (2011) reflections about the critics, and contradictions, towards Ethnomathematics by the academic community regarding its epistemological approach and, what interests us most here, regarding educational implications.

We completely agree with Pais (2011) that behind every criticism there is a different belief of the role of school and mathematics in relation to society. For example, some critics come from a more or less explicit idea of mathematics as a value-free and cultural-free knowledge; others come from a necessity to *uniformize* knowledge in response to the requirement of a global market-oriented society, where equality is more important than equity. Consequently, Ethnomathematics becomes a means of exclusion.

Our contribution is that the following critics are governed by the lack of dialogue between the etic and emic positions. Sometimes the critics are made from one perspective ignoring the other, while sometimes the educational activities are designed without considering one of the perspectives.

When local knowledge is used as folkloric curiosities to reach scholarly mathematics, it means that there is a lack of awareness of the principle of the emic perspective, because generally we cannot just apply concepts of scholarly mathematics to the ethnomathematics or QRS systems we found in the cultural heritage. In fact, these QRS systems may conceptualize space or quantities in different ways of knowing and it is important to detect and valorize these differences.

When the intent to bring Ethnomathematics into the classroom seems forced and disconnected from school mathematics, it generally depends on the conviction that local knowledge is incommensurable with school mathematics, so the relation is missing, in this case, a lack of interest in the principle of the etic perspective is shown. While usually there is a way to build bridges—maybe thought metaphors—between scholarly mathematics and ethnomathematics or QRS systems (let's recall the examples in one of the previous sections).

Often, Ethnomathematics is criticized to be the cause of a strengthening of social disparities between the majority and minority cultural groups.[4] According to our criteria, a critique in these terms is not justified because it would mean that the educational actions criticized are planned only from an emic point of view, in which only the vision of the minority cultural group is kept; while, if the actions took into consideration also an etic perspective, a confrontation with the dominant culture would be included to facilitate the learning of concepts. Let's just think about the Costa Rican indigenous pupils accustomed to counting with numeral classifiers, and the difficulties produced by the first approach with numbers in Spanish, the language that is used in their mathematical education. The awareness of cultural differences allows teachers more culture friendly teaching.

---

[4]This criticism was very strong during the years of the apartheid in South Africa.

So, as researched and educational actions designers, we should be careful with the planning of activities, always taking into account both perspective and so avoid possible future inconsistence with our purposes.

13.7 Conclusion

To sum up, in this document we propose two views of Ethnomathematics:

(1) the *recognition* of *mathematics in cultural practices*,
(2) the *discovery* of different *ways of thinking*.

The first view implies recognizing mathematics in the practices of cultural groups, and this research is performed by the categories and schemes of thinking of the researcher's culture. The second view implies discovering different ways of knowing quantity, space and relation aspects of human experience (QRS systems) considering a broader concept of (ethno)mathematics where the categories and schemes of thinking of the studied cultural group are taken into account.

The path in the literature allows us to identify features of this broad concept of mathematics, named ethnomathematics, in terms of a systematized structure of knowledge dealing with quantity, space and relation that can be oral and intuitive, without involving many symbols, but still having a way to justify actions. Likewise this literature review gives clues about how to face the methodological dilemma of studying the culture of the Other with or without the influence of one's own culture: a balance of the two perspectives proposed on one hand helps to avoid the strong influence of one's own culture, yet justifies the usefulness of some influence of this type.

We have pointed out which implications for Mathematics Education these two views suggest. Ethnomathematics as the study of mathematics in cultural practices encourages building tasks that consider a cultural contextualized environment and stimulate the reflection on certain mathematical concepts. Ethnomathematics as the study of different ways of knowing fosters epistemological reflection on the nature of mathematical knowledge: the concept of mathematics that Ethnomathematics introduces is focused on the idea of mathematics as a tool to understand and control reality (*practical dimension*), as a system of rules and symbols shared in a community to communicate (*social dimension*), and finally as a part of cultural heritage (*cultural dimension*), different for each population. Stressing this last idea of cultural relativism we have summarized the thought of many ethnomathematicians who indicate Wittgenstein's philosophy of a language game as the theoretical foundation of the field.

In addition, we have emphasized the necessity to integrate both of them in research as well as in educational actions. We also provided some examples to illustrate how it can be done. In terms of research, we have shown how the emic perspective allows us to find new conceptualizations of mathematical notions (geometrical figures and numbers) and methods (to find a midpoint of a segment), and in the same examples the etic perspective enables us to understand these new concepts in mathematical words or with metaphors connecting them to other mathematical notions. In our own research we studied two braiding crafts and the two perspectives guided us during the ethnographic work: first we looked for mathematics in the craft and we recognized a model that implies graphs. Then we ended up observing that in one craft the model was not evident in the artisans' thinking. Thus we discovered another model employed by these artisans based on patterns and numerical relation.

With respect to educational implications we have picked from our own work an example of how the etic perspective supports the design of contextualized tasks about geometrical figures based on the choreographic steps of an Argentinian folk dance. What is more, we have mentioned an example of how a workshop about models of braiding crafts in an emic perspective could lead to a reflection on the changes in the nature of mathematics that the Ethnomathematics program introduces, systematizing these changes in three dimensions: mathematics as a tool to control and manage reality, mathematics as a system of symbols and rules shared by a community, and mathematics as a part of a cultural heritage.

Nevertheless, as we have seen, the conceptions described present relevant differences, so it is worth remembering how important it is to be clear about the theoretical bases and the conceptions on which every Ethnomathematical research is built, in order to avoid inconsistency and/or contradictions. Moreover we have highlighted the necessity to integrate both of them into research as well as educational actions. For example, we believe that claiming to value cultural knowledge and then merely contextualizing mathematical tasks without reflecting on the nuances between scholarly concepts and cultural concepts means missing an opportunity and losing consistency (forgetting that there are different ways of knowing). At the same time, we think that claiming the incommensurability of various ethnomathematics does not mean respecting the culture, but rather not making a way to seek correspondence (forgetting that there are mathematics in cultural practices).

Finally, we would like to clarify that this proposal does not presume to be complete in terms of the theoretical foundations of Ethnomathematics, not even in the implications for mathematics education. However we believe that this contribution may help to understand why many research projects framed in the Ethnomathematical program seem to have such different objectives and such different implications for education, and this happens because of the different contexts of research and necessity of the sociocultural environment. These two interpretations of Ethnomathematics provide a possible way to integrate the differences and take advantage of them.

References

Adam, S., Alangui, W., & Barton, B. (2003). A comment on: Rowlands and Carson "Where would formal, academic mathematics stand in a curriculum informed by Ethnomathematics? A critical review". *Educational Studies in Mathematics, 52,* 327–335.

Albertí, M. (2007). *Interpretación matemática situada de una práctica artesanal.* Ph.D. Thesis. Barcelona: Universidad Autónoma de Barcelona.

Albanese, V. (2014). *Etnomatemáticas en artesanías de trenzado y concepciones sobre las Matemáticas en la formación docente* [Ethnomathematics in braid crafts and conceptions of mathematics in teacher education]. Ph.D. Thesis in Education. Universidad de Granada, Granada. Retrieved from: http://digibug.ugr.es/handle/10481/35195

Albanese, V. (2015a). Etnomatemáticas en artesanías de trenzado y concepciones sobre las Matemáticas en la formación docente [Ethnomathematics in braid crafts and conceptions of mathematics in teacher education]. *Enseñanza de las Ciencias, 33*(1), 277–278.

Albanese, V. (2015b). Desarrollo de una tesis doctoral en Etnomatemática: Construcción de una investigación emergente [Development of a Ph.D. thesis in Ethnomathematics: Construction of an emergent research]. *Revista Latinoamericana de Etnomatemática, 8*(2), 381–397.

Albanese, V. (2015c). Etnomodelos de trenzado, etnomatemática de una artesanía argentina [Ethnomathematics of an Argentine craft: Identifying ethnomodels of braid]. *Bolema - Boletim de Educação Matemática, 29*(52), 493–507.

Albanese, V., & Perales, F. J. (2014a). Pensar Matemáticamente: Una Visión Etnomatemática de la Práctica Artesanal Soguera [Thinking mathematically: An ethnomathematics view of the art and craft practices of Sogueria]. *RELIME - Revista latinoamericana de investigación en matemática educativa, 17*(3), 261–288. doi:10.12802/relime.13.1731

Albanese, V., & Perales, F. J. (2014b). Microproyectos etnomatemáticos sobre danzas folclóricas: Aprender matemática desde el contexto [Ethnomatematical Microproyect on folk dances: Learning mathematics from the context in teacher education]. *Profesorado, revista de currículo y formación de profesorado, 18*(3), 457–472.

Albanese, V., & Perales, F. J. (2015a). Ethnomathematics dimensions to analyse teachers' conceptions about mathematics. In: K. Krainer & N. Vondrová (Eds.), *Proceedings of the Ninth Congress of the European Society for Research in Mathematics Education* (pp. 1539–1543). Prague: European society of Research in Mathematics Education and Charles University.

Albanese, V., & Perales, F. J. (2015b). Enculturation with ethnomathematical microprojects: From culture to mathematics. *Journal of Mathematic and Culture, 9*(1), 1–11.

Albanese, V., Oliveras, M. L., & Perales, F. J. (2014). Etnomatemáticas en Artesanías de Trenzado: Aplicación de un Modelo Metodológico elaborado [Ethnomathematics in Braiding Crafts: Application of a methodological model]. *Bolema - Boletim de Educação Matemática, 28*(48), 1–20.

Albanese, V., Perales, F. J., & Oliveras, M. L. (2016a). Matemática y lenguaje: Concepciones de los profesores sobre las matemáticas, una mirada desde la Etnomatemática [Mathematics and language: Teacher perceptions from an ethnomathematics approach]. *Perfiles Educativos, 38* (152), 31–50.

Albanese, V., Adamuz-Povedano, N., & Bracho-López, R. (2016b, July). *A reinterpretation of the Ethnomathematical Program.* Communication presented in the 3rd Mathematics Education and Contemporary Theory (MECT3), Manchester, United Kingdom.

Barton, B. (1996a). Making sense of ethnomathematics: Ethnomathematics is making sense. *Educational Studies in Mathematics, 31*(1), 201–233.

Barton, B. (1996b). *Ethnomathematics: Exploring cultural diversity in mathematics.* Ph.D. Thesis. University of Auckland, New Zealand. Retrieved from https://researchspace.auckland.ac.nz/handle/2292/2332

Barton, B. (2004). Mathematics and mathematical practices: Where to draw the line? *For the Learning of Mathematics, 24*(1), 22–24.

Barton, B. (2008). *The language of mathematics: Telling mathematical tales. Melbourne*. Australia: Springer.
Barton, B. (2012). Ethnomathematics and philosophy. In H. Forgasz & F. D. Rivera (Eds.), *Towards equity in mathematics education: Gender, culture, and diversity* (pp. 231–240). Berlin, Germany: Springer.
Bishop, A. J. (1991). *Mathematical enculturation: A cultural perspective on mathematics education*. Dordrecht, The Netherlands: Kluwer Academic Publishers.
Davis, P. J., & Hersh, R. (1988). *Experiencia matemática [Mathematical experience]*. Barcelona: Editorial Labor y Ministerio de Educación y Ciencia.
D'Ambrosio, U. (1985). Ethnomathematics and its place in the history and pedagogy of mathematics. *For the Learning of Mathematics, 5*(1), 44–48.
D'Ambrosio, U. (2008). *Etnomatemática: eslabón entre las tradiciones y la modernidad [Ethnomathematics Link between Traditions and Modernity]*. Ciudad de México, México: Limusa.
D'Ambrosio, U. (2012). The program rthnomathematics: Theoretical basis and the dynamics of cultural encounters. *Cosmopolis, a Review of Cosmopolitics, 3–4*, 13–41.
François, K. (2016, July). *Wittgestein's late philosophy as a philosophical foundation for Ethnomathematics*. Communication presented in the 13th International Congress on Mathematical Education, Hamburg, Germany.
Gavarrete, M. E. (2012). *Modelo de aplicación de etnomatemáticas en la formación de profesores indígenas de Costa Rica* [A model for applying ethnomathematics in indigenous teacher education in Costa Rica]. Ph.D. Thesis. Universidad de Granada, Granada. Retrieved from http://www.redalyc.org/pdf/335/33524579005.pdf
Gavarrete, M. E. (2013). La Etnomatemática como campo de investigación y acción didáctica: su evolución y recursos para la formación de profesores desde la equidad [Ethnomathematics as a field of research and pedagogical action: Its evolution and resources for teacher training towards equity]. *Revista Latinoamericana de Etnomatemática, 6*(1), 127–149.
Gerdes, P. (1988). On culture, geometrical thinking and mathematics education. *Educational Studies in Mathematics, 19*(2), 137–162.
Gerdes, P. (1998). On culture and mathematics teacher education. *Journal of Mathematics Teacher Education, 1*(1), 33–53.
Gerdes, P. (2001). Symmetry aspects of Mavuku baskets among the Makhuwa (Mozambique). *Symmetry: Culture and Science, 12*(1–2), 87–114.
Knijnik, G. (2012). Differentially positioned language games: Ethnomathematics from a philosophical perspective. *Educational Studies in Mathematics, 80*(1–2), 87–100.
Miarka, R. (2011). *Etnomatemática: do ôntico ao ontológico* [Ethnomatematics: From the ontic to the ontological]. Ph.D. Thesis. Universidade Estadual Paulista, Rio Claro, Brasil.
Miarka, R., & Viggiani, M. A. (2012). Matemática e/na/ou Etnomatemática? [Mathematics in/and/or Ethnomathematics?]. *Revista Latinoamericana de Etnomatemática, 5*(1), 149–158.
Oliveras, M. L. (1996). *Etnomatemáticas. Formación de profesores e innovación curricular* [Ethnomathematics. Teacher education and curricular innovation]. Granada: Comares.
Oliveras, M. L., & Albanese, V. (2012). Etnomatemáticas en Artesanías de Trenzado: un modelo metodológico para investigación [Ethnomathematics in Braiding Crafts: A methodological model for research]. *Bolema- Boletim de Educação Matemática, 26*(44), 1295–1324.
Pais, A. (2011). Criticisms and contradictions of Ethnomathematics. *Educational Studies in Mathematics, 76*(2), 209–230.
Pais, A. (2013). Ethomathematics and the limits of culture. *For the Learning of Mathematics, 33* (3), 2–6.
Pais, A., & Mesquita, M. (2013). Ethnomathematics in non-formal educational settings: The Urban Boundaries project. *Revista Latinoamericana de Etnomatemática, 6*(3), 134–144.
Rosa, M., & Orey, D. C. (2012). The field of research in ethnomodeling: Emic, etic, and dialectical approaches. *Educação e Pesquisa, 38*(4), 865–879.

Rosa, M., & Orey, D. C. (2013). Ethnomodeling as a research theoretical framework on ethnomathematics and mathematical modeling. *Journal of Urban Mathematics Education, 6* (2), 62–80.

Sebastiani, E. (1991). Por uma teoria da etnomatemática [For a theory of ethnomathematics]. *BOLEMA, 6*(7), 30–35.

Sebastiani, E. (1997). *Etnomatemática: uma proposta metodológica [Ethnomathematics: A methodological proposal]*. Río de Janeiro, RJ: Universidade Santa Úrsula.

Vilela, D. S. (2010). Discussing a philosophical background for the ethnomathematical program. *Educational Studies in Mathematics, 75*(3), 345–358.

Chapter 14
The Critical-Reflective Dimension of Ethnomodelling

Daniel Clark Orey

The images of peace are ephemeral. The language of peace is subtle. The reasons for peace, the definitions of peace, the very idea of peace have to be invented, and invented again. Children, everybody, here's what we do during war: In a time of destruction, create something. A poem. A parade. A community. A school. A vow. A moral principle. One peaceful moment
Maxine Hong Kingston—The Fifth Book of Peace

Abstract This discussion is related to critical-reflective dimensions of ethnomodelling, which is the creation of indeed the exploration of contexts that develop creativity in students in order to enable them to solve problems they face in their own sociocultural contexts. The critical-reflective dimensions of ethnomodelling allows learners the opportunity to develop a sense of purpose and their own potential by using mathematics to examine and solve problems they themselves choose and deem important. In this chapter, the author also discusses the importance to use both *emic* and *etic* approaches in the implementation of ethnomodelling in investigations. Emic approach is essential for an intuitive and empathic understanding of mathematical ideas, procedures, and practices developed by the members of distinct cultural groups while etic approach is important for cross-cultural comparisons that demand standardized categories to facilitate communication. This context enables ethnomodelling to bring cultural elements to the modelling process because it is the study of mathematical phenomena within a culture. If mathematics is considered as a universal language, then the process of the elaboration of an ethnomodel is the act of creating a mathematical poem. Thus, working with learners in assisting them in this creation is the act of ethnomodelling.

Keywords Critical-reflective dimension · Emic approach · Ethnomodelling · Etic approach

---

D.C. Orey (✉)
Universidade Federal de Ouro Preto, Ouro Preto, Brazil
e-mail: oreydcead@gmail.com

M. Rosa et al. (eds.), *Ethnomathematics and its Diverse Approaches for Mathematics Education*, ICME-13 Monographs,
DOI 10.1007/978-3-319-59220-6_14

14.1 Introduction

Most critical-reflective dimensions of ethnomodelling occur when educators create supportive environments that allow learners to practice and develop critical mathematical abilities and to practice the use of the tools that enable them to solve increasingly complex problems faced in their own sociocultural context. This allows them to develop the competencies, abilities, and the skills that encourage them to focus on the data and problems they are motivated to explore, value and are engaged in. It is hoped that they develop the opportunity to see mathematics as worthy of deeper exploration.

In the past, the process of teaching and learning mathematics was a top-down phenomenon that has contributed to the development of traditional curricula and is related to an educational system that creates cynical, unmotivated, disengaged and passive learners. The purpose of ethnomodelling, as outlined here, is to propose a pedagogical action that engages both educators and students to become active, critical, and reflective participants in resolving problems they face in their own community or cultural context.

14.2 Conceptualizing Critical-Reflective Efficiency

This researcher was initially introduced to general curriculum and instructional issues with professors and colleagues who operated within the paradigm that was initially developed by Tyler (1949). The brilliance of Tyler's model is that it was one of the very first curriculum models and survives to this day as a simple and to me, elegant model. Tyler's theory was based on four fundamental questions that became known as the *Tyler Rationale*:

1. What educational purposes should the school seek to attain?
2. What educational experiences can be provided that are likely to attain these purposes?
3. How can these educational experiences be effectively organized?
4. How can we determine whether these purposes are being attained?

Tyler's model deeply influenced my early teaching and work. And so, when I first came across ethnomathematics, this writer naturally looked at it through the Tyler paradigm. There are certainly problems, and not a little controversy with Tyler's paradigm, there is not space to go into it here, but the four questions remain fundamental to my thinking and filter what I find interesting to my work. This model means that the community must participate in and be fully present in creating school experiences.

This writer is, after all, a former public school teacher trained in the western United States, who still wants to know how *this* might affect *my* practice, *my* students, and *my* community and the practices of my pre-service teachers as well.

Some may find it strange, but it still forms my core or my base as I work with ethnomathematics and modelling, and it provides a foundation to what later I found as a *critical* mathematics education. It also has consistently reminded me to return to and look at the data, the proof for me as a mathematics educator, that things work (or do not work) in classrooms, and in our communities.

On the other hand, one of the more important characteristics of a critical educator has become the ability to develop processes that allow for the analysis of data by both educators and learners that are related to ongoing and day-to-day phenomena. Critical perspectives in relation to the social conditions that affect the experiences of our students help to develop workable strategies needed to solve problems.

This is a transformational form of learning, and is based on the context and previous experiences of our learners. In this context, educators create conditions that help learners to challenge predominant opinions. By using their own context and their experiences while learning to apply a data-based critical reflection to these problems, students are able to develop a rational discourse in order "to create meanings necessary for the structural transformation of society" (D'Ambrosio 1990).

Rational discourse then becomes the form of dialogue in which all parties have the same rights, responsibilities and duties to engage in, claim, and test the validity of their observations in an environment free of prejudice, fear, and social and political domination. It develops sound arguments for action and allows us to develop plans that allow learners to transcend simple, often prejudicial opinions not based on data. It gives learners the opportunity to enter into dialogue, resolve conflicts, and collaborate.

In this type of discourse, intellectual honesty, the elimination of prejudices, and the use of critical analysis based on data and facts, are important aspects that allow for dialogue to happen rationally and freely (Rosa and Orey 2007). In this educational environment, the processes of discourse, conscious or mindful work, intuition, creativity, criticality, and emotion are important elements that combine to help students develop their own critical-reflectiveness as they learn to move away from emotion-laden opinions and focus on the data related to the problem.

In the context of ethnomodelling, *ethno* refers to differences in culture, which are combinations of, and are influenced by, geography, climate, language, history, religion, customs, institutions and on the subjective self-identification of people. These differences also include distinctions based on racial and religious oppression or nationality. Individuals from distinct cultural groups have diverse worldviews that are a product of centuries; they are far more fundamental than differences among political ideologies including the social, economic, and cultural backgrounds that define civilization as a cultural entity (Huntington 1993).

In addition, culture is expanded to include also the cultures of differing professional groups and age classes as well as social classes and gender (D'Ambrosio 1996). Culture is also defined as a social group's design for surviving in and adapting to its environment (Bullivant 1993). Despite these definitions, culture is considered as the ideations, symbols, behaviors, values, knowledge and beliefs that are shared by a community (Banks and Banks 1997).

The essence of our many diverse cultural groups is related to our many cultural artifacts, tools, or other tangible cultural elements as well as the ways we have come to interpret, use, and perceive them. Artifacts are used in distinct cultures in different ways as well as for diverse purposes. Thus, there is certainly a need for the development for a broader definition of culture to ensure that the mathematical knowledge developed by the members of distinct cultural groups is included in this process. Most of the most interesting problems we may encounter, are thought to be unsolvable, though they can be resolved using a different kind of heuristics. Differently from the traditional constrained view of modelling, ethnomodelling uses a combination of:

1. Organized structures and ethnomodels[1] to represent sociocultural information.
2. Diverse ways of manipulating organized information (algorithms).
3. The mechanical and linguistic realizations of these ways, ethnomodels, and structures.
4. The application of these ways, ethnomodels, and structures in different cultural groups and society.

The aim of ethnomodelling is not to question the foundations of modelling, but to understand the way that mathematical ideas, notions, procedures, and practices are elaborated, presented, and communicated among members of distinct cultural groups. It also aims at encouraging the search for cultural ideas, procedures, and mathematical practices, their examination, and adoption into the mathematics curriculum.

The implications for ethnomodelling are that modelling of cultural constructs may be considered as a symbol system organized by an internal logic of the member of distinct cultural groups. Thus, these members develop models of an ongoing set of mathematical ideas, procedures, and relations of one individual to another in order to identify or examine patterns and their interactions in relation to their own cultural contexts.

14.3 Emic and Etic Approaches to Mathematical Knowledge

When researchers investigate the mathematical practices used by members from distinct cultural groups, they often find distinctive, often curious characteristics of mathematical ideas, notions, procedures, and practices that are possible to label,

[1]Ethnomodels are cultural representations that facilitate the understanding of systems taken from the reality of the members of distinct cultural groups. They can be considered as external representations that are precise and consistent with the scientific and mathematical knowledge that is socially constructed and shared by these members (Rosa and Orey 2010).

explain or translate with ethnomodelling. However, an outsider's understanding of these *cultural traits*[2] is an understanding that emphasizes unimportant cultural features and might misinterpret essential concepts. The challenge that arises from this perspective is how to extract the culturally bound mathematical ideas; notions, procedures, and practices of others without letting the culture of the investigator interfere with the data or the culture of the members of a cultural group under study or observation. This happens when cultural group members have their own interpretation of their culture (emic) opposed by or simply misread by an outsider's interpretation (etic) of it.

There are two approaches to be considered in order to investigate and study what can often be thought of as novel, new or *different* mathematical ideas, concepts, procedures, and practices developed by the members of any given cultural group. By becoming mindful of these approaches, one can learn that there is more to diverse ways and we refrain from relegating profound mathematical patterns, ideas and ways of doing to the *folkloric* or *exotic*:

1. *Etic* approaches may be defined as the external or outsiders' view on beliefs, customs, behaviors, and scientific and mathematical knowledge of the members of a given cultural group. This approach relies upon the extrinsic concepts and categories that have meaning for scientific and researcher observers who are the sole judges of the validity of an etic account. It equates culture to "the collective programming of the mind which distinguishes the members of one group or category of people from another" (Hofstede 1997, p. 5). The focus of this approach is the comparison of one culture to another. Researchers that follow the etic approach in cross-cultural research generally look for universal or culture-free concepts and theories. They search for variables and constructs common to all cultures, which are compared in order to discover how they are different from or similar to each other. It is equated with the objective explanation of sociocultural phenomena from the external point of view. These individuals are considered as *culturally universal* (Sue and Sue 2003).
2. *Emic* approaches may be defined as the internal or the insiders' view about their own customs, beliefs, behaviors, and scientific and mathematical knowledge. Hence, it is an approach of understand behaviors, beliefs, customs as well as ideas, procedures and scientific and mathematical practices. This approach focuses on the intrinsic cultural distinctions that are meaningful to these members whether the natural world is distinguished from the supernatural realm in the worldview of that specific culture (Pike 1954). This approach is identified with the sympathetic comprehension of subjective experience from the internal

[2]The terms cultural traits are used for simple behavior patterns that are transmitted by a society and to which the society gives recognition and meaning. They are learned meaning system that consists of patterns of traditions, beliefs, values, norms, meanings and symbols that are passed on from one generation to the next and are shared to varying degrees by interacting members of distinct cultural groups (Ting-Toomey and Chung 2011).

point of view because it focuses upon understanding issues from the viewpoint of the subjects being studied. In this context, culture may be defined as "the *lens* through which all phenomena are seen. It determines how these phenomena are apprehended and assimilated. (…) it is the *blueprint* of human activity. It determines the coordinates of social action and productive activity, specifying the behaviors and objects that issue from both" (McCracken 1988, p. 73). Consequently, the emic approach investigates how local people think, how they perceive and categorize the world, their rules for behavior, what has meaning for them, and how they imagine and explain things (Kottak 2006). Thus, the members of distinct cultures are the sole judges of the validity of the descriptions of their own cultural ideas, notions, procedures, and practices. These individuals are considered as *culturally specific* (Sue and Sue 2003).

The terms *emic* and *etic* were coined by anthropologist Pike (1943, 1954; Headland et al. 1990) who stated that there are two approaches in the study of cultural group systems, which contain the point of view of either the *insiders* or the *outsiders*. Unfortunately, misunderstandings that arise may imply that two dichotomous approaches to cultures have arisen. It is often presupposed that this methodological dichotomy corresponds to the opposition between behaviors of members of distinct cultural groups.

Ethnomodelling helps to engage researchers and educators in a dialogue between emic and etic approaches. The goal is for the emic approach is to be on par with, indeed respected equally along with that of the etic. For example, the results of research conducted by Eglash et al. (2006) show that mathematical ideas, procedures, and practices are merely analyzed from an *etic* approach such as in the application of the symmetrical classifications from crystallography to textile designs and patterns. Rosa and Orey (2010) and Eglash et al. (2006) affirmed that ethnomathematical researchers often use the term translation to describe the process of modelling by using the application of etic academic mathematical representations to translate mathematical phenomena.

In some cases, the *translation* from *emic* to *etic* mathematical knowledge is direct and simple such as studies related to counting systems and calendars. In other cases, the mathematics as *embedded* in processes such as iteration in beadwork and in Eulerian paths in sand drawings in which the act of translation could be considered as a process of mathematical modelling (Eglash et al. 2006).

However, it is necessary to be cautious in this process because it is easier to strictly use numeric systems and counting procedures rather than to look at the embedded mathematical knowledge that is used in architecture and crafts that requires the application of modelling (Eglash et al. 2006). This tells us that culture, is essential to how we think, apply and use and even develop new forms of mathematics. Ethnomodelling can help us to approach this goal.

14.4 The Emphasis of Ethnomodelling

It is our experience that modelling relates to the mathematical ideas, notions, procedures, and practices that describe *systems* taken from the reality of the members of distinct cultural groups that help them to describe their world by using small units of information named ethnomodels that compose a representation for outsiders' understanding. In this regard, they learn to develop and retain the necessary information needed to solve any given problem they face daily and share how this occurs with others.

Therefore, the emphasis of ethnomodelling is related to the processes that help the construction and development of mathematical knowledge, which includes important aspects of collectivity, creativity, and inventiveness. Thus, it is impossible to imprison mathematical concepts in registers of univocal[3] representation of reality and its universal explanation (Craig 1998). Thus, it is not possible to conceive mathematics as a universal language because its principles, concepts, and foundations are not the same everywhere around the world (Rosa and Orey 2007).

In this regard, the "choice among equivalent systems of representation can only be founded on considerations of simplicity, for no other consideration can adjudicate between equivalent systems that univocally designate reality" (Craig 1998, p. 540). This means that the processes of production of mathematical ideas, notions, procedures, and practices operate in the register of interpretative singularities regarding the possibilities for the symbolic construction of the mathematical knowledge developed by the members of distinct cultural groups.

In this context, the etic approach claims that knowledge developed by the members of any given cultural group has no necessary priority over its competing emic claims. Therefore, it is necessary to depend on acts of translation between emic and etic approaches (Eglash et al. 2006). In this regard, cultural specificity may be better understood with an attempt toward comprehending and respecting the background of contextualized methods independent of the subjectivity of the observers.

From an ethnomodelling perspective, the emic constructs are accounts and descriptions of behaviors, beliefs, customs, and mathematical ideas, procedures, and practices that are expressed and analyzed in terms of the conceptual schemes and categories that are regarded as meaningful (consciously and unconsciously) by the members of distinct cultural groups. This means that these constructs must be in accordance with the perceptions and understandings deemed appropriate by the insider's culture. The validation of emic knowledge comes with a matter of consensus of the local people who must agree that these constructs match the shared perceptions that portray the main characteristics of their culture. Emic knowledge can be obtained through either an elicitation or an observation. In this regard,

[3]Univocal representations provide only one meaning or interpretation of reality, which leads individuals to develop only one explanation or conclusion.

external observers may infer local perceptions, while being careful and respectful of their perspective, context, and setting.

Etic constructs are the accounts and descriptions of behaviors, beliefs, customs, and mathematical ideas, procedures, and practices that are expressed and analyzed in terms of the conceptual schemes and categories that are meaningful and appropriate to the community of scientific observers. An *etic* construct must be precise, logical, comprehensive, replicable, and observer-researcher independent. The validation of *etic* knowledge thus becomes a matter of logical and empirical analysis, in particular, the logical analysis of whether the construct meets the standards of comprehensiveness and logical consistency. *Etic* knowledge may also be obtained through elicitation as well as observation because local people possess scientifically mathematical valid knowledge.

One of the primary issues raised in mathematics education is concerned with the actual position of researchers, educators, and students in relation to the *etic* (culturally universal) and *emic* (culturally specific) approaches. Most researchers and educators may operate from an *etic* position because they believe that their own mathematical ideas, concepts, procedures, and practices occur in the same way in every culture (Rosa and Orey 2013). This, of course, is not quite right since we are in danger of basing our beliefs on what is often considered *superior* Western ideas and ignoring alternative solutions, perspectives and experiences. As a consequence, it is important to argue here that educators should not use the etic approach to acquire the emic approach because of the danger of colonization of the local knowledge (emic) by the western knowledge (etic).

Once again, this writer cannot emphasize enough the importance that members of every cultural group have come to construct, develop, acquire, accumulate, and diffuse mathematical knowledge. It may be different, it may not launch a rocket to the moon, or navigate an airliner from A to B, but it exists, no matter how humble, unusual or technical. In this regard, minimal modifications in our pedagogical practices are required because they consider mathematical knowledge universal and equally applicable across cultures. Therefore, if the assumption regarding the universal origin, process, and manifestation of mathematical knowledge is similar across cultures, then general guidelines and strategies for the pedagogical work would appear to be appropriate to apply in all cultural groups.

Many researchers and educators who take on an emic approach often believe that many factors such as cultural values, morals, and lifestyle come into play when mathematical ideas, notions, procedures, and practices are developed in regards to the cultural backgrounds of the members of distinct cultural groups. Since students come from different cultures, they have developed different ways of doing mathematics in order to understand and comprehend their own cultural, social, political, economic, and natural environments.

Given this context, our students may come to operate from an emic rather than an etic approach. According to this assertion, it is paramount that educators and teachers alike acknowledge that our lifestyles, cultural values, and different worldviews influence the development of the students' mathematical knowledge because its development arises from cultural contexts. This is one of the most important educational issues currently confronting researchers and educators because it is pointing directly at how worldwide current guidelines and standards for mathematical instruction are culturally bound (Rosa 2010).

14.5 Emic and Etic Approaches in Ethnomodelling

Considering both research and educational fields, which approach must be applied? Should researchers and educators be based on the culturally universal or culturally specific? These questions allow us to argue that some professionals believe in a "cultural universality", which focuses on similarities and minimizes diverse factors that lead to alternative views, while others take on techniques and beliefs of cultural specificity, which focus on cultural differences. Then, the question is whether it is necessary to understand a cultural specificity that requires a specific theoretical basis and concepts (emic) against the background of universal and generic theories and methods (etic).

Over the last three decades, there has emerged a demand for local psychologies besides the Western or Euro-American psychology. Therefore, it is necessary to support the valorization and use of emic approaches in investigations in order to show that many theories and methods seem to be susceptible to the differences in distinct cultures and, consequently, demand cultural contextualization. Conversely, it is most certainly naive to state that the members of distinct cultural groups do not share universal mathematical characteristics. Globalization, the Internet, cultural interchange, even tourism has encouraged a *mudding of the waters*, so to speak. This is not a bad thing, as this exchange, causes new ideas to emerge.

For example, Bishop (1991) stated that many of our everyday activities as well as those of diverse members of cultural groups, involve a substantial amount of mathematical application. In this regard, there are six universal activities practiced by the members of any cultural group. These activities are: (1) Counting, (2) Measuring, (3) Designing, (4) Locating, (5) Explaining, and (6) Playing.

They provide the fundamental facets used to probe the traditional or daily living practices of all members of the human species. These universals are inseparably intertwined with other aspects of the daily life of the members of any cultural group. Through a study of these applications, it is possible to glimpse the wonder of how humans have came to resolve problems, communicate and maybe see how their early experiences incorporated mathematics. However, even though these activities may be universal, it is important to recognize that they are merely generalized perspectives in relation to those who may or may not share similar cultural-historical characteristics and perspectives. Once again, it is naive to believe that mathematical concepts do not reflect the cultural values and lifestyles of the members of any given cultural group.

In this regard, perhaps a better approach to opposing views may be to try to understand the universality of mathematical ideas, notions, procedures, and practices, which are relevant to researchers and educators and to consider them linguistically, that is as different accents or regional dialects. It is also necessary to state that these approaches may take into consideration the relationship between cultural norms, values, attitudes, and the manifestation of mathematical knowledge in different educational fields.

Researchers and educators must also take into account their own worldviews. If they become more aware of themselves and their worldviews and values, then they can become increasingly more open to apply ethnomodelling in their pedagogical practices. This may lead them to a clearer decision between the two approaches. Many scholars have come to believe mathematics activity as highly cultural (D'Ambrosio 1990; Eglash 1997; Rosa and Orey 2007). For example, mathematicians do not agree on the nature of mathematics, debating whether this subject is culturally bound (internalists) or culture-free (externalists) (Dossey 1992).

Internalists such as Bishop (1988) and D'Ambrosio (1985) have shown how mathematics is a cultural product, which is developed as a result of various activities such as counting, locating, measuring, designing, playing, and modelling. Other mathematicians such as Kline (1980) are considered externalists because they believe mathematics is a culturally free activity. Thus, they do not believe in the connection between mathematics and culture. For example, the results of the study conducted by Rosa and Orey (2013) showed that the majority of teachers possess an externalist view of mathematics, which means that they perceive mathematics as culture-free while few of these professional possess an internalist view of mathematics because they perceive mathematics as a cultural product.

Much of the research in ethnomathematics provides strong challenges to the primitivist view that local societies have developed only simplistic or folkloric technologies and associated mathematics (Ascher 2002; D'Ambrosio 1990; Eglash 1997; Gerdes 1991; Horne 2000 Rosa and Orey 2010; Zaslavsky 1999). It is crucial to show that the members of these societies developed sophisticated mathematical practices,[4] not just trivial, exotic or folkloric examples by directly challenging the epistemological stereotypes most damaging to their mathematical knowledge. These practices include diverse economic practices; geometric principles in craftwork, architecture, and the arts; numeric relations found in measuring, calculation, games, divination, navigation, and astronomy; and a wide variety of other artifacts and procedures (Eglash et al. 2006).

Ethnomodelling often uses the term *translation* to describe the process of modelling local systems with a *Western* academic mathematical representations. However, as with all translations, the success is always partial, and intentionality is one of the areas in which the process is particularly tricky or difficult to share. Often

[4]Sophisticated mathematical practices often reflect deep design themes that provide cohesive structures to many of the important local knowledge systems such as cosmological, spiritual, and medical for the members of distinct cultural groups. Examples of sophisticated mathematical practices include the pervasive use of fractal geometry in African design and the prevalence of fourfold symmetry in Native American design (Eglash et al. 2006).

local designs are merely analyzed from a Western view, such as the application of symmetry classifications from crystallography to local textile patterns (Eglash et al. 2006).

Ethnomathematics also makes use of modelling, but here it attempts to use modelling to establish and translate the relations between the local conceptual framework and the mathematics embedded in local designs, traditions, or contexts. Therefore, ethnomathematically speaking, mathematics can be seen as arising from emic rather than etic origins. In some cases, this process of *translation* to Western mathematics is direct and simple: as with counting systems and calendars. In other cases, the mathematics is *embedded* in a process such as iteration in beadwork, and in Eulerian paths in sand drawings, on in counting and measuring (Eglash et al. 2006).

According to this context, ethnomodelling can be considered as the application of "ethnomathematical techniques and the tools of mathematical modelling [that] allows us to see a different reality and give us insight into science done in a different way" (Orey 2000, p. 250). In order to solve problems, students need to understand alternative mathematical systems and they also need to be able to understand more about the role that mathematics plays in a societal context (Rosa and Orey 2007). This aspect promotes a deeper understanding between diverse mathematical systems and traditions by using mathematical modelling. Which is, to repeat the process of the translation and elaboration of problems and questions taken from systems that are part of the reality of the students or community (Rosa and Orey 2013).

As early as 1993, D'Ambrosio defined a *system* as a part of reality, which is to be integrally considered. In this regard, a system is a set of items taken from a certain student context and/or reality. The study of a system considers the study of all its components and the relationship between them. Mathematical modelling is a pedagogical strategy used to motivate students to work on the mathematics content and helps them to construct bridges between the mathematics of school and the mathematical concepts they use in their own reality.

For example, D'Ambrosio (2002) commented about an ethnomathematical example that naturally comes across as having a mathematical modelling methodology. In the 1989–1990 school year, a group of Brazilian teachers studied the cultivation of grape vines that were brought to Southern Brazil by Italian immigrants in the early twentieth century. This was investigated because of the cultivation of grapes and the production of wine is directly linked to the culture of the people in that region. Both Bassanezzi (2002) and D'Ambrosio (2010) believed that this particular case study is an excellent example of the connection between ethnomathematics and mathematical modelling by applying ethnomodelling (Rosa and Orey 2007).

Once again, educators and teachers should search for problems taken from students' reality that translate their deepened understanding of real-life situations through the application of culturally relevant activities. This process enables students to take a position such as sociocultural, political, environmental, and economic in relation to the system under study. According to Rosa (2010), the main objective of this pedagogical approach is to rehearse the established mathematical context that allows students to see the world as consisting of opportunities to employ mathematical knowledge that helps them to make sense of any given situation.

14.6 Distinctions Between Emic and Etic in Ethnomodelling

Our ongoing development of ethnomodelling serves as a vehicle to transfer meaning and value from the culturally constituted world around us to mathematical procedures. The communications of these ideas are represented in the effect of our cultural outlook, or lens, on mathematics concepts developed by distinct cultural groups—either ours, or the *other*. When considering this, it is worth noting that from an emic perspective, culture may not be seen as a construct apart from and causing the development of, mathematical practices.

Answers to the most fundamental anthropological questions, including the origins of humanity, the characteristics of human nature, and the form and function of human social systems are part of the worldview of every culture. Mathematicians and anthropologists have been acculturated to some particular worldview, and they therefore need a means of distinguishing between the answers they derive as acculturated individuals and the answers they derive as anthropological observers/researchers. Defining *emics* and *etics* in epistemological terms provides a reliable means of making that distinction.

From an applied perspective, definitions of culture from emic and etic approaches might be considered as two sides of the same coin. Emic knowledge is essential for an intuitive and empathetic understanding of a culture, and it is essential for conducting effective ethnographic fieldwork. It is important to state that emic approaches do not intend to directly compare two or more differing cultures, but to promote a complete understanding of the cultural group under study. The methods used in conducting emic research do not provide *culture-free* measures that can be directly compared. Instead, they provide *culture-rich* information. The choice of emic versus etic approaches depends on several important factors, including the nature of the research question, the investigators' resources, and the purpose of the study.

In accordance to this context, traditional mathematical models do not fully take into account the implications of the cultural aspect of human social systems. This cultural component is critical because its accounts "emphasize the unity of culture, viewing culture as a coherent whole, a bundle of practices and values" (Pollak and Watkins 1993, p. 490) that are incompatible with the rationality of mathematical models. However, in the greater context of mathematical knowledge that defines mathematics as more than that taught from a western-academic sense of mathematics, and includes a historical-cultural perspective of time and place as described above, the cultural component then encourages us to respect the wide variation of mathematics found word-wide.

Again, this comes from viewing mathematical practices as socially learned and diffused to members of distinct cultural groups and includes the mathematical practices with an internal logic. Then, it is the process by which transmission takes place from one person to another that is central to elucidating the role of culture in the development of mathematical knowledge. Culture then comes to play a far-reaching and constructive role with respect to teaching/learning mathematical practices.

If mathematical knowledge consists of socially learned, transmitted, and diffused mathematical practices, then, in accordance to Read (2004), the cognitive aspects of the human brain play a relatively minor role when constructing models of behaviors of sociocultural systems. The cognitive aspect of the brain that is needed in this framework is primarily related to the decision processes by which members of distinct cultural groups either accept or reject a conduct as part of one's own repertoire of behaviors. In regards to mathematical knowledge, it appears that there are two ways in which we recognize, represent, and make sense of mathematical phenomena.

First, there is a level of cognition that we share, to varying degrees, with the members of distinct cultural groups. For example, we all locate, we all judge distance and time, we all count and order objects, we all distinguish larger for smaller amounts or objects, and we all seem to enjoy patterns. This level would include cognitive modelling that we may do at a non-conscious level that serves to provide an internal organization of external mathematical phenomena and provides the basis upon which a mathematical practice takes place.

Second, there are many culturally constructed representations of external mathematical phenomena that provide internal organization for this phenomena, but where the form of the representation arises or differs is through formulating an abstract, conceptual structure that provides form and organization for external phenomena in a variety and diversity that need not be consistent with the form and patterning of those phenomena as external phenomena; that is, the cultural construct provides a constructed reality.

The implications for mathematical modelling of human systems are that our modelling of a cultural construct may be based on our symbol systems that are organized by our own internal logic or *grammar*. There is modelling of the ongoing set of behaviors and relationships of one individual to another, such as the networks we use to identify an actual pattern of interactions of individuals along one or more dimensions is deemed to be relevant for the organizational form of individuals making up a social unit.

According to this discussion, in the emic approach, the information and observations reflect the target population's own vocabulary, scientific and mathematical knowledge, conceptual categories, language of expression, religion, and cultural belief system. The word contrasts with etic, that refers to the information collected in terms of a conceptual system and the categories used by researchers or *other* outsiders. It is often necessary to use the local language or dialect and gather information in a very open-ended, nondirective way. The participants' perspective on the phenomenon of interest should unfold as the participants perceive it (emic) not as the investigators view it (etic approach).

For example, when we use the pile sort technique and ask informants to "group the food items in any groups they wish to or any way that they happen to think of", the resulting groups are *emic* categories. In the etic approach, observations and data are constructed in the researcher's system of categories and definitions. In other words, models and ethnography should be mutually reinforcing to each other. Ethnographers, if not blinded by prior theory and ideology, should come with an informed sense for the differences that make a distinction from the point of view of people being modeled. They should, in the end, be able to tell outsiders what matters most to insiders in a critical and reflective manner.

14.7 Theoretical Basis for a Critical-Reflective Dimension in Ethnomodelling

Critically reflective teaching places both educators and learners at the center of teaching and learning processes. These classrooms become active laboratories where educators coach students to develop intuitive, creative and a databased criticality by applying pedagogical approaches to real life situations. The act of teaching becomes a social and cultural activity that introduces students to the process of knowledge creation instead of passively receiving information (Freire 2000).

Currently, the debate between what are often conflicting teaching approaches continues mostly due to the over emphasis on ranking, evaluation and testing; where teaching is more or less centered on the memorization and the testing of content in preparation for further testing. Simply put, I reject this approach. The need to elaborate a mathematics curriculum that promotes critical analysis, active participation, and reflection on social transformation by students (Rosa and Orey 2007) is critical as humankind engages in resolving serious political, environmental and social problems that threaten all of us, everywhere.

It is important to call here for a curriculum change that prepares, indeed encourages, educators and learners to become critical, reflective, and responsible citizens, not as a by-product of a standard curriculum and instructional practices, but as a fundamental goal for all learners. This aims to find practical solutions to the many real problems faced by society that can actively engage learners to apply and value mathematics. It is impossible to teach mathematics or other curricular subjects in a way that is not neutral or insensitive to the experiences of students (Fasheh 1997).

Ethnomodelling focuses on critical-reflective efficiencies that engage students in relevant and contextualized activities allowing them to be involved in the construction of their own mathematical knowledge. In this context, Rosa and Orey (2007) argue that the theoretical basis for critical-reflective dimensions of ethnomodelling has its foundations in both Sociocultural Theory and the Critical Theory of Knowledge.

It involves the mathematical practices developed, used, practiced, and presented in diverse situations in the daily life of the members of these groups. This context is holistic and allows those engaged in this process to study mathematics as a system or an ecology taken from their own contextual reality in which there is an equal effort to create an understanding of all components of these systems as well as the interrelationship among them (Rosa and Orey 2007).

Ethnomodelling takes in consideration the essential processes found in the construction and development of mathematical knowledge, which includes often curious and unique aspects of collection, creativity, and invention. It is the study of mathematical phenomena within a culture since it is a social construction that is culturally bound, which applies cultural elements to the modelling process.

14.7.1 Sociocultural Theory

A fundamental aspect for this is through socialization, where knowledge is best constructed when students work in groups and learn to work cooperatively. Construction of knowledge is connected to other knowledge areas in an interdisciplinary manner. It is through real social interactions among students from diverse socio-cultural groups that learning is initiated and built (Vygotsky 1986).

In ethnomodelling processes, diverse socio-cultural environments greatly influence student cognition. Collaborative work among educators and learners makes learning more effective because it generates higher levels of engagement in mathematical thinking through the use of socially and culturally relevant activities, and this makes use of *dialogical constructivism* because the source of knowledge is based on social interactions between students and environments in which cognition is the result of the use of *cultural artifacts* in these interactions. These artifacts act as vehicles allowing students to understand problems they face in their own community (Rosa and Orey 2007).

14.7.2 Critical Theory of Knowledge

Studies of Habermas' *Critical Theory of Knowledge* reinforce the importance of social contexts in the teaching and learning process. Habermas demonstrates how critical consciousness in learners is increased as they analyze social forces around them. This occurs through the use of strategies such as interpersonal communication, dialogue, discourse, critical questioning, and the use of problems taken from their own reality. The effects of social structure influence distinct knowledge areas taken by individuals from their social context. As well there are three generic knowledge domains: *technical*, *practical*, and *emancipatory* (Habermas 1971).

14.7.3 Knowledge or Prediction

The ability to predict, using technical knowledge, is determined by how individuals manipulate their environmental contexts. It is gained through working with diverse empirical investigations and is governed by technical rules. In ethnomodelling processes, students learn to apply this as they collect data by coming to observe and document attributes of specific phenomena, verify if a specific outcome can be produced and reproduced, and know how to use rules to select different and efficient variables to manipulate and elaborate mathematical models.

In this process, students improve their ability to communicate by using data-based hermeneutics (written, verbal, and non-verbal communication) to verify social actions/norms modified by communication. It is here that meaning and interpretation of communicative patterns interact to construct and elaborate understandings that serve to outline agreements in social performance.

14.7.4 Knowledge or Criticism and Liberation

The process of gaining insight emancipates individuals from institutional forces that often limit and control their lives and is equally important to the development of critical-reflective dimensions. How we come to determine our own unique opinions, answers and solutions to the social condition around us forms an essential objective. Knowledge is used to liberate individuals from outdated, often oppressive modes of social domination by developing the tools needed to exercise databased decision-making. In the ethnomodelling process, these insights are gained through mathematical modelling.

Ethnomodelling encourages learners to recognize what is needed to solve problems. Knowledge is gained by reflecting on the data that they themselves observe, collect, and analyze. Learning is linked to the growing technical knowledge of learners and their experiences gained in conjunction with the social and cultural aspects around them through dialogical activity that enables understanding of data collected. In the ethnomodelling process, this approach helps students to learn to take responsibility and/or ownership of their own knowledge and processes. Knowledge and ideas are then translated in interdisciplinary and dialogical ways as modelers begin to focus on the data used as instruments for social transformation.

14.8 Critical-Reflective Dimensions of Ethnomodelling

Currently, there is little consensus for specific epistemologies in the critical-reflective dimensions of ethnomodelling. However, Rosa and Orey (2010) state that the main objectives of these dimensions are to:

1. Provide learners with mathematical tools necessary to study, act on, modify, change and transform their own reality.
2. Teach that deep learning of mathematics starts from the social and cultural contexts of students by providing them with opportunities to develop logical reasoning and creativity.
3. Facilitate the learning of mathematical concepts that help students build knowledge in mathematics so that they are able to understand their social, historical and cultural contexts.

Ethnomodelling encourages learners to describe, inquire about, and investigate problems coming from their immediate sociocultural context, where they work with real problems that they have a personal connection to, and learn to use mathematics as a language for translating, understanding, simplifying, and solving problems. Even this process seems to be similar to the development of mathematical modelling, ethnomodelling is the process in which members of distinct cultural groups come up with different mathematical tools that help them to organize, analyze, solve, and model specific problems located in their our own sociocultural contexts. These tools allow theses members to identify and describe specific mathematical ideas, techniques, and procedures by schematizing, formulating and visualizing problems in different ways, and discovering relations and regularities in their own cultural practices (Rosa and Orey 2010).

Modelers critically intervene in their own reality by obtaining a mathematical representation of the situation by means of reflective discussions related to the development and elaboration of the findings in their own models (Rosa and Orey 2007). As they come to use mathematics to describe a setting, opinion, or problem-situation, ethnomodelling is akin to writing a mathematical poem. Critical-reflective dimensions of ethnomodelling are based on:

1. An increasing comprehension and understanding of sociocultural context in which learners live through reflection, critical analysis and actions based on data.
2. Learners borrow existing systems they study in the context of symbolic, systematic, analytical, and critical aspects.
3. Starting from given problem-situations taken from their own sociocultural context, learners make hypotheses, test them, fix and improve upon them, draw inferences, generalize, analyze, conclude, and make decisions about the object under study.

From this perspective, ethnomodelling makes use of three gradually more complex mathematical modelling pedagogical phases based on the research developed by Barbosa (2001). I would like to take time now to share a couple of cases, or steps that we move students through as we engage in developing ethnomodels.

14.8.1 Case 1: Educators Choose a Problem

In this type of practice, the teacher chooses the situation and then describes it for students, more often than not by using preselected textbook or classic examples. This is in accordance to curriculum content where the teacher provides students with the appropriate mathematical tools needed in the elaboration of classic mathematical models. In our experience, this is often the first step as learners learn to integrate ethnomodelling strategies in their experience (Rosa and Orey 2007). These first experiences in many ways are related to Halpern's (1996) *critical thinking* that involves a range of thinking skills that leads toward desirable outcomes; and Dewey's (1933) *reflective thinking* that focuses on the process of making judgments about what has happened or was observed.

This approach allows students to solve classic problems, such as the study of the calculation of the circumference of the earth by Eratosthenes. By being shown how to describe a problem, organize variables and data, set up related equations to enable them to translate real situations into mathematical terms; they see how to make observation of patterns, testing of conjectures, and the eventual estimation of results. These are important processes for later stages as they become increasingly more and more autonomous and sophisticated in their ability to model. This most certainly introduces them to the process of mathematization, and allows students to construct and look at classic mathematical models.

14.8.2 Case 2: Educators Suggest and Elaborate the Initial Problem

At this stage, the teacher gives a common theme to the students, such as repaving a certain street, river pollution or transportation fares, and then modelers investigate the problem by collecting data, formulating independent hypotheses, and making necessary modifications in order to develop their mathematical models, that they share between groups. Students themselves are given more autonomy to participate in the activities proposed as they develop their own modelling awareness. One of the most important stages of this modelling process is related to the elaboration of a set of assumptions, which aims to simplify and solve the model to be developed (Rosa et al. 2012).

14.8.3 Case 3: Educators Facilitate the Modelling Process and Engage in Ethnomodelling

Educators at this stage facilitate the modelling process by allowing students to choose and justify their own theme. The *ethnomodelers* are encouraged to develop a project in which they are responsible for all stages of the process: from justification for and formulation of the problem to its validation and final presentation. At this stage, supervision by the teacher is akin to coaching, and involves encouragement in the modelling process. It also means the teacher, may teach short lessons to the groups that enable them to learn and understand new mathematical content needed to elaborate their model. This enables a vital critical-reflective engagement in proposed activities by encouraging modelers to develop and justify their own hypotheses and opinions based on their data and research.

Once again, this is very much like writing a poem or essay, at some point the learners must be freed from traditional limitations and the rote learning practices of mathematical grammar and prescribed textbook problems in order to practice developing their own prose and mathematical poems. During the development of ethnomodels, problems are chosen and suggested by the modelers themselves and are used to reflect critically on the aspects involved in the situation modeled. These are related to:

1. The diverse interdisciplinary connections they encounter;
2. Access and uses of many forms of technology; and
3. The discussion of environmental, economic, political, cultural, and social issues.

The use of mathematical content in this process is directed towards a critical databased analysis of problems faced by the members of their own community.

14.8.4 One Example of Ethnomodelling

The results from a conversation during a morning walk with students along a street in Ouro Preto encouraged exploration and development of a simple model that explored the relationship between mathematical ideas, procedures and practices that developed connections between community members and formal academic mathematics.

By observing the architecture along the wall of one of the schools in Ouro Preto, this professor and his students were able to converse and explore and eventually determine ways to relate functions of three types of curves: exponential, parabolic, and catenary to the patterns found on its wall (Rosa and Orey 2013). In this case I wonder aloud if the pattern shown in Fig. 14.1 was a series of parabolas catenaries or exponential functions.

After examining the data collected they measured various curves on the wall of the school and attempted to fit them to functions (exponential and quadratic) through mathematical models and came to the conclusion that the curves on the wall of the school closely approximated that of a catenary curve function.

Fig. 14.1 Curves on the wall of the school. *Source* Photo by the author

14.8.5 An Emancipatory Approach in the Critical-Reflective Dimension of Ethnomodelling

Because these pedagogical practices offer an open activity that allows us to apply multiple perspectives to solve a given problem, this is to my mind, related to the emancipatory aspect of mathematics. However, the *open* nature of the (ethno) modelling activity may be difficult for students to establish and for them to develop models that satisfactorily represent the problem under study (Barbosa 2001). Thus, the coaching, dialogical, and mediator roles of educators are vital during the ethnomodelling process; this is why this approach is considered an extension of critical theories of knowledge, and forms the emancipatory aspects by addressing social-political issues.

According to the Brazilian National Curriculum for Mathematics (Brasil 1998), all students must develop their own autonomous ability to gather data, solve problems, make decisions, work collaboratively, and communicate effectively. This cannot be done in environments that focus on uniform curriculum and testing. The approach as outlined here is based on developing *emancipatory powers*, where students are encouraged to become flexible, adaptive, reflective, critical, and creative citizens.

This perspective is related to sociocultural dimensions of mathematics, and is directly associated with the political dimension of the ethnomathematics program as it develops "actions that guide students in transition processes from subordination to autonomy in order to guide them towards a broader command of their rights as citizens" (Rosa and Orey 2016). It emphasizes the role of mathematics they see in their own context, by developing the ability to analyze a problem in relation to critical and reflective thinking aspects. As well, the role of modelling processes are used can be used to solve everyday challenges present in contemporary society.

Ethnomodelling may be understood as a language to study, understand, and comprehend problems faced daily by the community used to develop their own mathematical prose. For example, this approach is used to analyze, simplify, and solve daily phenomena in order to predict results or modify the characteristics of

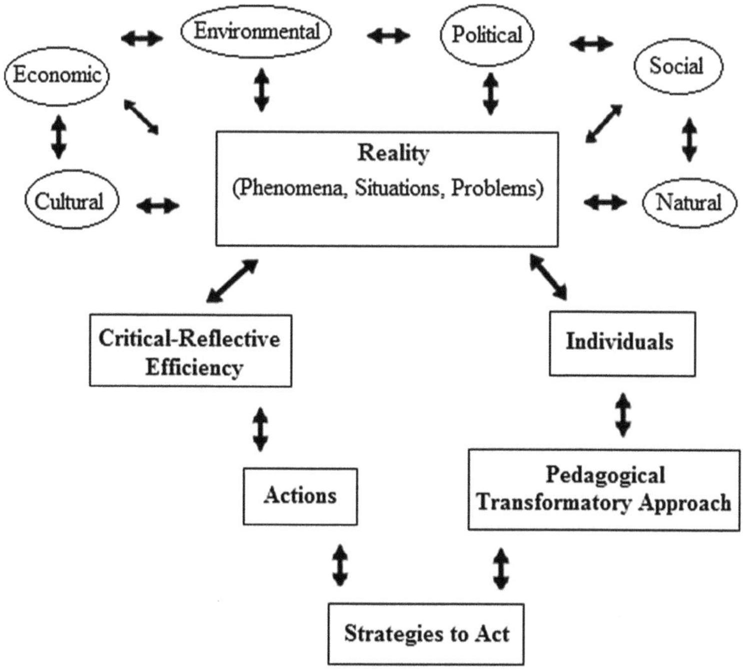

Fig. 14.2 Critical-reflective ethnomodelling cycle. *Source* Rosa and Orey (2015)

these phenomena (Bassanezzi 2002). Figure 14.2 demonstrates this *Critical-Reflective Ethnomodelling Cycle.*

Reflections on the reality modeled become a transforming action (Rosa and Orey 2007), and this system allows modelers to make their own representations by using strategies that enable them to explain, understand, manage, analyze, and reflect their own unique problem-situation.

The application of critically-reflective dimensions of modelling allows mathematics to be seen as a dynamic language used to explain phenomena and used to resolve important conflicts and problems; and is related to the reality of students as when they are passionately involved in sports or video gaming; where gamers deal with more and more abstractions, increasing in difficulty, and the development of and the creation of new tools, where the formulation of new concepts and theories allows them to solve (beat) the problem.

14.9 Final Considerations

Ethnomodelling enables students to question ideas and themes that are develop to help them to understand, explain, and make predictions related to the phenomena under study through the elaboration of ethnomodels that represent these situations

(Rosa and Orey 2007). In this context, ethnomodels are related to processes that critically check parameters selected for the solution of models gleaned from holistic interconnected contexts that the modelers themselves find valuable.

It is not possible to explain, know, understand, manage, and cope with reality outside of realistic, interconnected and holistic contexts (D'Ambrosio 1990). In the critically reflective dimensions found in ethnomodelling, it is impossible to work only with isolated theories or techniques that facilitate models to be memorized, tested upon and quickly forgotten. Ethnomodelling gives access to creativity, conceptual elaboration, and the development of logical, reflective, and critical thinking and empowers modelers to connect what they learn to what they experience outside of school.

Fundamental characteristics for critical-reflective ethnomodelling are based on opinions stemming or that may emerge from databased analysis of problems the students found meaningful. For example, Rosa and Orey (2015) discuss a typical exercise given to students in trigonometry classes: *From the top of a cliff, whose height is 100 m, a person sees a ship under a depression angle of 30°. Approximately, how far is the ship from the cliff?* They argue that this exercise does not provide the development of the critical-reflective dimension of ethnomodelling because students can apply a simple mathematical model related to the tangent function, $tg30° = 100/d$, in order to determine the distance from the base of the cliff to the ship.

In the process of modelling this problem, it is important to discuss with students the assumptions that have been previously established as a critical analysis of its solution since some generalized simplifications of reality were established such as the assumptions that the ocean is flat, the cliff is perfectly vertical to the straight line chosen to represent the distance from the base of the cliff to the ship, a straight line can reasonably approximate the distance from the base of the cliff to the ship, and the curvature of the Earth is ignored. These assumptions are not critically discussed nor reflected with the students.

The critical perspective of students in relation to social conditions that affect their own experience helps them to identify common problems and develop strategies used to solve them (D'Ambrosio 1990). Thus, the analysis of sociocultural phenomena focuses on the development of critical-reflective dimensions of ethnomodelling by using contextualized situations in the mathematics teaching and learning process (Rosa and Orey 2007). Thus, ethnomodelling is based on the ongoing development, comprehension and understanding of reality by the students themselves, and makes use of the reflections, analysis, and critical action they take. When students begin to study the symbolic, systematic, analytical and critical aspects, ethnomodelling can then explain different ways of working with reality.

Critically-reflecting becomes a transformational action that reduces complexity of the reality by allowing students to explain it, understand it, manage it, and find solutions to the problems they themselves have developed. There is a certain intuitive aspect to this application of mathematics that in this time of destruction allows both teachers and students to create something beautiful such as a mathematical poem. Why we are teaching learners to create *mathematical poems*?

Learning to triangulate their opinion by using data gleaned from an emic-etic approach reminds me of a story shared with me by Allen (1992):

The Fable of the Roasted Pig

Once upon a time, a forest where some pigs lived caught fire, and all the pigs were roasted. People, who at that time were in the habit of eating raw meat only, tasted the roasted pigs and found them delicious. From that time on, whenever men wanted roasted pork they set a forest on fire.

Due to the many bad points of *the system*, complaints grew at an increasing rate, as the system expanded to involve more and more people. It was obvious that "the system" should be drastically changed. Thus every year there were a number of conventions, and congresses, and a considerable amount of time and effort was spent on research to find a solution. But apparently no way of improving the system was ever found, for the next year and the year after and the year after that there were more and more conventions and congresses and conferences. And this went on and on and on (…).

Those who were experts on the subject put down the failure of the system to a lack of discipline on the part of the pigs, who would not stay where they should in the forests; or to the inconstant nature of fire, which was hard to control; or to the trees, which were too green to burn well; or the dampness of the earth; or the official method of setting the woods on fire or (…) or (…).

There were men who worked at setting the woods on fire (firemen). Some were specialists in setting fires by night, others by day. There were also the wind specialists, the *anemotechnicians*. There were huge compounds to keep the pigs in, before the fire broke out in the forest, and new methods were being tested on how to let the pigs out at just the right moment.

There were technicians in pig feeding, experts in building pig pens, professors in charge of training experts in pig pen construction, universities that prepared professors to be in charge of training experts in pig pen construction, research specialists who bequeathed their discoveries to the universities that prepared professors to be in charge of training experts in pig pen construction, and…

One day a fireman named John Commonsense said that the problem was really very simple and easily solved. Only four steps need to be followed: (1) the chosen pig had to be killed, (2) cleaned, (3) placed in the proper utensil, and (4) placed over the fire so that it would be cooked by the effect of the heat and not by the effect of the flames.

The director general of roasting himself came to hear of the Commonsense proposal, and sent for John Commonsense. He asked what Commonsense had to say about the problem, and after hearing the four-point idea he said:

"What you say is absolutely right–in theory, but it won't work in practice. It's wasteful. What would we do with our technicians, for instance?"

"I don't know," answered John.

"Or the specialists in seeds, in timber? And the builders of seven-story pig pens, now equipped with new cleaning machines and automatic scenters?"

"I don't know."

"Can't you see that yours is not the solution we need? Don't you know that if everything was as simple as all that, then the problem would have been solved long ago by our specialists? Tell me, where are the authorities who support your suggestion? Who are the authors who say what you say? Do you think I can tell the engineers in the fire division that it is only a question of using embers without a flame? And what shall be done with the forests that are ready to be burned - forests of the right kind of trees needed to produce the right kind of fire, trees that have neither fruit, nor leaves enough for shade, so that they are good only for burning? What shall be done with them? Tell me!"

"I don't know."

What you must bring, are realistic solutions, methods to train better wind technicians; to make pig sties eight stories high or more, instead of the seven stories we now have. We have to improve what we have; we cannot ignore history. So bring me a plan, for example, that will show me how to design the crucial experiment, which will yield a solution to the problem of roast reform. That is what we need. You are lacking in good judgment, Commonsense! Tell me, if your plan is adopted, what would I do with such experts as the president of the committee to study the integral use of the remnants of the ex-forests?"

"I'm really perplexed," said John.

"Well, now, since you know what the problem is, don't go around telling everybody you can fix everything. You must realize the problem is serious and complicated; it is not so simple as you had supposed it to be. An outsider says, "I can fix everything."

But you have to be inside to know the problems and the difficulties. Now, just between you and me, I advise you not mention your idea to anyone, for your own good, because I understand your plan. But, you know, you may come across another boss not so capable of understanding as I am. You know what that's like, don't you, Eh?"

Poor John Commonsense didn't utter a word. Without so much as saying goodbye, stupefied with fright and puzzled by the barriers put in front of him, he went away and was never seen again. It was never known where he went. That is why it is often said that when it comes to reforming the system, Commonsense is missing.

Anonymous

We know that we need to give learners more time to explore, to build new ideas, to express themselves mathematically, yet many of our school systems are more concerned or worried about testing, and less about true, or deeper learning. Ethnomodelling engages our educators and students in the process of teaching and learning mathematics. Many of our students remark that it was the first time they were able to use, explore and engage in mathematics in any real, free, or honest fashion, and they were motivated to continue studying more mathematics, because they could see its power, its elegance, it worth.

It is clear that mathematics provides the foundation of the technological, industrial, military, economic, and political systems and that in turn mathematics relies on these systems for the material bases of its continuing progress. It is important to question the role of mathematics and mathematics education in arriving at the present global predicaments of humankind (D'Ambrosio 2010). So many of our learners learn mathematics, yet rarely have the opportunity to see for themselves how it is used, how it can resolve conflict, how it connects to their lives, how it allowed them to turn a conflict into one peaceful, yet empowered moment by writing their own mathematics poem.

References

Allen, D. W. (1992). *Schools for a new century: A conservative approach to radical school reform*. Santa Barbara, CA: Praeger.

Ascher, M. (2002). *Mathematics elsewhere: An exploration of ideas across cultures*. Princeton, NJ: Princeton University Press.

Banks, J. A., & Banks, C. A. M. (1997). *Multicultural education issues and prospectives*. Needham Heights, MA: Allyn and Bacon.

Barbosa, J. C. (2001). *Modelagem matemática*: concepções e experiências de futuros professores [Mathematical modelling: Conceptions and experiences of prospective teachers]. Doctorate thesis in Mathematics Education. Rio Claro, SP: UNESP.

Bassanezzi, R. C. (2002). *Ensino-aprendizagem com modelagem matemática* [Teaching and learning with mathematical modelling]. São Paulo, SP: Contexto.

Bishop, A. (1988). *Mathematical enculturation: A cultural perpective on mathematics education*. Dordtrecht: Kluwer.

Bishop, A. (1991). *Mathematical enculturation: A cultural perspective on mathematics education*. Dordrecht, The Netherlands: Springer.

Brasil. (1998). *Parâmetros curriculares nacionais (PCN):* matemática [National curricular parameters prospects] (pp. 29–47) Boston, MA: Allyn and Bacon.

Craig, E. (1998). *Routledge encyclopedia of philosophy: Questions to sociobiology* (Vol. 8). Now York, NY: Routledge.

Bullivant, B. (1993). Culture: Its nature and meaning for educators. In J. A. Banks & A. M. Banks (Eds), *Multicultural education issues and à prática* [Mathematics education: From theory to practice]. Campinas, SP: Papirus.

D'Ambrosio, U. (1985). Ethnomathematics and its place in the history and pedagogy of mathematics. *For the Learning of Mathematics, 5*(1), 44–48.

D'Ambrosio, U. (1990). *Etnomatemática* [Ethnomathematics]. São Paulo, SP: Editora Ática.

D'Ambrosio, U. (1996). *Educação matemática: da teoria à prática* [Ethnomathematics: From theory to practice]. Campinas, SP: Papirus.

D'Ambrosio, U. (2002). *Etnomatemática: elo entre as tradições e a modernidade* [Ethnomathematics: Links between tradition and modernity]. Belo Horizonte, MG: Autêntica.

D'Ambrosio, U. (2010). Mathematics education and survival with dignity. In H. Alro, O. Ravn, & P. Valero (Eds.), *Critical mathematics education: Past, present and future* (pp. 51–63). Rotterdam, The Netherlands: Sense Publishers.

Dewey, J. (1933). *How we think*. Lexington, MA: D. C. Heath and Company.

Dossey, J. (1992). The nature of mathematics: Its role and its influence. In D. A. Grouws (Ed.), *Handbook of research in mathematics teaching and learning* (pp. 39–48). New York, NY: Macmillan.

Eglash, R. (1997). When math worlds collide: Intention and invention in ethnomathematics. *Science, Technology and Human Values, 22*(1), 79–97.

Eglash, R., Bennett, A., O'Donnell, C., Jennings, S., & Contorino, M. (2006). Culturally situated design tools: Ethnocomputing from field site to classroom. *American Anthropologist, 108*(2), 347–362.

Fasheh, M. (1997). Is math in classroom neutral or dead? A view from Palestine. *For the Learning of Mathematics, 17*(20), 24–27.

Freire, P. (2000). *Pedagogy of the oppressed*. New York, NY: Continuum.

Gerdes, P. (1991). *Etnomatemática: cultura, matemática, educação* [Ethnomathematics: Culture, mathematics, education]. Maputo, Moçambique: Lulu.

Habermas, J. (1971). *Knowledge and human interests*. Boston, MA: Beacon Press.

Halpern, D. F. (1996). *Thought and knowledge: An introduction to critical thinking*. Mahwah, NJ: L. Erlbaum Associates.

Headland, T. N., Pike, K. L., & Harris, M. (1990). *Emic and etics: The insider/outsider debate*. Newbury Park, CA: Sage.

Hofstede, G. (1997). *Culture's consequences: International differences in work-related values.* Newbury Park, CA: Sage.
Horne, C. (2000). *Geometric symmetry in patterns and tilings.* Oxford, UK: Woodhead Publishing.
Huntington, S. P. (1993). The clash of civilizations? *Foreign Affairs, 72*(3), 22–49.
Kline, M. (1980). *Mathematics: The loss of certainty.* New York, NY: Oxford University Press.
Kottak, C. (2006). *Mirror for humanity.* New York, NY: McGraw Hill.
McCracken, G. (1988). *The long interview: Qualitative research methods.* Newbury Park, CA: Sage.
Orey, D. C. (2000). The ethnomathematics of the Sioux tipi and cone. In H. Selin (Ed.), *Mathematics across culture: The history of non-western mathematics* (pp. 239–252). Dordrecht, The Netherlands: Kluwer Academic Publishers.
Pike, K. L. (1943). *Phonetics: A critical analysis of phonetic theory and a technic for the practical description of sounds.* Ann Arbor, MI: University of Michigan.
Pike, K. L. (1954). *Emic and etic standpoints for the description of behavior.* Glendale, IL: Summer Institute of Linguistics.
Pollak, R., & Watkins, S. (1993). Cultural and economic approaches to fertility: Proper marriage or mésalliance? *Population and Development Review, 19,* 467–496.
Read, D. (2004). Mathematical modelling issues in analytical representations of human societies. *Cybernetics and Systems, 35*(2–3), 163–172.
Rosa, M. (2010). *A mixed-methods study to understand the perceptions of high school leader about English language learners (ELL): The case of mathematics.* Doctorate dissertation. College of Education. Sacramento, CA: CSUS.
Rosa, M. & Orey, D. C. (2007). A dimensão crítica da modelagem matemática: ensinando para a eficiência sociocrítica [Critical dimension of mathematical modelling: Teaching for social-critical efficiency]. *Horizontes, 25*(2), 197–206.
Rosa, M., & Orey, D. C. (2010). Ethnomodelling: A pedagogical action for uncovering ethnomathematical practices. *Journal of Mathematical Modelling and Application, 1*(3), 58–67.
Rosa, M., & Orey, D. C. (2013). The mathematics of the curves on the wall of the Colégio Arquidiocesano and its mathematical models: A case for ethnomodelling. *Journal of Mathematical Modelling and Application, 1*(8), 42–62.
Rosa, M. & Orey, D. C. (2015). Social-critical dimension of mathematical modelling. In G. A. Stillman, W. Blum, & M. S. Biembengut. (Orgs.). *Mathematical modelling in education research and practice: Cultural, social, and cognitive influences* (pp. 385–395). NewYork, NY: USA: Springer.
Rosa, M. & Orey, D. C. (2016). State of the art in ethnomathematics. In M. Rosa, U. D'Ambrosio, D. C. Orey, L. Shirley, W. V. Alangui, P. Palhares, & M. E. Gavarrete (Eds.) *Current and future perspectives of ethnomathematics as a program* (pp. 18–23). ICME13 Topical Surveys. London, England: SpringerOpen.
Rosa, M., Reis, F. S. & Orey, D. C. (2012). A modelagem matemática crítica nos cursos de formação de professores de matemática [Critical mathematical modelling in the mathematics teacher education program]. *Acta Scientiae, 14*(2), 159–184.
Sue, D. W., & Sue, D. (2003). *Counseling the culturally diverse: Theory and practice.* New York, NY: Wiley.
Ting-Toomey, S., & Chung, L. C. (2011). *Understanding intercultural communication.* New York, NY: Oxford University Press.
Tyler, R. (1949). *Basic principles of curriculum and instruction.* Chicago, IL: University of Chicago Press.
Vygotsky, L. (1986). *Thought and language.* Cambridge, MA: The MIT Press.
Zaslavsky, C. (1999). *Africa counts: Number and pattern in African cultures.* Chicago, IL: Lawrence Hill Books Chicago Review Press.

Part VI
Conclusions

Chapter 15
Some Conclusions About Ethnomathematics: Looking Ahead

Wilfredo V. Alangui and Lawrence Shirley

Abstract This final chapter reviews the 13 contributions in the volume, organizing them using the four categories of *mathematics of cultural groups*, *classroom applications*, *cross-cultural situations*, and *theoretical basis of ethnoamthematics*. Taken together, the featured works provide examples of new trajectories for research, and they offer a glimpse of future directions in the field of ethnomathematics, especially its role in advancing inclusive and culturally relevant mathematics education.

Keywords Mathematics of cultural groups · Classroom applications · Cross-cultural situations · Theoretical basis of ethnomathematics

15.1 Looking Back

This volume contains thirteen chapters on ethnomathematics research, projects, and thoughts. The authors worked independently unaware of each other's contributions, only following the general topic requirement of *Ethnomathematics and Mathematics Education*. Collectively, they offer a small, but fairly representative sample of current work in ethnomathematics and offer suggestions for further work. To consider them as a group, to see what we can learn from them about ethnomathematics, it is valuable to fit them into a broader context of theories, applications, and data of the field.

W.V. Alangui
University of the Philippines Baguio, Baguio City, Philippines
e-mail: wvalangui@up.edu.ph

L. Shirley (✉)
Towson University, Towson, MD, USA
e-mail: LShirley@towson.edu

M. Rosa et al. (eds.), *Ethnomathematics and its Diverse Approaches for Mathematics Education*, ICME-13 Monographs,
DOI 10.1007/978-3-319-59220-6_15

15.2 Four Work Areas of Ethnomathematics

In the early 1990s, the *International Study Group on Ethnomathematics* (ISGEm) used a structure of four interest groups to help sort out the various kinds of work that was happening. Sometimes, ISGEm meetings even broke into interest subgroups to review projects in more detail. It has been suggested (Shirley 2012), that this four-way categorization can still be useful today, even though some projects can be fitted into more than one category. Sometimes this organization can also point out gaps and areas where additional research is called for. That is also true for the chapters of this volume. Let us review the four categories in general and then look at the placements of the thirteen chapters as we review them.

The first area is basic *collection of field data*. This is fundamental to a research program that investigates the "mathematics of cultural groups" or the interplay between "mathematical knowledge and knowledge embedded in cultural practice" (Alangui 2010, p. 84). In some sense, this is similar to anthropological data collection. The researcher discusses activities with the subjects of the research, seeking and hoping to recognize instances of mathematical thinking being used in an activity. Earlier, much of this data collection was done by outside researchers, who had to learn about the culture itself before seeking mathematical activity. More recently, the researcher may be a member of the culture, or at least one who is close to and familiar with the societal activities (Alangui 2010). In either case, the research needs to follow the processes of the activity, constantly on the alert to spot mathematics, which may not even resemble mathematics in the Western sense. Ascher (1991) tried to use the term *mathematical thinking* in cases that seemed very different from Western mathematics.

In some cases, the research is organized with the researcher serving as a liaison to Western mathematics, comparing notes with the cultural representatives and educators or, in some cases (Adam 2010), with academic mathematicians. It seems that both sides learn and gain from this exchange. The data results may be numerical or geometrical; they may be linguistic in defining terms used in mathematical applications; they may describe a cultural structure or an engineering-type process of completing a cultural activity. Whatever information is found needs to be analyzed and reported to show the mathematics being used and, usually, relate them to other mathematics.

The second area was called *classroom applications*. This can be traced to school classrooms even before there was a term called *ethnomathematics*. Teachers have often sought unusual examples to bring to their instruction, often highlighting the curriculum content in unexpected ways for better student attention, understanding, and involvement. When those examples come from mathematics in cultural groups, this is an instance of the classroom application work area. A classic example from the 1960s and 1970s was the New York secondary mathematics teacher, Claudia Zaslavsky. Her son was working in Tanzania, and she turned a visit with him into an expedition of gathering classroom teaching materials of mathematics in East African cultures. She later broadened her search and eventually put it together in a book entitled *Africa Counts* (Zaslavsky 1973); for nearly three more decades, she published several other books on mathematics from cultures around the world. This was the initial use of classroom applications, usually in Western classrooms.

However, a more direct classroom application has been to use cultural mathematics examples directly in the curriculum of that same culture. This serves the purpose of helping to localize school curricula, especially in cases where a non-indigenous curriculum may be irrelevant or even oppressive (Knijnik 1997). In this case, local pride can be developed and cultural traditions are acknowledged and honored. Generally, it helps provide an education that is appropriate and useful for the students' situation. Often, this kind of project is carried out with the joint assistance of cultural experts and school authorities.

The third work area in ethnomathematics is working in *cross-cultural situations*. As the world becomes increasingly interdependent, there is a growing number of situations where two or more cultures are interacting, in particular, involving mathematics and/or mathematics education. This may be a teacher from one culture working in another; it may involve curriculum development work incorporating Western and local content or trying to make local content fit into external standards; it may be people of several cultures cooperating in an educational project. In many cases, this area is getting work done rather than writing papers about the investigations, but it may be the clearest application of the ideals of ethnomathematics.

The final area of ethnomathematics work is more abstract. It reviews the theoretical basis of ethnomathematics, drawing from history, sociology, and philosophy, often based on early writing of D'Ambrosio (1985) and others. It may include thoughts on the new directions of work, political issues or controversies arising from applications, or reasoning on why certain studies may be useful. Perhaps this essay itself fits into that category.

15.3 Review of the Chapters

Now, let us review the chapters of this volume with the four-category structure in mind and seeking ideas for future work. The 13 chapters in this volume represent important work in the field of ethnomathematics that is going on many parts of the world. While these may be classified according to the four-category structure, most of the chapters fall into more than one category. This suggests two things: firstly, while the four-category classification is useful in understanding the kind of work that is being done in the field, it also somewhat simplifies the nature of our field; and secondly, that our work in the field of ethnomathematics is multi-faceted, complex, and exciting.

Five of the chapters are works that investigate the *mathematics* in cultural groups or practice. The work of Miriam Amit and Fouze Abu Qouder among Bedouin students looks at the *mathematical elements* in the day-to-day activities of the people, especially how they measure length and weight in their traditional manner. Jaya Bishu Pradhan's work *uncovers* the mathematical knowledge embedded in the artifacts and other wooden products of the Chundaras. Toyanath Sharma and Daniel C. Orey discuss the mathematics *found* on the Rai cultural artifact *dhol* in Nepal, referring to this as *contextualized mathematics*; Tod Shockey and John Bear Mitchell do the same in their work on Penobscot hemispherical lodge, and Tony Trinick, Tamsin Meaney and Uenuku Fairhall in their search for cultural and mathematical symmetry inside *Rauru*, a Maori meeting place located in Hamburg. These works highlight a fundamental principle guiding ethnomathematical work—that some cultural practices of diverse cultures, particularly those that are highly systematized, may lead us to new concepts and alternative ways of thinking and doing mathematics (Barton 1996).

Five chapters may be classified under *classroom applications*, the second category. Both the works among the Bedouin and the Rai peoples may again fall under this work area. In the first, mathematical elements are used to help address Bedouin students' persistent difficulties in the subject, while in the second, the idea is to connect school mathematics with the home cultures of the Rai students in Nepal, with the aim of helping improve teaching and learning practices for teachers and students. Wilfredo V. Alangui's work highlights the development of culturally relevant mathematics lessons for Indigenous Mangyan students in the Philippines. Daniel C. Orey's chapter on ethnomodelling is shown to allow students to use mathematics in examining and solving problems they themselves choose and deem important. Charoula Stathopoulou on the one hand, shows how a critical ethnomathematical perspective can help develop a bottom up curriculum for Roma students. Overall, these chapters show how an ethnomathematical approach or perspective can help promote inclusive and culturally relevant mathematics education especially among historically marginalized groups.

It can be argued that all the 13 ethnomathematical projects featured in this volume are works that are located in *cross-cultural situations*, the third category, and rightly so. As an area that investigates the interplay between mathematics and culture, ethnomathematics provides an avenue for cross cultural research and understanding between ethnomathematics researchers and academics and members of the cultures that they work with. The works of Alangui, Amit and Abu Quoder, Kay Owens, Pradhan, Sharma and Orey, Shockey and Mitchell, Stathopoulou, and Trinick, Meaney and Fairhall involve Indigenous and ethnic groups. For example, Owens immerses herself among the Papua New Guinean Indigenous cultures in order to show the power of spatial thinking and visual reasoning.

Artisans are featured in the work of Maria Cecilia Fantinato and Jose Ricardo e Souza Mafra where they investigate the production of *cuia* crafts by a group of rural women in Santarem in the Amazon region in Brazil. Mogege Mosimege argues for the importance of familiarising with the language (an identity marker) of the culture being considered in the ethnomathematical research, an imperative that certainly requires a cross-cultural undertaking. Even in instances when the researcher or academic investigates her or his own culture (in this volume, Trinick and Fairhall), one can argue that one's culture as a university researcher or academic is relatively different from the culture of the community that is featured in the ethnomathematical project. Thus, cross-cultural learning necessarily occurs between the academic/researcher and the partner cultural group even if they belong to the same culture. Orey's work on ethnomodelling with Brazilian students uses the knowledge of their own culture by connecting ethnomathematics and modelling.

Finally, the works of Veronica Albanese, Natividad Adamuz-Povedano, and Rafael Bracho-Lopez, and Ubiratan D'Ambrosio and Milton Rosa are discourses about the theory, but their ruminations focus on fundamental challenges in the field: how to work with diverse cultures, and how to work towards a cross- and multi-cultural interdependent world free from inequity and injustice.

Another important point has to be made. Of the 13 projects in this volume, 7 have multiple authors. One can argue that individually, these authors bring in their personal histories, values, and perspectives in the shaping of their project. For the other 6 single-authored chapters, most if not all of these projects involved working with cultures different from their own. What these projects show is that our work in ethnomathematics necessarily entails cross-cultural encounters, and that collaborative work is an important feature of our field, not only between the researchers/academics and our partner communities, but also among and between us ethnomathematicians.

The last area of work deals with the theory of ethnomathematics, or theories about ethnomathematics. D'Ambrosio and Rosa reiterate a vision of an egalitarian world that motivates ethnomathematicians to continue the work that they/we are doing; Albanese, Adamuz-Povedano and Bracho-Lopez attempt to resolve a long-standing tension in the field (what they referred to as two views of ethnomathematics), and their chapter would certainly help not only those who are newly initiated in ethnomathematics but also those who continue to struggle with this tension; Orey details a new theory that could arguably be considered as a subfield of ethnomathematics as it focuses on a mathematical tool, modelling, but in *context*. Alangui argues for the use of ethnomathematics as a theory grounding Indigenous peoples' education, especially in providing culturally relevant mathematics education to Indigenous students.

All the other works contribute to better understanding of the theoretical basis of ethnomathematics as a research programme. They give clear indications of how our work should proceed when collaborating with Indigenous groups, ethnic groups, and groups of artisans especially when we highlight culturally valued practices in our research. The work of Stathopoulou reminds us to be mindful of how power/power relations operate within the cultures that we investigate. Other important cultural aspects like language and visuospatial reasoning prompt us to be sensitive to other ways of thinking, doing and talking about mathematics when working with diverse cultures.

15.4 Looking Ahead: Possibilities and Future Directions

The chapters that compose this book show the necessity to debate about issues regarding Mathematics Education, classroom knowledge, knowledge of a specific cultural group, and the nature of mathematical knowledge. The discussions surrounding these issues, as illustrated by the different chapters, affirm the importance of the field of ethnomathematics and the possibilities it offers as an instrument to improve mathematics education, as a theory that helps clarify the nature of mathematical knowledge and of knowledge in general, and as research programme that helps build a just and inclusive society.

Taken together, the works featured in the volume provide examples of new trajectories for research, and they offer a glimpse of future directions in the field of ethnomathematics, especially its role in advancing inclusive and culturally relevant mathematics education. It is an encouraging and inspiring view of the future.

References

Adam, N. (2010). *Weaving culture and mathematics: An evaluation of mutual interrogation as a methodological process in ethnomathematical research*. Unpublished doctoral dissertation. Auckland, New Zealand: University of Auckland.

Alangui, W. (2010). *Stone walls and water flows: Interrogating cultural practice and mathematics*. Unpublished doctoral dissertation. Auckland, New Zealand: University of Auckland.

Ascher, M. (1991). *Ethnomathematics: A multicultural view of mathematical ideas*. Pacific Grove, CA: Brooks-Cole Publishing Company.

Barton, B. (1996). *Ethnomathematics: Exploring cultural diversity in mathematics*. Unpublished doctoral dissertation. Auckland, New Zealand: University of Auckland.

D'Ambrosio, U. (1985). Ethnomathematics and its place in the history and pedagogy of mathematics. *For the Learning of Mathematics, 5*(1), 44–48.

Knijnik, G. (1997). An ethnomathematical approach in mathematical education: A matter of political power. In A. B. Powell & M. Frankenstein (Eds.), *Ethnomathematics: Challenging Eurocentrism in mathematics education* (pp. 403–410). Albany, New York, NY: State University of New York Press.

Shirley, L. (2012). *The work of ethnomathematics*. Quarta Conferência Brasileira em Etnomatemática—CBEm4 (Fourth Brazilian Conference on Ethnomathematics). Belém, PA, Brazil: UFPA.

Zaslavsky, C. (1973). *Africa counts: Number and pattern in African culture*. Boston, MA: Prindle, Weber, and Schmidt.

Author Index

M. Rosa et al. (eds.), *Ethnomathematics and its Diverse Approaches for Mathematics Education*, ICME-13 Monographs,
DOI 10.1007/978-3-319-59220-6

Subject Index

M. Rosa et al. (eds.), *Ethnomathematics and its Diverse Approaches for Mathematics Education*, ICME-13 Monographs,
DOI 10.1007/978-3-319-59220-6

MIX
Papier aus verantwortungsvollen Quellen
Paper from responsible sources
FSC® C105338

If you have any concerns about our products,
you can contact us on
ProductSafety@springernature.com

In case Publisher is established outside the EU,
the EU authorized representative is:
Springer Nature Customer Service Center GmbH
Europaplatz 3, 69115 Heidelberg, Germany

Printed by Libri Plureos GmbH
in Hamburg, Germany